Construction Scheduling
with Primavera Project Planner®

Construction Scheduling
with Primavera Project Planner®

David A. Marchman

Delmar Publishers

I Ⓣ P™   An International Thomson Publishing Company

Albany • Bonn • Boston • Cincinnati • Detroit • London • Madrid • Melbourne
Mexico City • New York • Pacific Grove • Paris • San Francisco
Singapore • Tokyo •Toronto • Washington

Delmar Staff

Publisher: Alar Elken
Senior Administrative Editor: John Anderson
Developmental Editor: Michelle Ruelos Cannistraci
Production Manager: Mary Ellen Black
Art and Design Coordinator: Cheri Plasse
Project Editor: Cori Filson

COPYRIGHT © 1998
Delmar is a division of Thomson Learning. The Thomson Learning logo is a registered trademark used herein under license.

Printed in the United States of America
2 3 4 5 6 7 8 9 10 XXX 03 02 01 00 99

For more information, contact Delmar, 3 Columbia Circle, PO Box 15015, Albany, NY 12212-0515; or find us on the World Wide Web at http://www.delmar.com

International Division List

Japan:
Thomson Learning
Palaceside Building 5F
1-1-1 Hitotsubashi, Chiyoda-ku
Tokyo 100 0003 Japan
Tel: 813 5218 6544
Fax: 813 5218 6551

Australia/New Zealand:
Nelson/Thomson Learning
102 Dodds Street
South Melbourne, Victoria 3205
Australia
Tel: 61 39 685 4111
Fax: 61 39 685 4199

UK/Europe/Middle East:
Thomson Learning
Berkshire House
168-173 High Holborn
London
WC1V 7AA United Kingdom
Tel: 44 171 497 1422
Fax: 44 171 497 1426

Latin America:
Thomson Learning
Seneca, 53
Colonia Polanco
11560 Mexico D.F. Mexico
Tel: 525-281-2906
Fax: 525-281-2656

Canada:
Nelson/Thomson Learning
1120 Birchmount Road
Scarborough, Ontario
Canada M1K 5G4
Tel: 416-752-9100
Fax: 416-752-8102

Asia:
Thomson Learning
60 Albert Street, #15-01
Albert Complex
Singapore 189969
Tel: 65 336 6411
Fax: 65 336 7411

Library of Congress Cataloging-in-Publication Data:

Marchman, David A.
   Construction scheduling with primavera project planner / David A. Marchman.
     p.   cm.
   Includes index.
   ISBN 0-8273-7086-5
   1. Building—Superintendence. 2. Production scheduling. I. Title.
TH438.4.M37   1997
690′.068—dc21   97-10619
                  CIP

Contents

SECTION 2   Scheduling   **55**

CHAPTER 3   Schedule Calculations   **56**

CHAPTER 4   P3 Schedule Preparation   **70**

Preface

COMPUTERS IN CONSTRUCTION SCHEDULING

The Advantages of Using Computers in Construction Scheduling

One of the primary advantages of using computers for construction scheduling is that the mathematical computations are instantaneous and error free. The speed and accuracy of the mathematical scheduling computations and the analysis of the information produced make computerized scheduling an invaluable tool for construction project controls. The low cost and superb on-screen and hard copy graphics make it an effective communication and project control tool for owners, contractors, subcontractors, suppliers, and vendors.

The Advantages of *P3*

Primavera Systems, Inc.'s most powerful project management/scheduling software package is *Primavera Project Planner* or *P3*. This book is written for use with *P3*. The advantages of *P3* are:

- Between *P3* and its sister package, *SureTrak*, Primavera Systems, Inc. controls between 70% and 80% of the market for construction scheduling software in the United States.
- Primavera Systems, Inc. is among the top fifty computer software vendors in the United States.
- Primavera Systems, Inc. offers excellent training and customer support in the use of their software.
- Unlike some other construction scheduling software companies, Primavera Systems, Inc.'s total market is construction project management software.

P3 for Windows System Requirements

- 486 or better computer
- 16 MB total memory
- 58 MB of free disk space for a complete installation or 32 MB of free disk space for a minimal installation
- DOS 5.0 or later (running Share or Vshare)
- *Windows 3.1* (enhanced mode), *Windows 95*, or *Windows NT*
- VGA or SuperVGA monitor
- Mouse

Version 2.0 Enhancements

This book is based upon *P3 for Windows*, Version 2.0, which includes:

- PERT (*Primavera*'s Easy Relationship Tracing)—PERT is an interactive pure logic view of activities and relationships, making it much easier to evaluate the critical path of a schedule.
- Advanced resource management—*P3* now has resource calendars; users can split, stretch, and/or crunch activities based on resource availability.
- Web Publishing Wizard
- E-mail integration
- *Primavera* Post Office
- RA: *P3*'s OLE automation server

Help Topics

The on-line help in *P3* is extensive. Help Topics can be accessed three ways for ease in use. They are:

- **Contents**. Displays *P3* help contents by category (books).
- **Index**. Displays the **Search** dialog box that enables you to find help information using keywords.
- **Find**. Enables you to search for specific words and phrases in Help Topics instead of searching by category.

Another helpful feature of *P3* is that by depressing the F1 key (first function key), the Help Topics screen for that field can be pulled up from almost any field within *P3*. The tutorial is a very helpful walk-through of the different functions of the product.

ABOUT THIS BOOK

This book is a graphic, step-by-step, introduction to good construction project control techniques. This book depicts construction scheduling as it has not been presented before. It includes the traditional theory on planning, scheduling, and controlling construction projects. Topics included are schedule development, activity definition, relationships, calculations, resources, costs, and monitoring, documenting, and controlling change. The difference between this and other scheduling books is that this book takes the student through all the above topics with an example construction project schedule using *P3*. *P3* is generally recognized as the premier construction scheduling software on the market today. The student is not only exposed to on-screen images, but also is shown how to manipulate hard copy prints for exceptional communication and demonstration results.

The step-by-step tutorials are the strength of this book. This book is designed for the classroom, but will work very well for people in industry that need to learn *P3* by themselves.

WHO SHOULD USE THIS BOOK

This book was written for students of construction, construction technology, and for working professionals who wish to gain a better understanding of *P3*, the scheduling software system by Primavera Systems, Inc.

USING THIS BOOK

This book can be used either as a stand-alone book or in conjunction with another more traditional planning/scheduling book. Students should be able to use this book as a step-by-step guide while they are using *P3*. The strength of this book lies in its detailed examples of a sample schedule that uses scheduling information in numerous ways. This book can be used in conjunction with *P3*'s Help Topics and tutorial for an efficient introduction to planning and scheduling and the use of *P3*. Scheduling is presented primarily from the general contractor's point of view, but most of this information can be applied to the subcontractor, fabricator, owner, and construction manager.

USING THE EXERCISES

A sample problem and exercises at the end of each chapter make this text useful in a classroom setting. A student who completes the exercises by using *P3* will have a sound basis for producing construction schedules.

ABOUT THE AUTHOR

David Marchman teaches estimating, scheduling, project management, and other construction courses in the School of Engineering Technology at the University of Southern Mississippi. His teaching career is supported by extensive experience in the construction industry (residential, commercial, and industrial), including seven years with Brown & Root. He continues to consult with numerous companies on estimating, scheduling, and cost accounting. Mr. Marchman has spent summers with Bill Harbert International Construction, Hensel Phelps Construction, and Brice Building Company. He also has taught numerous seminars and workshops on construction-related topics.

ACKNOWLEDGMENTS

I would like to thank my friend Jim Goodwin of Technical Construction Systems in Birmingham, Alabama. Jim's guidance and assistance was the inspiration for originally undertaking the writing of this book. His help on this and other projects has been immeasurable. My friend Leo Ney (Houston, Texas) was a great support with his proofreading and ideas. Michelle Cannistraci, my editor with Delmar, was heaven sent. Her patience and support made this a manageable project.

I also would like to thank the following reviewers for providing feedback and guidance:

Arch Alexander,
  Purdue University,
  West Lafayette, IN

William R. Doar,
  East Carolina University,
  Greenville, NC

Kweku K. Bentil, Ph.D.,
  University of Washington,
  Seattle, WA

Gary B. Gehrig,
  Colorado State University,
  Fort Collins, CO

Gregg Corley,
  Auburn University,
  Auburn, AL

David Goodloe,
  Clemson University,
  Clemson, SC

 Section 1

Planning

Introduction

Objectives

Upon completion of this chapter, the reader should be able to:

- Explain the necessity for good scheduling
- Identify project scheduling needs
- Identify and name activities
- Recognize different types of schedules

HISTORY OF SCHEDULING

Need for Structures

Construction is as old as civilization. Humans have always needed shelter, bridges, roads, and buildings to store water and the food they gathered, killed, or cultivated as well as for public meetings. Of course, the earliest structures were simple and rudimentary in nature.

Human Construction Capabilities

As civilization evolved, so did the buildings to house increasingly sophisticated political, religious, economic, and cultural activities. We marvel still at the skills of the builders of such ancient monuments as the Pyramids of Egypt, the temples of Ancient Greece, and the great European cathedrals. These projects required vast resources of materials, labor, and equipment, not to mention wealth. The Pyramids of Egypt were built by a workforce of thousands over many years. The huge cut stones comprising these structures were hauled many miles by water and land, then lifted and placed with great effort. The commitment to build such enduring structures and the dedication of wealth, time, and other resources to the task reflect the greatness of a society.

Process

Scheduling has always been an essential part of construction. Resources and time to complete projects are always limited. Users of every structure anxiously await its completion. Scheduling techniques have advanced over the years to meet the demand for more advanced structures at the least cost. A scheduling feat that preceded the advent of the computer and modern scheduling techniques was construction of the famous Empire State Building in New York City, the tallest building in the world at the time. The site was midtown Manhattan, so there was tremendous traffic congestion. Without space on site to temporarily store construction materials, the builders had to bring in nightly the structural steel to be used the next day. Compounding the site-related difficulties were labor problems, and the owner wanted the use of the building as soon as possible. This clearly was a project demanding good scheduling techniques. Remarkably, because of efficient scheduling, the project averaged the erection of one floor every three days!

Progress

The size and nature of projects modern society constructs display the ability of people to dream, plan, and marshal their resources as did the Pyramids and cathedrals of old. Construction projects are getting larger and more complex all the time. An ambitious recent example is the tunnel built under the English Channel. Whether of such great scope or much more modest, no construction project can be accomplished without adequate planning and scheduling.

NECESSITY FOR GOOD SCHEDULING

Resources

A resource is a source of supply or an available means. Any type of construction project (whether residential, commercial, industrial, or small or large) requires the gathering of resources of specialized labor, materials (fabricated both on and off site), construction equipment, and management for a particular site. These resources are expended over the time needed to complete the project. Planning and controlling resources and the pace of building to meet contractual deadlines and quality requirements are crucial to the success of any construction project.

Time: The Owner's Viewpoint   The old adage "Time is money" is particularly true in construction. When a project is finished, and the owner has use of the facility, the owner starts to receive a return on investment from the use of the building to repay funds used in the construction of the project. The return on investment is the monies left after the product or service is sold and the costs to produce the product have been deducted. The cost of the facility used to produce a product or service is usually a large part of the investment of production.

Since the owner wants to produce a return as quickly as possible, another old adage applies: "Time is of the essence." This means that the project completion date is of concern to the owner, and any delay could result in damages paid by the contractor to the owner for losses incurred. With most construction contracts, whether lump-sum or negotiated project, time is as important to the contract as project cost. In a lump-sum project the owner signs a contract with the constructor to produce a project for a set dollar value within a specified time. In a negotiated project the owner negotiates with the constructor to act as his agent to produce the project for a set fee.

Duration: Contractor's Viewpoint Typically the contractor also would like to minimize a project's duration. Once a project is complete, the contractor's indirect costs stop. Indirect costs of a project depend directly on its duration. They include the cost of temporary facilities and utilities as well as field supervision and construction equipment. The longer these resources are kept at the job site, the greater the cost. A crane might cost $1,000 per day to have at a job site. If the crane is needed for only 10 days rather than the 20 days scheduled, the project is saved $10,000. The same work is accomplished, but through better scheduling or allocation of resources, monies are saved. Such savings reduce total project costs. (Of course if a schedule is shortened by means of overtime, overall costs may not be reduced at all.)

Service

The construction industry is a service industry. The contractor is selling his or her firm's management ability to put the owner's project in place. For a traditional, lump-sum project, the owner furnishes the money, the architect and engineer furnish the contract documents, and the contractor furnishes the construction expertise. Any time the construction management team saves in meeting a project schedule can be spent on the next project. Saving time thus increases the volume of work and, potentially, profit. The efficient construction company can produce more work per year with the same management staff. Time is money!

Project Control

Good project management is characterized by good "project controls," which prevent a project from getting out of control. All projects have two primary control documents: the estimate and the schedule. The estimate defines the scope of the project in terms of quantities of materials, work hours of labor, and hours of equipment. Since all of these resources can be measured in dollars, the estimate also serves as the budget for cost.

The schedule is produced to control the expenditure of another resource: time. Since Time is money, the resources and the time frame over which they are expended are interrelated, and the elements of the estimate and the schedule are interrelated The information from one document feeds the other: the estimate computes resources needed over the specified time, while the schedule defines the time needed given the specified resources. These documents are the vehicle the project team uses to build the project on paper and in their minds before the actual construction begins. This is the essence of planning.

Surprises   Surprises are unwelcome at the construction job site. Estimating and, even more so, scheduling, provide the project organization with an efficient mechanism for anticipating and solving problems during the planning phase before resources are expended rather than on the job site during the construction phase. The process of planning is intended to make the construction process run smoothly, minimize surprises, and prevent expending costly resources with poor results.

Communication

Another key to avoiding unwelcome and costly surprises is communication. Only the very smallest of construction projects is completed by a single individual. Most projects require the services of many different experts, various contractual relationships, and numerous functions. The project manager needs to give and receive certain information at every stage. The same is true for superintendent, foremen, subcontractors, owners, and supervisors of the equipment and personnel departments. In order for a project to run smoothly, all parties need to be on the "same page at the same time." This requires continual communication so that all participants will know what the plan is and what their responsibilities, requirements, and expectations are. Keeping everyone informed is one of the critical functions of schedulers.

PROJECT MANAGEMENT

The purpose of project management is to achieve a project's goals and objectives through the planned expenditure of resources that meet the project's quality, cost, and time requirements.

Quality

Good management assures that a project attains the level of quality defined in the contract documents. Not only do quality and pride in workmanship go hand in hand and improve both worker morale and project outcome, but they also lead to profitability through repeat business, referrals, and negotiated works.

Cost

The ability to estimate and then to complete a project within the budget is at the very heart of good construction project management. This is true whether the estimate is lump sum or negotiated. The owner looks to the contractor for the efficient expenditure of resources to get the most results for his or her dollar.

Time

"Time is money"! "Time is of the essence"! Time is usually just as critical to the contract and the owner's needs as quality and cost constraints. The ability to determine a schedule and then to complete the project within the time frame also goes to the very essence of being a construction project manager. Many times an owner will choose a contractor based on his or her ability to marshal forces to complete a project in a timely manner, rather than basing the decision solely on the lowest cost. Getting the project in service and generating a return on investment as soon as possible is critically important to the owner. Tax or market considerations may concern the owner just as much as cost and quality.

Safety

Another integral component of good project management is safety. Accidents can be extremely costly, not only in their direct costs, but also their influence on the Workmen's Compensation Modifier Rate. That rating of the construction company's safety record affects the amount the company must pay for insurance coverage for employees. Accidents also affect worker productivity, morale, and other direct and indirect costs. Good safety practices are an indicator of good project management.

Risk

Among other important objectives, a constructor is in business to be successful and make money. This is accomplished through estimates and schedules that are fair, honest, and reasonable. Financial risks can be controlled and minimized to produce a successful project.

SCHEDULING FOR DESIGN-BID-BUILD PROCESS

Linear Construction

In "linear" construction the processes are consecutive rather than concurrent. There are breaks between the design, bid, and construction functions. The owner contracts with the architect and engineer to produce the project contract documents, which define the project scope. Next, bids for construction services are accepted from contractors. The owner/architect selects the successful contractor's bid, awards a contract, and construction commences. Thus, the contractor knows the scope of the work he is bidding on, and the owner knows the price and duration of the work before the contract is awarded. For the constructor, scheduling lump-sum linear projects deals primarily with scheduling field construction activities. The preconstruction activities of architecture/engineering design and definition of project scope and the postconstruction activities of maintenance and operation are not part of the contractor's responsibility and therefore are not scheduled.

Field Construction

In controlling the field construction, the contractor is concerned with controlling time, resources, labor, materials, subcontractors, vendors, equipment, and money. Influences to the schedule to be controlled are the:

- relationship of the activities to each other (Which activities precede which others, and which can occur at the same time?)
- size of crews
- availability of labor
- methods of operation
- types of construction equipment
- work schedule (number of hours per day, shifts, weekends, holidays)
- material deliveries
- inspections
- payment schedules

All these factors must be efficiently organized, sequenced, and controlled in order to optimize the plan and maximize efficiencies.

Timing

Typically on a lump-sum project, detailed scheduling begins once the contractor has signed the contract with the owner to construct the project. Although some scheduling takes place during the estimating/bidding phase, this is usually general in nature for example, determining how long a superintendent will be needed on the job site.

Since producing the detailed schedule can cost a substantial amount, it is not begun until the contract is awarded, and it is completed as soon as possible. There are three primary reasons for preparing the schedule immediately. First, the contractor wants the schedule to control field operations. Second, the planning and scheduling process is a preplanning tool to help point out and solve problems before they arise in the field. The third reason is that the owner/contractor contract may require presenting a project schedule to the owner/architect for use in monitoring the project.

The scheduling process used in setting up project controls for lump-sum linear construction involves four phases. They are planning, scheduling, monitoring, and controlling.

Planning

Decision Making Prior to starting construction, the management team must plan the execution of the project. Planning is a form of decision making since it involves choosing among alternative courses of action.

Information Gathering Planning entails defining the activities necessary to construct the project and establishing the relationships between the activities. Communication is required in order to gather data from many persons and places. For example, questions might include: When and from what source is the right crane available? Are enough skilled masons available to complete the brickwork on time? What will be the impact on the project schedule of long lead-time items, such as the elevator for the project that will require special fabrication at a shop where there is a backlog? Activity relationships are defined with information being derived from the contractor's management team, the accounting department, the equipment department, the plans and specs, a visit to the site, the nature of the work, the owner, the banker, the subcontractors, the governmental agencies involved and their requirements, and the suppliers.

Identifying/Defining Activities The team preparing the schedule must create a rough diagram identifying and defining activities and their relationships to other activities. Essentially, the project is built on paper with activities and interrelationships as building blocks. The entire project is first constructed in the minds of the scheduling team, then put on paper. The relationships between activities, building methods, problem solving, and communications that define the plan take place in this fertile environment that defines the project in terms of the schedule.

Creativity   Good planning requires being creative and not being bound by preconceived notions. Just because a company has always tackled a certain construction sequence or segment of work a particular way, it needn't assume that way is best. By trying to break away from established "paradigms" or models or false constraints, planners can incorporate improved methods into the schedule. One method to incorporate new ideas and methods is "brainstorming" with key participants to solve a specific problem. Say the question is finding the best method to pour the elevated concrete column on a specific project. Participants in a brainstorming session would each state the first method or solution to the problem that comes to mind, no matter how far-fetched or nontraditional the approach. This and other methods to expand and change the way team members look at the project are very useful in the planning process.

Flexibility   Good planning also requires flexibility. It is the natural tendency of first-time schedulers to build a long chain of activities, one after another, with no branches and with only one thing happening at a time. The most efficient way to shorten project duration is to have as many activities going on concurrently as possible without hindering each other's progress. Instead of having single crews working consecutively at the job site, there could be five crews working concurrently though not interfering with each other's progress. In Figure 1-1, activity Rough Framing Walls comes after Pour Slab. They are scheduled consecutively, or one must be complete before the other can begin. Activities Rough Plbg, Ext Fin Carp, Rough Elect, Rough HVAC, and Inst Wall Insul are scheduled concurrently, or they can be going on at the same time.

Interrelationships   During the planning stage, no durations and usually no resources are applied to the schedule. Developing activities necessitates some consideration of resources. For example, an activity might be cast-in-place concrete or precast. Thus, some resource decision has

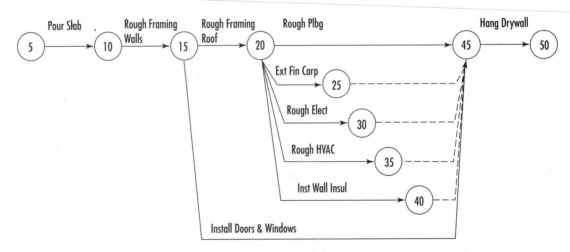

Figure 1-1   Activity-on-Arrow Diagram—Example of Consecutive and Concurrent Activities

already been made. But the activities are usually only defined along with their associated interrelationships and constraints. This is the phase of the scheduling process with the most potential for being creative and developing new approaches, systems, or methods for putting the work together. When the project has been defined on paper in the form of the rough schedule, the planning stage gives way to the scheduling phase.

Scheduling

Durations   The second phase of the overall scheduling process is filling in more precise estimates of time and other resources in the rough schedule produced in the planning phase. The other resources might include labor, equipment needs, and division of responsibility (for example, crew versus subcontractor). By calculating the estimated duration of each activity, the scheduler can calculate project duration. If the first pass yields a project duration that is too long, resource constraints can be adjusted; for example, by increasing crew size to complete an activity; or sequences of activities can be modified or activities made concurrent. The schedule is continually fine tuned until it becomes satisfactory to all parties, and it is then accepted as the project schedule.

Evolution   Development of the schedule from the rough stage to the project schedule is an evolutionary process that requires communication and approval of all parties involved—contractor's management team, subcontractors, major suppliers, and, of course, the owner. Take care at this stage. The contractor must receive input from these parties and wants each of the parties to accept or "buy off" on the schedule. This process makes it "our schedule" instead of "your schedule." Typically the contract documents make the contractor the keeper of the schedule for the construction project. It is the contractor's responsibility to resolve scheduling disputes and conflicts between the different subcontractors and the contractor's own forces. Thus the process includes soliciting input from the different parties, drafting a rough schedule, circulating it for feedback, making modifications, resolving disputes, and reaching agreements. This process produces a project schedule that all parties can accept.

Monitoring

The third phase in the scheduling process is monitoring. The "data date" is the date on which a schedule is updated with current information. Typically at fixed intervals throughout the project, usually at the end of each month, progress is determined as of that date. This is the cutoff date for comparing the actual project progress to the planned progress. The data date is usually the first day of the month. Monitoring involves determining the physical progress in the field and inputting the progress of each activity. First, it is necessary to establish a database of the actual expenditure of resources for each activity. This information is compared to planned expenditure to determine the percent complete for each activity. The progress of each individual activity is established using the information from the database. Then the schedule is recalculated with the updated information to determine if each activity is ahead or behind schedule when compared with the plan (baseline) and if the overall project is ahead or behind schedule. This process is called updating the schedule.

Controlling

Change   The fourth and last phase in the scheduling process is the controlling phase. Controlling usually involves the documenting and communicating of change to the plan and schedule. As projects develop, the sequence of activities originally planned may change. The reason may be updating schedule, changes in scope, material delays, lower or higher productivity, or other reasons. An example of this type of change may be the decision to use a different erection method or system, which essentially alters the schedule. The new revised schedule then becomes the current project schedule. If a project is behind schedule, the contractor may have to "crash" or accelerate the schedule to make up for lost time by adding shifts, having laborers work overtime, or adding craftsmen to a crew. Keeping the schedule relevant and useful requires redrawing the plan to incorporate such changes in the relationships. The revised schedule will be better, since it is based on more current information.

Progress   Usually, schedule progress is the basis of monthly meetings with the owner/architect to determine the contractor's compliance with the contract and invoice for payment. The owner is presented with an updated schedule showing the data date (the date up to which progress determination was made), progress during the last 30 days, and the forecast for the next 60 days. The schedule is a critical document for determining whether the project is progressing according to the original plan. By using the current (progressed) schedule, along with the target (original budgeted) schedule, it is easy to spot which activities are in trouble and whether the project itself is on schedule or in trouble. This concept is known as "management by exception." Management and usually the owner want to know which parts of the schedule are in trouble so they can determine which activities to spend time on and which activities are the most likely candidates for making up lost time. It doesn't do any good to put extra resources on an activity that is not critical to the completion of the project schedule. Controlling is the process of constantly modifying the schedule to make sure it is current with the latest plan for how the project will be constructed.

Documentation

When changes are made, the construction team must document or record the changes for historical information and project backup. The historical information will be used as a database for reference for future projects. The project backup is necessary for use in possible legal claims or settlement of project disputes.

Scheduling for Negotiated, Phased, or Fast-Track Construction

Integration

Fast-track or phased construction involves the integration of detailed design and construction. On larger projects, significant time can be saved by overlapping or concurrently designing and building the project. The primary disadvantage is that since the design is not complete, the complete scope of the work is not known before field work commences. These projects are therefore negotiated contracts with the owner accepting a greater portion of the risk for project cost increases.

Team Approach

Another advantage of fast-track construction is the team approach to the design/construction process. Teamwork among owner, designer, and constructor contrasts with the confrontational relationship that lump-sum construction typically leads to. The constructor typically works as the owner's agent, is paid a fee for services, and is part of a team. Under the lump-sum arrangement, the constructor is paid according to bid and receives any money saved. This leads to the construction firm looking out for its own best interest and not that of the owner. Under the team concept, if the constructor's fee is fixed, money saved returns to the owner, and the constructor is looking out for the best interest of the owner.

Design

Another advantage of fast-track construction is that the constructor is involved in the design process enhancements. Significant cost savings can be derived by improving "constructability"—designing for ease and economy of construction. Another way to save is by "value engineering," a systematic approach to evaluating a number of cost alternatives to choose the best for the project.

A good rule of thumb is that 80 to 90 percent of a project's cost is fixed by the time conceptual design (the sketching phase) is completed. The project cost is fixed to a large extent before the first detailed drawing is ever completed. The reason for this is that when certain project parameters are defined, the scope and, therefore, the cost of the project become fixed. Once the function, location, size, vertical or horizontal orientation, type of construction, and type of environmental controls are fixed, the project cost is essentially determined. Since the constructor's knowledge of cost and how the project goes together is likely to be greater than that of other members of the team, the constructor's input during the conceptual design phase is invaluable in controlling project cost and schedule.

Entire Project

The four phases of scheduling (planning, scheduling, controlling, and monitoring) are the same for negotiated and lump-sum projects. The primary difference is that the scheduling effort in the lump-sum project focuses on the field construction effort, whereas a negotiated schedule integrates the conceptual design, detailed design, project management, procurement, expediting, documentation, field construction, and possibly maintenance/operations functions. The entire project is looked at as a whole with break-out schedules for each of the areas, such as design, construction, or project management.

SCHEDULING LEVELS

Project Schedule

During the control phase of the schedule process, most contractors consider the project schedule established at the beginning of the project as the general starting point. Good scheduling requires more detailed preparation as a particular activity gets closer to actual installation.

10-Day Look-Ahead

It is good practice to prepare a written 10-day (2-week schedule) look-ahead or a preplanning sheet (short-interval schedule) of all upcoming activities. Depending on the size and complexity of the project, the preplanning sheets are organized by some combination of the project engineer/superintendent/foreman. This short-term schedule is usually a field function used for communication with the crews. Good preplanning techniques require that all necessary sketches or shop drawings be complete, the plan be in writing, and materials be available for the work to commence. The preplanning process is essentially a problem-solving exercise to save time and money by averting problems that might arise in the field. Preplans should be in writing to reduce mistakes and rework. A major cause of rework in the construction industry is the heavy reliance on oral communication rather than written communication. Workers try to put in place what they *think* their supervisors want. However, since they do not always understand exactly what the supervisors want, they sometimes complete the wrong work. Precise written plans can prevent such miscommunication.

ACTIVITIES

Activity Defined

One of the first steps in putting any schedule together is identifying the activities or tasks that must be completed to attain the project goals and objectives of the project team. Since this book is based on Primavera Systems, Inc.'s most powerful project management/scheduling software package, *Primavera Project Planner*, abbreviated *P3*, activity definitions and other relationships used in this book will reflect the use of this software. This book is based on *P3 for Windows*, Version 2.0.

Construction projects are made up of a number of individual activities that must be accomplished in order to complete the project. There are numerous types of activities within *Primavera*, but the primary activity type is the task activity, which requires time and usually resources to complete. Task activities have five specific characteristics. An activity

- Consumes time
- Consumes resources
- Has a definable start and finish
- Is assignable
- Is measurable

Techniques used to help define activities are:

- The estimate
- Historical information
- Work breakdown structure (WBS)
- Experience

Other types of *Primavera* activities are:

- Milestone Activity—a major event, phase, or any other important point in the project
- Meeting Activity—used where resources cannot work independently
- Independent Activity—used to make resources work according to their own resource calendars

Time Consumed   The task activity breaks the schedule down into more easily estimated smaller components. A task activity consumes weeks, days, or hours. Duration is a function of the scope of work for the activity and resources assigned to accomplish the work.

Resources Consumed   Usually resources must be expended to complete a task activity. (The assigned resources should be scheduled according to the activity's base calendar.) Labor is expended to install material resources, and permanent equipment is expended to complete the finished structure. The quantity or scope of work, as defined by the estimate of labor, materials, and equipment, determines the duration of the activity.

Definable Start and Finish   Task activities consume time and resources and are tied to related activities by relationships. These relationships determine which activities must be complete before the activity in question can begin. The scheduler determines duration of the activity from the estimated quantities of materials to be placed and the size of the crew to place the materials. By knowing the relationships between the activities and their durations, the scheduler can determine the planned start and finish dates of the activity. Therefore, each activity is definable in terms of its planned start time, duration, and planned finish.

Assignability   Determining responsibility for each activity is critical to any scheduling effort, since any construction project involves bringing together many crafts, subcontractors, suppliers, and others to attain project completion. Task activities should be defined so that the responsibility for activity completion is clear and assignable to a single party. If the activity is to be completed by the contractor's own forces, it should be defined by crew to identify the proper superintendent or foreman. If the activity is to be completed by a subcontractor or vendor, responsibility should be assigned to that party, so that as the project progresses, communication with the parties can be expedited using the schedule as the baseline indicator of the performance. The plan is something to measure against to determine performance.

Measurability   The duration and resources assigned to an activity must be measurable to determine whether or not the budget for the activity duration was met. How many days were actually spent? What was the actual physical progress in terms of days? How many resources were actually expended? What was the actual physical progress in terms of resources? The duration and resources are measured as of a particular date, and an evaluation is made about how the project is going in terms of the original plan. Is the project ahead or behind the original schedule with respect to time? Is the project ahead or behind the original schedule in terms of the expenditure of resources? Answering these questions is essential for monitoring and controlling the project. Since the activities are assigned, the party responsible for controlling the duration and the expenditure of resources is identified by activity, and communications to resolve project control problems are enhanced.

Activity Codes/Categories

It is usually not enough to identify only how activities fit into the over-all logic of the project. Responsibility for completion of an activity can also be assigned. Good management requires defining authority and responsibility by activity so that communications about the activity and the management chain of command are clear. This is clearly a choice for using activity data and is not mandatory. When responsibility is assigned by activity, specialty sorts of information can be requested using *P3* to give the responsible party copies of only their activities so that they won't have to dig through mountains of data to find the information they require.

P3 has the built-in capability to sort activities defined by project requirements. The usual sorting parameters of department, responsibility, phase, work breakdown structure, or other sort requirements can be defined to the system.

Establishing project activity categories/codes and assigning related activities gives the project management team the ability to sort the schedule by category. The party responsible for that portion of the schedule can concentrate on the activities over which he or she has control.

Engineering   The engineering category of activities usually relates to construction permits, documentation control (tracking shop drawings, etc.), payment, and inspection. Often these activities or events fall into the category of a milestone activity, an event that enables other activities to continue or commence. For example, the visit by the off-site inspector enables the rest of the activities to begin. Even though the inspection had no cost or resources connected to it, it can be considered a milestone activity.

Mobilization   The mobilization category of activities relates to moving onto the job site. Besides bringing in construction equipment and temporary materials, it also includes installation of temporary facilities and utilities.

Procurement   The procurement category of activities includes the identification, procurement, expediting, delivery, and control of bulk materials, fabricated materials, and permanent equipment to be used on the job site.

Construction with Own Forces   Work in this category is performed by craftsmen who report directly to the contractor and not to a subcontractor.

Here the contractor sorts the activities by the foreman responsible for carrying out the work. The labor resources expended are sorted by type, such as rough carpentry, finish carpentry, concrete finishing, etc.

Construction with Subcontract   It is critical to establish subcontractor's responsibility for the activity in this category. It is a good practice to document the labor to be expended according to subcontractor's craft classification. The contractor is not responsible for the direct supervision of the craftsmen. But, by comparing the expenditure of resources according to the subcontractor's original plan, the contractor can determine if the sub is within budget. Comparing actual with anticipated expenditure of resources also enables the contractor to assess physical progress and evaluate payments.

Start-up   The start-up category of activities relates to testing, punch list, and start-up of the facility. This is the point when the owner takes possession of the facility to use for the intended purpose. There are many activities related to the testing, accepting, and starting of the facility and also activities relating to the owner taking responsibility for insurance, utilities, security, and possession of the structure.

Demobilization   The demobilization category of activities relates to moving off the job site. This includes the removal of temporary facilities, construction equipment, and temporary materials.

Activity Identification

The activity identification (activity ID) is the way the activity logic is identified to the computer. It is used by Primavera (and most other scheduling software vendors) to give each activity a short name or identifier that can be used for sort functions. *P3* maintains a database of information about each activity, and the key or primary field used to sort the information is the activity ID field.

The activity ID itself can be used as a convenient means to sort activities. Take care when developing the naming format to make the activity ID a sortable field. Be consistent in naming all project activities to make the sort possible.

Activity Detail

Activity detail defines the appropriate level of information breakdown needed to meet the project needs. Usually, a daily unit is appropriate, and units less than a day should be consolidated if possible.

Other considerations besides the daily unit should be considered. They are:

- Who is going to use the schedule, and what are their needs?
- Complexity. What communication is needed?
- Division of responsibility. Who is doing the work (division of responsibility)?
- What is the contractor's management philosophy?
- Will more or less information affect the usefulness of the schedule?
- Will short-term scheduling be the appropriate place for more detailed scheduling?

Activity Description

The activity description is a *P3* field that is longer than the activity ID field and is used to describe the activity. An important communication tool, it must be clear, concise, and have the same meaning to all parties using the schedule (contractor's forces, subcontractors, owner, and the architect/engineer). This allows all parties to read and understand the schedule.

The short description of each activity (48 characters in *P3*) must communicate the scope and location of the portion of the work that activity encompasses. Since much information is communicated in a small space, descriptions must be constant in format, so abbreviations are typically used. Abbreviations and procedures for naming activities should be consistent throughout the project. Consistency will make the schedule much easier to use. Whenever possible, standard industry abbreviations should be used, such as "Ftg" for footing and "Conc" for concrete.

Activity Relationships

The relationships between activities determine which other activities must come before, come after, or can be going on at the same time as the activity being defined.

BAR CHARTS

The bar chart (Gantt chart) is a convenient and easy-to-read method of viewing the schedule. Henry L. Gantt and Fredrick W. Taylor popularized this graphical representation in the early 1900s. Simply put, the horizontal axis represents a timescale of the project and the vertical scale lists the general activities necessary to put the project together. These activity descriptions can be as broad or as narrow as the author needs to adequately describe the project. Figure 1-2 is an example of a bar chart.

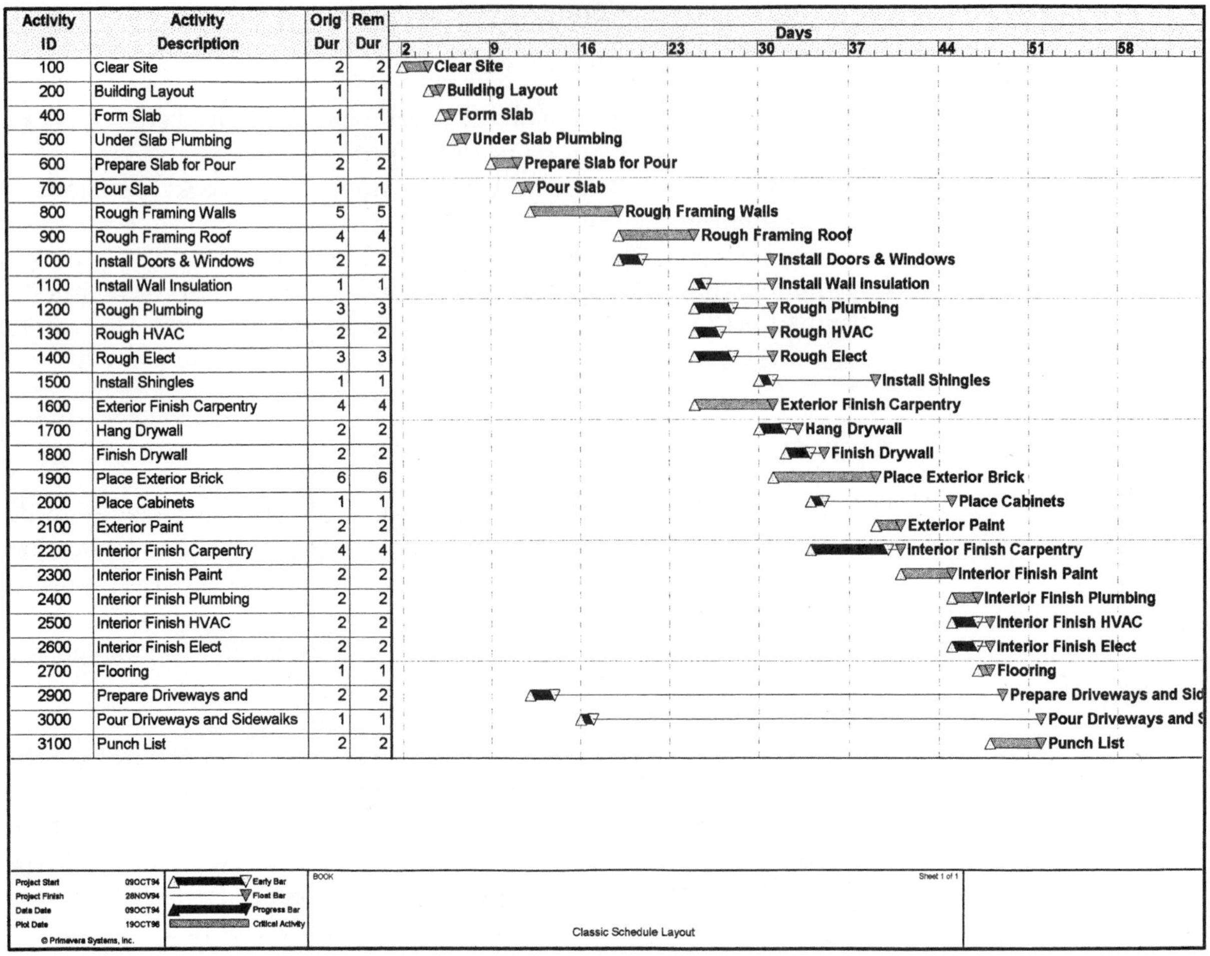

Activity ID	Activity Description	Orig Dur	Rem Dur
100	Clear Site	2	2
200	Building Layout	1	1
400	Form Slab	1	1
500	Under Slab Plumbing	1	1
600	Prepare Slab for Pour	2	2
700	Pour Slab	1	1
800	Rough Framing Walls	5	5
900	Rough Framing Roof	4	4
1000	Install Doors & Windows	2	2
1100	Install Wall Insulation	1	1
1200	Rough Plumbing	3	3
1300	Rough HVAC	2	2
1400	Rough Elect	3	3
1500	Install Shingles	1	1
1600	Exterior Finish Carpentry	4	4
1700	Hang Drywall	2	2
1800	Finish Drywall	2	2
1900	Place Exterior Brick	6	6
2000	Place Cabinets	1	1
2100	Exterior Paint	2	2
2200	Interior Finish Carpentry	4	4
2300	Interior Finish Paint	2	2
2400	Interior Finish Plumbing	2	2
2500	Interior Finish HVAC	2	2
2600	Interior Finish Elect	2	2
2700	Flooring	1	1
2900	Prepare Driveways and	2	2
3000	Pour Driveways and Sidewalks	1	1
3100	Punch List	2	2

Project Start	09OCT94		Early Bar
Project Finish	28NOV94		Float Bar
Data Date	09OCT94		Progress Bar
Plot Date	19OCT98		Critical Activity

© Primavera Systems, Inc.

BOOK   Sheet 1 of 1

Classic Schedule Layout

Figure 1-2 Bar Chart—Typical House Construction

Advantages

There are three primary advantages to using the bar chart. First, it is usually easy to read. It is easy to interpret when the activities should take place. Anyone involved in the construction process—owner, architect, banker, bonding agent, contractor, subcontractor, supplier—can interpret this simple document. Another advantage is that, due to its simplicity, it is a great communications tool. The third advantage is that it is easy to update the project to show progress. However, progress can be shown in many different formats, some more viable and visual than others.

Disadvantages

The primary disadvantage of the bar chart is that it does not show the interrelationship of various activities. What happens if one of the activities is late being completed? How is the rest of the project delayed? The impact of the delay can be evaluated by analysis of the bar chart, but since the relationship between the activities is not shown, the conclusions are open to debate. The logic of interrelated activities may be very formalized, but it is not clearly and completely conveyed to the user of the bar chart. A great majority of the construction claims relating to schedules are lost by contractors because the contractor cannot prove the impact of schedule delays. A construction claim requires proof through documentation. For example, in residential construction, when the drywall subcontractor is late in hanging the drywall, typically all interior work is delayed. This would have an impact on finishing the drywall, placing cabinets, interior finish carpentry, and the rest of the schedule. But because the bar chart does not show the direct impact on these activities, there is room for argument and the claim of the contractor is difficult to prove. A better tool for proving impact cost or ripple damages is the network logic diagram. The activity-on-node and activity-on-arrow diagrams are discussed below.

Format

The bar chart format (along with the pure logic and the timescaled logic diagrams) is one of *P3*'s primary hard copy graphical print formats. Once the scheduler has input the activity information and the associated relationships into *P3*, any of the three print formats can be requested. The bar chart is a convenient vehicle for confirmation and dissemination of the information used for more complex formats. Figure 1-2 (p. 20) is an example of the *P3* hard copy print of a bar chart for a typical residential construction.

ACTIVITY-ON-NODE DIAGRAMS

The graphics for logic diagrams within *P3* are organized for activity-on-node diagrams. In these, the nodes (rectangles in Figure 1-3a and b) are the activities and are connected by arrows, which show relationships between activities. The nodes contain possible information about the activities. The activity-on-node diagrams originated as PERT (Program Evaluation Review Technique) diagrams. PERT diagrams were developed for projects where activity duration and scope of work could not be determined with great accuracy. These were for new types of projects that had never been built before. The full extent of the work, or the relationships between the activities, was not understood. PERT diagramming was developed by the Special Projects Office of the Navy Bureau of Ordnance in the late fifties and early sixties and was used in the development of the Polaris missile project.

The node or box is the activity, and the arrows connecting the boxes show the relationships between the activities. Compare Figure 1-3a and b to Figure 1-1 (p. 10) to see the difference between activity-on-node and activity-on-arrow diagramming. The activity-on-node diagram is *P3*'s pure logic diagram. Figure 1-3a and b is an example of a hard copy print of *P3*'s pure logic diagram.

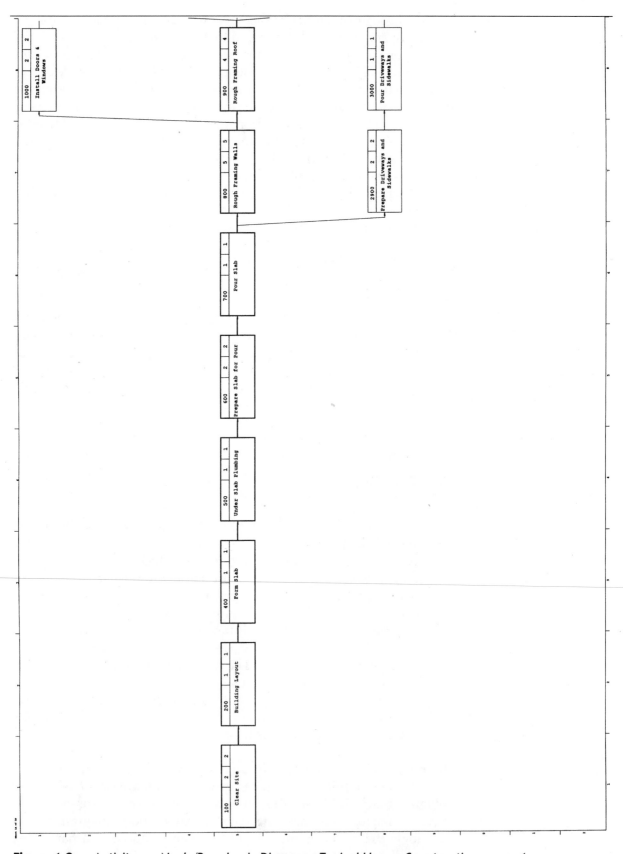

Figure 1-3a   Activity-on-Node/Pure Logic Diagram—Typical House Construction, page 1

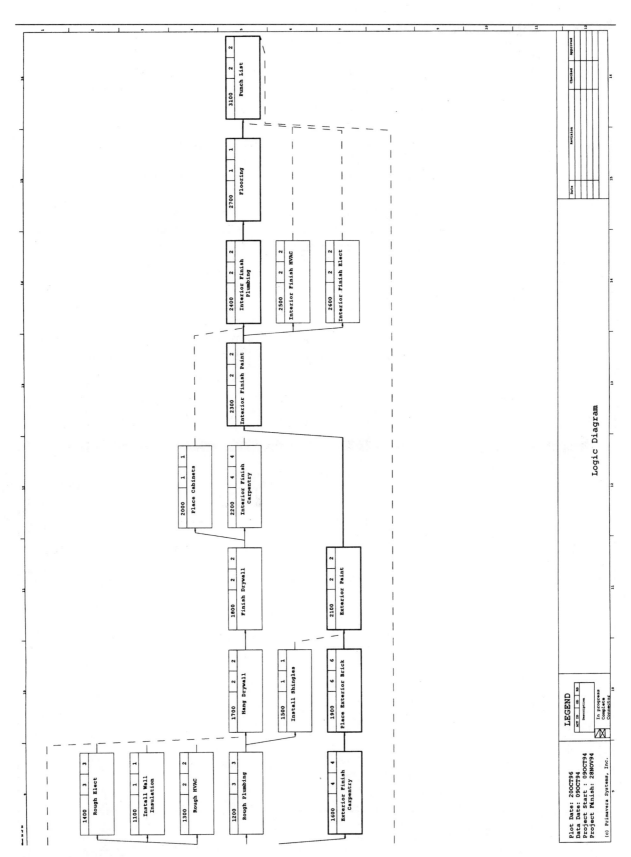

Figure 1-3b   Activity-on-Node/Pure Logic Diagram—Typical House Construction, page 2

ACTIVITY-ON-ARROW DIAGRAMS

P3 for Windows does not support activity-on-arrow notation to the same level that it supports activity-on-node notation. The activity-on-node diagram is neater and more efficient from a graphical point of view. The reason is that it is much easier to design graphics around putting information into a box, rather than associating it with a line of unknown length with which you are also trying to show logical relationships. Because of this, activity-on-arrow diagrams are falling into disuse.

The activity-on-arrow diagramming method for scheduling was developed by the E. I. du Pont de Nemours Company in the late fifties and was called the critical path method (CPM) (Figure 1-1, p. 24).

TIMESCALED LOGIC DIAGRAMS

P3's timescaled logic diagram (Figure 1-4a and b) combines the advantages of the bar (Gantt) chart and the pure logic diagramming methods

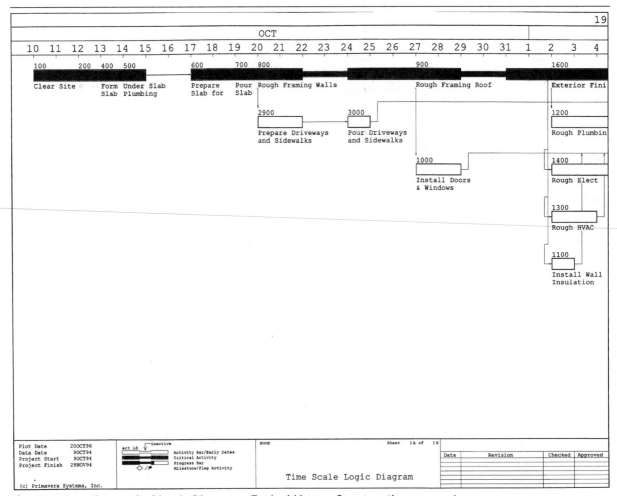

Figure 1-4a   Timescaled Logic Diagram—Typical House Construction, page 1

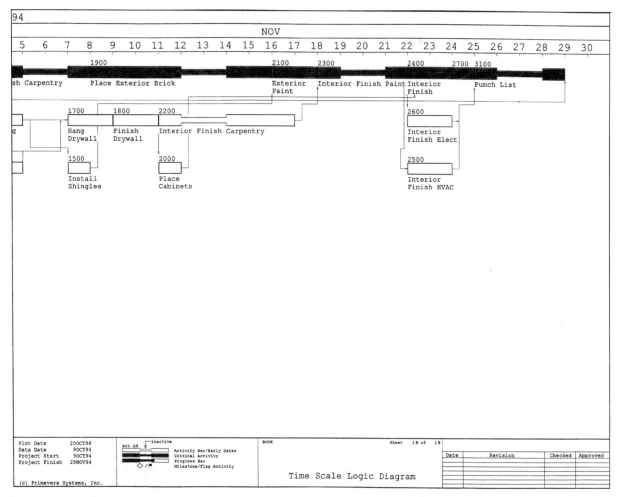

Figure 1-4b   Timescaled Logic Diagram—Typical House Construction, page 2

(PERT). Like the bar chart, it shows the activities' relationship to time (either workdays or calendar days). It also shows the relationships among activities.

The advantages of the timescaled logic diagrams are that they are easy to understand, like the bar chart, and they define the logic. For updating and documentation purposes, this is a tremendous advantage. Figure 1-4a and b is an example of a hard copy print of *P3*'s timescaled logic diagram.

TABULAR REPORTS

Sometimes a table or tabular report is the easiest way to communicate information, or to update information. Figure 1-5 is an example of a tabular report, which lists the data associated with each activity.

```
----------------------------------------------------------------------------------------------------
Southern Constructors                    PRIMAVERA PROJECT PLANNER

REPORT DATE 20OCT96  RUN NO.   19                           START DATE  9OCT94  FIN DATE 28NOV94
               9:44
Classic Schedule Report - Sort by ES, TF                    DATA DATE   9OCT94  PAGE NO.    1
```

ACTIVITY ID	ORIG DUR	REM DUR	%	ACTIVITY DESCRIPTION	EARLY START	EARLY FINISH	LATE START	LATE FINISH	TOTAL FLOAT
100	2	2	0	Clear Site	10OCT94	11OCT94	10OCT94	11OCT94	0
200	1	1	0	Building Layout	12OCT94	12OCT94	12OCT94	12OCT94	0
400	1	1	0	Form Slab	13OCT94	13OCT94	13OCT94	13OCT94	0
500	1	1	0	Under Slab Plumbing	14OCT94	14OCT94	14OCT94	14OCT94	0
600	2	2	0	Prepare Slab for Pour	17OCT94	18OCT94	17OCT94	18OCT94	0
700	1	1	0	Pour Slab	19OCT94	19OCT94	19OCT94	19OCT94	0
800	5	5	0	Rough Framing Walls	20OCT94	26OCT94	20OCT94	26OCT94	0
2900	2	2	0	Prepare Driveways and Sidewalks	20OCT94	21OCT94	24NOV94	25NOV94	25
3000	1	1	0	Pour Driveways and Sidewalks	24OCT94	24OCT94	28NOV94	28NOV94	25
900	4	4	0	Rough Framing Roof	27OCT94	1NOV94	27OCT94	1NOV94	0
1000	2	2	0	Install Doors & Windows	27OCT94	28OCT94	4NOV94	7NOV94	6
1600	4	4	0	Exterior Finish Carpentry	2NOV94	7NOV94	2NOV94	7NOV94	0
1200	3	3	0	Rough Plumbing	2NOV94	4NOV94	3NOV94	7NOV94	1
1400	3	3	0	Rough Elect	2NOV94	4NOV94	3NOV94	7NOV94	1
1300	2	2	0	Rough HVAC	2NOV94	3NOV94	4NOV94	7NOV94	2
1100	1	1	0	Install Wall Insulation	2NOV94	2NOV94	7NOV94	7NOV94	3
1700	2	2	0	Hang Drywall	7NOV94	8NOV94	8NOV94	9NOV94	1
1500	1	1	0	Install Shingles	7NOV94	7NOV94	15NOV94	15NOV94	6
1900	6	6	0	Place Exterior Brick	8NOV94	15NOV94	8NOV94	15NOV94	0
1800	2	2	0	Finish Drywall	9NOV94	10NOV94	10NOV94	11NOV94	1
2200	4	4	0	Interior Finish Carpentry	11NOV94	16NOV94	14NOV94	17NOV94	1
2000	1	1	0	Place Cabinets	11NOV94	11NOV94	21NOV94	21NOV94	6
2100	2	2	0	Exterior Paint	16NOV94	17NOV94	16NOV94	17NOV94	0
2300	2	2	0	Interior Finish Paint	18NOV94	21NOV94	18NOV94	21NOV94	0
2400	2	2	0	Interior Finish Plumbing	22NOV94	23NOV94	22NOV94	23NOV94	0
2500	2	2	0	Interior Finish HVAC	22NOV94	23NOV94	23NOV94	24NOV94	1
2600	2	2	0	Interior Finish Elect	22NOV94	23NOV94	23NOV94	24NOV94	1
2700	1	1	0	Flooring	24NOV94	24NOV94	24NOV94	24NOV94	0
3100	2	2	0	Punch List	25NOV94	28NOV94	25NOV94	28NOV94	0

Figure 1-5 Tabular Report—Typical House Construction

EXAMPLE PROBLEM: Getting Ready for Work

Table 1-1 is a list of activities for getting ready for work. Figures 1-6 to 1-8 are completed samples of a bar chart, pure logic diagram, and a timescaled logic diagram using the list of activities from Table 1-1.

	Activity Name	Duration (Minutes)		Activity Name	Duration (Minutes)
1.	Turn Off Alarm	1	12.	Place Underwear	1
2.	Get Out of Bed	1	13.	Place Shoes	1
3.	Remove Pajamas	1	14.	Place Shirt	1
4.	Brush Teeth	1	15.	Place Pants	1
5.	Take Shower	5	16.	Place Tie	1
6.	Wash Hair	2	17.	Take Vitamins	1
7.	Make Coffee	1	18.	Fix Cereal	1
8.	Perk Coffee	5	19.	Eat Cereal	3
9.	Drink Coffee	10	20.	Make Bed	2
10.	Dry Hair	3	21.	Leave for Work	1
11.	Comb Hair	1			

Table 1-1   Activity List—Getting Ready for Work

Activities	1	2	3	4	5	6	7	8	9	10	11	12	13	14	15	16	17	18	19	20	21	22	23	24	25	26
Turn Off Alarm	X																									
Get Out of Bed		X																								
Remove Pajamas				X																						
Brush Teeth														X												
Take Shower					X	X	X	X	X																	
Wash Hair								X	X																	
Make Coffee			X																							
Perk Coffee				X	X	X	X	X																		
Drink Coffee											X	X	X	X	X	X	X	X	X	X						
Dry Hair											X	X	X													
Comb Hair														X												
Place Underwear														X												
Place Shoes																	X									
Place Shirt																X										
Place Pants															X											
Place Tie																		X								
Take Vitamins													X													
Fix Cereal																				X						
Eat Cereal																					X	X	X			
Make Bed																								X	X	
Leave for Work																										X

Figure 1-6   Example Problem—Bar Chart Format

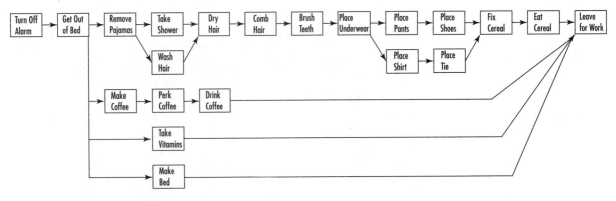

Figure 1-7   Example Problem—Pure Logic Diagram Format

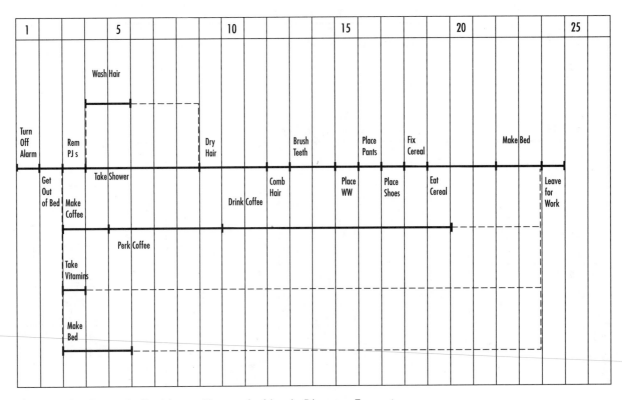

Figure 1-8   Example Problem—Timescaled Logic Diagram Format

EXERCISES

1. Building a Shed

The objective of this exercise is to organize and sequence activities. Using the tabular list of activities from Table 1-2, produce a bar chart and a pure logic diagram. Refer to the previous example problem for the format of each of these schedule types. The preparer has to assume the precedence relationships.

	Activity Name	Duration (Days)			Activity Name	Duration (Days)
1.	Clear Site	1		11.	Place Exterior Paneling	1
2.	Remove Topsoil	1		12.	Ext Trim Carpentry	2
3.	Form Slab	2		13.	Install Overhead Door	1
4.	Place Rebar/Embeds	1		14.	Rough Electrical	1
5.	Pour Slab	1		15.	Finish Electrical	1
6.	Prefab Wood Walls	2		16.	Place Shingles	2
7.	Erect Wood Walls	1		17.	Install Finish Carpentry	1
8.	Install Siding	2		18.	Place Topsoil/Grade	1
9.	Place Trusses	1		19.	Landscape	1
10.	Place Roof Sheathing	2				

Table 1-2   Activity List—Building a Shed

2. Purchasing a New Automobile

The objective of this exercise is to organize and sequence activities. Using the tabular list of activities from Table 1-3, produce a bar chart, and a pure logic diagram. Refer to the previous example problem for the format of each of these schedule types. The preparer has to assume the precedence relationships.

	Activity Name	Duration (Days)
1.	Decision—Type Car	5
2.	10 Models—Make List	1
3.	10 Models—Obtain Consumer Ratings	1
4.	10 Models—Obtain Pricing Publication	1
5.	10 Models—Talk to Vehicle Owners	1
6.	3 Models—Decision	2
7.	3 Models—Test Drive	1
8.	3 Models—Obtain Information	1
9.	3 Models—Compare Lease/Purchase Options	1
10.	1 Model—Decision	2
11.	1 Model—Negotiate Purchase Contract	1
12.	3 Institutions—Compare Financing	1
13.	Decide on Institution	1
14.	Obtain Financing	1
15.	Money for Down Payment	1
16.	Obtain Insurance	1
17.	Obtain Tag	1
18.	Drive Away with Purchased Vehicle	1

Table 1-3   Activity List—Purchasing a New Automobile

2 Rough Diagram Preparation—An Overview

Objectives

Upon completion of this chapter, the reader should be able to:

- Enumerate project phases
- Decide which schedule format to use
- Determine schedule information needed
- Run schedule meetings
- Estimate activity durations

ROUGH DIAGRAM

The four steps of project control are *planning, scheduling, monitoring,* and *controlling.* The very first step of planning is preparation of a rough diagram. A rough diagram builds the project on paper. It defines activities and relationships (logic) between them and the time to complete them. Once accepted, developed, and refined, the rough diagram becomes the project plan. Durations and resources are then applied in scheduling. It is during the rough diagram phase of planning that the approach to placing the project takes shape. Decisions are made about construction and management.

Construction decisions concern:

- Construction methods
- Flow of materials
- Prefab vs. on-site assembly of materials
- Types of construction equipment
- Crew size and balance
- Productivity
- Subcontractor definition

Management decisions include:

- Work breakdown structure (WBS)
- Division of responsibility

- Division of authority
- Software to be used
- Level and distribution of reports
- Interface with other functions (payroll, accounting, etc.)

APPROACHES TO ROUGH DIAGRAMMING

The four most common approaches to handling the rough diagramming meeting are using a tape recorder, making a list of activities, drawing the actual diagram, and using a software integrator.

Tape Recorder

Some schedulers prefer to tape-record the initial meeting of the key project participants. These persons verbally walk through the project from beginning to end, identifying activities and their sequence and interrelationships. The scheduler then prepares the graphical schedule manually or by computer. The group meets again to review the first rough activity diagram. When agreement is reached on activities and logic, the scheduler obtains from the owner, architect/engineer, subcontractors, equipment suppliers, and suppliers of long-lead items input that will affect the schedule. This information is incorporated into the rough diagram and reviewed. If there are conflicts between that of the contractor's team and the information provided by the other parties, some negotiation or conflict resolution is in order. It is better to resolve the differences in opinion before the project starts. When all conflicts are resolved, the plan is accepted (signed off). Then the rough diagram is ready for scheduling, the phase when durations, and possibly resources, are applied to the planned activities. During the scheduling stage, a certain amount of fine tuning is always necessary. Rough diagram planning must be reconciled with the estimate (cost and resource constraints) and time constraints (usually imposed by the contract documents).

List of Activities

In the second approach to the preparation of the rough diagram, the scheduler in the initial meeting makes a written list of activities instead of using a recording, remarking on the relationships between activities and other pertinent information.

Sketch for the Diagram

In the third approach, during the initial meeting, the scheduler simply sketches the diagram on a large, usually continuous, sheet of paper showing the activity names and logic. The disadvantage of this approach is that sometimes the scheduler gets so bogged down in diagramming that the flow of conversation and the effective use of meeting time is diminished. If the meeting has more than two people present, this method is usually not efficient.

Combination   A combination of listing the activities and sketching a diagram is popular. After identifying the activities and discussing their relationships in the initial meeting, the planners write the activity names on small stick-on notes. The notes can be moved around on a large sheet of paper by members of the scheduling team in refining the original logic to the accepted plan.

Reconciliation   In the first three approaches to creating a rough diagram, no reconciliation to the estimate is made until *after* the rough diagramming stage. Reconciliation is done during scheduling, when durations and resources are assigned to project activities. All costs input to the estimate have to be input a second time if they are to be used in the schedule. This extra step of having to reenter all the information prohibits many contractors from using labor or resource profiles and detailed cost analysis by activity.

Integrator

The fourth approach—using a software integrator—to produce the rough diagram differs fundamentally from the others. This is the most efficient method for a number of reasons. The estimate is the document that defines the bottom line for a project—the budget for resources and costs. Unfortunately the estimator and the scheduler look at the world differently. The estimator's goal is to turn out the greatest number of estimates with the best accuracy for the least cost. He or she will probably only get one out of every six to ten jobs estimated. The estimator is more concerned with defining cost rather than providing a breakdown of cost for sequencing purposes. For example, an estimate will have one entry for concrete grade beams of a particular type with breakout by the different materials making up the grade beam (concrete, formwork, rebar, embeds, finishing, waterproofing, rubbing, etc.). This is the most efficient way to estimate cost. The scheduler, however, is not interested in looking at the work that way. He or she needs to relate the work usually by type of crew (carpentry, finishing, etc.) to individual pours or groups of pours. Since labor crews are usually the resource to be maximized and kept working efficiently, the scheduler needs information sorted by type of crew.

Using an estimate/schedule software integrator for preparing the rough diagram avoids inputting the same information twice. Many of the largest selling estimating programs provide an interface program for the estimate to be "dumped" into the estimate/schedule integrator. While the information is in the integrator, individual line entries from the estimate can be either combined or split and attached to a named activity. The strength of the integrator is its ability to name the activities, then move any associated costs and resources in the estimate to the named activity. The named activity will retain the cost and resources when it is dumped to the scheduling software package. This exercise assures a correlation between the data in the estimate and that of the schedule. The beauty of using the integrator is that all the information from the estimate does not have to be reinput into the schedule. It simply has to be distributed when naming the activities. This simplification process helps ensure the contractor will actually use the management tools available through software packages such as *Primavera* to control cost and resource functions.

Markup of the Estimate   In using a software integrator to produce the rough diagram, the first step is for the project team to manually mark up the hard copy print of the estimate and define the activities. In the integrator, line entries from the estimate are either combined or split to form activities. When complete, the computer estimate file is "dumped" into the scheduling software package. All associated cost and resource information identified by activity are brought into the scheduling software package automatically when the dump occurs. The scheduling software package establishes relationships between the activities and calculates the schedule. This version of the rough schedule is presented to all key project participants for fine tuning, as are rough diagrams in the first three approaches. This approach to producing the rough diagram offers the advantage of reconciling the estimate and the schedule without having to input information twice. But care must be taken not to let reliance on the computer stifle the creativity and freedom of the planning stage.

ROUGH DIAGRAM PREPARATION

Stages

There are five main steps in developing a rough diagram. The stages for preparation of a construction-only schedule are as follows:

Contractor's Initial Meeting   The contractor's project team members have an initial planning meeting. At this time the estimator, project manager, possibly the superintendent, and the person putting the schedule together discuss the entire project from beginning to end. The team builds the project on paper, starting from the estimate. Either a tape recorder is used to record the conversation or the scheduler takes notes.

Rough Diagram   The scheduler uses the information from the initial meeting to draw a rough diagram on paper showing interrelationships between the activities. Some schedulers prefer to go straight to scheduling software at this stage rather than manually drawing the rough diagram on paper. Those who have software with an integrator function can use it to estimate the cost and resources needed for each activity at this stage.

Review   The entire project team must "buy off" the rough diagram so that it becomes "our" schedule. It is critical for them to agree as a team to the concept, methods, and procedures to be used in the construction of the project.

Major Subcontractors and Suppliers   Input from major subcontractors and suppliers of long-lead material and/or equipment items is critical. The contractor's plan must be ready for communication with these parties. With the typically used contract documents, the general contractor or construction manager has overall responsibility for coordination and scheduling of work. It is up to the contractor to coordinate all parties' work and to solve disputes. Input is usually required of the subcontractor in the contractor-subcontractor contract. Through discussions with the subcontractor/suppliers, the contractor modifies the plan to reconcile differences. Then the subcontractors/suppliers accept the contractor's plan as the project plan. Contractors need to make sure they have obtained information from any outside party that can impact the schedule.

Project Schedule   Once the contractor has reviewed the revised rough schedule with all subcontractors/suppliers, it becomes the official project schedule.

Information Gathering

Estimate   During the prebid or estimating phase of the project life cycle, the estimate becomes the focal point in the gathering of information relating to the project. Information must be gathered from many sources and incorporated into the estimate. The following is information that must be gathered and used:

- Owner time constraints and other input
- Scope definition
- Building methods and procedures to be used
- Productivity rates, crew balances, and crew sizes
- Labor availability and wage rates
- Construction equipment to be used
- Construction equipment availability and use rates
- Material availability and prices
- Subcontractor availability and prices
- Fabricator availability and prices
- Project organizational structure
- Rough preliminary schedule

- Temporary facilities requirements
- Permit, test requirements
- Tax requirements
- Insurance requirements

In-Depth Planning   If a contractor's bid wins the contract for a project, the estimate is used to generate the schedule. But now that the project is a real live job rather than just a proposal, more in-depth planning can take place. A logic diagram rather than a "quick-and-dirty" bar chart is used to fine tune the information already gathered in the estimating stage.

Parties   The schedule becomes the focal point of project information that is received from the following parties: project manager, superintendent, foremen, estimator, subcontractors, fabricators, suppliers/vendors, owner, and architect/engineer.

Meetings

Agenda   To keep meetings from degenerating into a general waste of time requires certain steps. First, distribute an agenda in advance of the meeting. It should document the date, time, and location of the meeting, information to be covered, what each participant should bring to the meeting, and points each participant should know before the meeting begins. Each member should already be familiar with the contract documents and the major parameters. It wastes everyone else's time if some project members come to the meeting unprepared.

Action Items   The immediate product of a scheduling meeting is a list of agreed-upon action items that need to be accomplished before the next meeting. The list should designate the person(s) responsible for each item and set a date for the next meeting. The way we schedule and carry out scheduling the schedule says a lot about our management abilities in scheduling and carrying out the project.

Communications

Project Schedule   It is frequently commented that the act of preparing the rough diagram is the most valuable part of the entire scheduling process. The reason is communication: everyone is aware of the "plan"; everyone knows everyone else's point of view regarding project goals. The project schedule that is accepted after all discussion can be used by all parties to organize their work.

Team   If all key parties feel that they are a part of the "team" with a chance to contribute their information and ideas, the project is more likely to succeed. Some of the best suggestions for improving project efficiency and lopping time off the schedule may come from unexpected sources. Communication leads to evaluation and reevaluation of the plan to streamline construction.

Brainstorming Alternatives   When a decision has to be made to choose a particular system or technique for some portion of the work, brainstorming is a good way to look at the alternatives. For example, to decide on the most effective way to pour the concrete columns in a multistory building, the group would start by simply tossing out the question during a scheduling meeting by asking: What is the most effective way to pour the concrete columns on this project? Each member of the group would reply by tossing out the first solution that comes to mind, no matter how ridiculous the solution might immediately sound. Try to get as many possible solutions before beginning any analysis of the solutions.

The object of this brainstorming exercise is to break established patterns or mind-sets and look for new and better ways to construct the project. Many times we get stuck in a rut doing things in the same old way simply because "that's the way we have always done it and it has always worked." Only by looking for new and more effective ways to "build the mouse trap" can we improve construction operation.

WHAT IS A PROJECT?

Profit Center

Most construction companies look upon each project as a profit center. The project represents a unique set of activities or actions that must take place to produce a unique product.

Schedule

Start   The lump-sum project is divided into actions that precede and follow the signing of the contract. Precontract activities relate to bidding. The contractor wants to invest as little as possible in the project until the client has signed a contract. Once the contract is formalized, detailed planning and scheduling proceed.

Ending   At the end of the project, the owner takes possession of the structure to use for its intended purpose, takes responsibility for utilities and insurances, and submits final payment to the contractor.

Criteria for Success

Each project has a definable start and end and a unique set of characteristics and activities that must be accomplished to fulfill a contract. Each project is judged as a success or failure in terms of usually preset criteria. The following are a list of criteria on which a project is judged:

Cost
Did the project come in under budget?
Did the project make any money?
Were changes controlled through change orders?

Time
Did the project come in on time?
Were changes controlled through change orders?

Quality
Did the project meet the requirements of the contract documents?
Were both client and employee satisfaction achieved?

Safety
What was the project safety record?

Resource allocation
Was this project an effective use of company resources?

Uniqueness of Each Project

No two construction projects are ever the same. Even if the plans and specifications are almost identical, building sites differ. The amount of time required to build different projects differs, and therefore projects are subject to different weather conditions. And the project team that builds the project is composed differently depending on the type and size of the project.

Experience

Despite the differences between projects, however, each project is a learning experience. Lessons learned about productivity, project layout, flow of materials, crew sizing, and communications can help improve the quality of estimates and schedules in future projects.

PROJECT PHASES

Time and the control of time relates to all phases of the construction project life cycle. The construction project life cycle can be broken into seven phases:

1. Conceptual design
2. Detailed design

3. Bidding
4. Construction
5. Commissioning
6. Closeout
7. Maintenance

Conceptual Design

In the conceptual design phase the design professional (architect/engineer) defines the owner's need in a conceptual project-scope document. This document defines the project in enough detail so that detail design can begin. The conceptual design information needed varies by type of project. An office building differs from a refinery, for example, but the general idea is the same. The conceptual design of a new building would contain sketches of the following:

- Site
- Footprint of the building on site
- General floor plan of the building by floor
- Major wall sections
- Major elevations
- General definition of traffic flows
- Major environmental considerations such as type of HVAC system
- Conceptual estimate (must include a projection of project duration)

Evaluating Alternative Designs   The conceptual design usually involves evaluating many different design scenarios from the standpoint of building cost, maintenance cost, and appearance. These cost studies play "What-if?" games to pick the best approach. Only sketches are drawn and general concepts defined at this stage. The great expense of producing detailed documents is not incurred until all major concepts have been agreed upon. A widely used rule of thumb states that 80 to 90 percent of the project's costs are fixed when the conceptual design is complete. This percentage is fixed before the detailed plans and specifications are even started and before the contractor has had a chance to bid on the project. The reason these costs are fixed is that by this stage the following cost-determining characteristics of the project have already been defined:

- Size of the project (number of square feet, number of floors)
- Type of building system (steel vs. concrete building, flat vs. waffle slab)
- General arrangement showing bathrooms etc.
- Geographical location

Economic Viability   The conceptual design phase is where the global, high-dollar decisions about the project are generally defined and refined. These ideas are estimated and compared to the owner's budget and calculations of return on invested capital. The conceptual design phase is when decisions about the economic viability of the project are made, before the owner invests in detailed plans and specifications and concrete. When participants believe that a viable project is attainable, detailed design can begin.

Detailed Design

Detailed design involves applying the broad concepts defined in the conceptual design to produce detailed plans and specifications and the rest of the contract documents. In order to be able to make a lump-sum bid specifying cost value and duration in days, a contractor has to know the scope of the project. The purpose of the detailed design is to produce the contract documents in enough detail such that they can be used for three primary reasons: bidding the project, building the project, and settling any claims. A clear, concise set of contract documents is a tremendous asset to the constructor during bidding, construction, and closeout phases of the project.

Bidding

In the bidding phase of the project life cycle the constructor prepares the estimate based on the contract documents prepared in the detailed design phase. The constructor quantifies the project in terms of material quantities, productivity rates, worker hours, wage rates, material dollars, subcontractor dollars, overheads, and indirects. A rough schedule, usually a bar chart, is prepared for use in the bidding process.

Construction

During building, the constructor marshals at the job site all the management, expertise, manpower, materials, equipment, subcontractors, temporary facilities, and financial wherewithal necessary to construct the project according to the contract documents. The goal is to complete the project within cost and time constraints in order to make money.

Commissioning

Commissioning is the process of testing and starting up systems. It includes the final inspection and preparation of a punch list of remaining items to be completed, modified, or repaired before final acceptance by the owner.

Closeout

The closing-out phase in the project life cycle involves completing the paperwork necessary for a contractor to receive final payment and be released from the project. Documents include:

- Affidavit of release of liens from suppliers, vendors, and subcontractors
- Maintenance/owner's manuals for equipment and systems
- As-built drawings
- Request for final payment/lien waiver

Maintenance

Construction projects are designed for a 20-, 30-, 50-year, or possibly even longer productive life. To last that long, they require continual maintenance. Even on a project with a 50-year anticipated life cycle, the roof may only have a 20-year anticipated life. The elevator and the HVAC system will require constant attention. Some contract proposal forms require from the contractor not only a bid for construction of the building, but also maintenance for a certain number of years of the building's life. However, maintenance is usually an ownership function.

PROJECT TEAM

No construction project of any size is ever built by an individual. It is constructed by a team. The concept of team crosses company boundaries. The makeup of the team preparing the rough diagram depends on the construction life cycle phases to be controlled with the schedule. Members of the lump-sum, construction-only scheduling team include the following:

Contractor's organization
  Project team
- Project manager
- Superintendent
- Foremen/craftsmen

  Home office support
- Estimating
- Scheduling
- Cost
- Purchasing
- Accounting
- Equipment management

Subcontractor's organization
Field team
- Project manager
- Foremen/craftsmen

Home office support
- Estimating
- Scheduling
- Cost
- Purchasing
- Accounting
- Equipment management

Vendors'/suppliers' organizations
Fabrication/storage yard team
- Project manager
- Foremen/craftsmen

Home office support
- Estimating
- Scheduling
- Cost
- Purchasing
- Accounting
- Equipment management

Owner
Construction representative

Architect/engineer
Project representative
- Engineering consultants

Government
Inspection/code enforcement
Labor law enforcement
Safety law enforcement
Tax enforcement

TEAMWORK

Successful projects require a mind-set of teamwork and problem solving. The goal has to be to work together to settle disputes at the lowest level for the mutual benefit of the entire group rather than fighting, finger pointing, and litigation. The modern concepts of "partnering" and the "team" cross company boundaries. Looking at the project from the perspective of a team rather than one dominated by company boundaries helps to control time and project duration. The contractor is not solely in control of all the variables required for successful project completion. Only through teamwork can the project schedule be met successfully.

DESIGN-BID-BUILD CONSTRUCTION

Consecutive Order

Design-bid-build construction carries out the construction life cycle phases in consecutive order rather than concurrently. Each phase is completed before the next begins. Building the project with this constraint takes longer. Since each step of the process is defined, finished, and usually paid for before the next begins and since this process is usually based on lump-sum contracts, the owner transfers the risk of cost overruns to other parties. The owner knows the cost of each phase before committing to pay for it. Another advantage to the owner is that, since the project can be canceled at any time, the owner is only at risk for the work that has been released.

No Input   Typically with a lump-sum contract (design-bid-build construction), the constructor's contract includes the construction, commissioning, and closeout phases of the construction life cycle but no input during the conceptual or detailed design phases. This is unfortunate, since who knows more about minimizing construction cost by efficient design, constructibility, materials selection, the use of prefabricated materials than someone who is involved in the actual building process every day?

Steps in the Lump-Sum Project   The following is a sequential listing of the steps typically followed with lump-sum construction in the United States:

1. Owner determines a need.
2. Owner contacts architect/engineer for conceptual design.
3. Conceptual design is completed.
4. Owner approves conceptual design or sends it back for modification.
5. Owner selects architect/engineer for detailed design contract documents.
6. Detailed design is completed.
7. Owner approves detailed design or sends it back for modification.
8. Owner puts contract documents out for bid.
9. Bids are received from contractors and negotiated.
10. Owner and contractor sign contract.
11. Construction proceeds.

12. Project is commissioned.
13. Project is closed out.

Negotiated, Fast-Track Construction

Contractor's Input in Planning

With a construction-management, fast-track-type project, the constructor is typically involved in all construction life cycle phases except possibly maintenance. The constructor has a negotiated contract with the owner and acts as the owner's agent in a fiduciary relationship. He or she is looking out for the owner's best interest. The constructor will have input into defining the owner's need during both the conceptual and detailed design phases primarily concerning minimizing construction cost by efficient design, constructibility, materials selection, and the use of prefabricated materials while the critical decisions about these factors are being made. The constructor will be the party producing the conceptual estimate and cost studies of different design scenarios. Having a member of the design team who is a cost-conscious, knowledgeable constructor can be a tremendous advantage to the owner in producing a successful project.

Steps in Fast-Track Construction

The following sequence of steps is typical of fast-track construction in the United States.

1. Owner has a need.
2. Owner contacts construction manager for services, which include control of design, estimating, and scheduling.
3. Construction manager works with architect/engineer to produce conceptual design.
4. Construction manager estimates costs to fine tune conceptual design until it meets the owner's return-on-investment requirements.
5. Owner/construction manager approves conceptual design.
6. Owner/construction manager contracts architect/engineer for detailed design contract documents, then breaks down design into packages or phases that can be completed and put out for construction bid.
7. Owner/construction manager approves detailed design one package at a time.
8. Owner/construction manager puts contract documents out for bid, one package at a time.
9. Owner and successful contractor sign contract for single package.

10. Owner/construction manager bring all the packages through the steps of detail design, bid, award, and construction.

11. Construction progresses simultaneously with further detail design. The construction manager monitors the schedule, controls multiple contractors, and approves pay requests. The construction manager acts as the general contractor at the job site.

12. Project commissioning is usually handled one package or system at a time.

13. Project closeout is handled by the construction manager in much the same way as a general contractor would handle it.

Concurrent Steps   With the fast-track approach to construction services, the owner assumes more of the risk, since construction is proceeding before the design documents are completed. The construction manager is the owner's agent and therefore typically does not sign a lump-sum contract with the owner. The real advantage to the owner is the time savings of producing the detailed design and construction concurrently rather than consecutively. This approach can cut in half the overall time needed to bring a project "on line," thus saving time and money.

Use of Schedules

A critical part of any planning effort is deciding how the schedule will be used to control a project. The following is a list of criteria to be defined before the rough diagram is prepared:

- Type of project
- Purpose of the schedule
- Software requirements
- Parties involved: owner, construction manager, architect/engineer, contractors, major subcontractors, minor subcontractors and sub-subcontractors, suppliers of long-lead items
- Authority/responsibility of involved parties (Who is "keeper of the schedule"?); responsibility for resolving disputers and interferences
- Needs of the involved parties from a scheduling point of view
- Type and depth of reports to be provided
- Schedule update requirements (What is the time frame for providing updated information, in what format, and who is to provide the information?)
- Project change requirements
- Resources to be controlled: labor, materials, subcontractors, construction equipment, field indirects
- Cash-flow requirements
- Payment (schedule of values) requirements

TIME UNITS

Days vs. Hours

The Day   The time unit used for most construction schedules is days. If days are chosen, the schedule can then be summarized in terms of weeks, months, quarters, or years. If a larger unit such as weeks is chosen as the base unit, daily detail is not shown.

In the typical construction schedule, with days as the unit of measure, when activities are being defined no block of work is assigned a duration of less than a day to complete. If it takes less than a day, it is either rounded to a day or combined with another task/activity.

The Hour   Sometimes the hour is a more convenient unit than the day. This is the usual unit of choice for a "turnaround" schedule as used in a completed paper mill, chemical plant, refinery, or other facility in operation. The facility must be brought down or "off line" for maintenance or to add or modify plant systems. Typically, the owner loses thousands of dollars per day while the plant is out of operation. Making the modifications as quickly as possible and getting the plant going again is essential to the owner. Each hour is critical.

CONCURRENT RATHER THAN CONSECUTIVE LOGIC

Besides trying to improve efficiencies, schedulers also need to improve the logic of activity interrelationships and sequences when putting the rough diagram together. Instead of having just one thing at a time happen at the job site, as many things/crews/work functions need to be happening as possible without interfering with each other and without risking exposure or damage by project components being in place too early. What is the shortest and most efficient way to accomplish the work?

ESTIMATING ACTIVITY DURATION

The duration of an activity is a function of the quantity of work to be done by the activity and the rate of production at which the work can be accomplished. The formula is:

$$\text{Activity Duration} = \frac{\text{Quantity of Work}}{\text{Productivity Rate}}$$

For example, a masonry activity with a quantity of work of 10,000 regular masonry blocks to be placed by a planned crew of three masons, two laborers, and a mixer can place a productivity rate of 800 blocks per day. Thus, duration = 10,000 blocks/800 blocks per day = 12.5 or 13 days.

RESOURCE AVAILABILITY

Critical to estimating activity durations is the availability of the resources needed, including labor, materials, equipment, subcontractors, and suppliers/fabricators.

Driving Resources

Most activities have certain driving resources that control activity duration. For example, in the masonry activity mentioned in the previous paragraph, the three masons are the driving resource. Their craft determines activity duration. Once the ratio of laborers and mixers has reached maximum efficiency for three masons, no matter how many more laborers and mixers are added, the performance of the three masons will not be increased.

Reduction of Duration

There are three ways to reduce the duration of 13 days for the masonry activity mentioned above. First, increase the productivity rate of the three masons to better than 800 blocks per day. This could be accomplished by an improved method or system for placing the blocks. The second way would be to add more masons, with enough laborers and mixers to ensure their full production. The third way would be to have the masons work overtime (more hours per day and/or more days per week).

Extended Scheduled Overtime

Extended scheduled overtime can have a negative impact because of lost productivity. Also the cost of overtime pay should be considered.

As a rule, the more resources that are available to put an activity in place, the less time it takes to place the activity. If any of the resources necessary to place the activity are missing or are not handled properly, the activity duration calculation is impacted.

QUANTITY OF WORK

Project Scope

The contract documents define the scope of the project. Using the plans and specifications, the estimators survey quantities of materials necessary to complete the project. The estimate may not provide a bill of materials in enough detail to buy all materials for the project, but it will provide enough detail to bid the project.

Quantity Survey

In producing a quantity survey, the estimators organize the takeoff by type of work. Building contractors usually use the Construction Specifications Institute's (CSI) cost code structure. The estimators assign work definition by spec division, by phase, and then by the item or unit of work within the phase.

Contractors usually produce a quantity survey only for work to be completed by the contractor's own forces. For work that is to be subcontracted, contractors usually depend on the market for the best price.

PRODUCTIVITY RATE

CSI Format

The contractor organizes the code of accounts and quantity survey into identifiable areas or phases, usually organized according to CSI format, that each type of crew will accomplish. The major CSI phase headings are:

01000	General Requirements	09000	Finishes
02000	Sitework	10000	Specialties
03000	Concrete	11000	Equipment
04000	Masonry	12000	Furnishings
05000	Metals	13000	Special Construction
06000	Wood and Plastics	14000	Conveying Systems
07000	Thermal and Moisture Protection	15000	Mechanical
08000	Doors and Windows	16000	Electrical

The cost code structure is organized by craft designation so that work boundaries are understood by all parties involved.

Rate by Activity

The quantity survey defines the amount of materials for a particular type of work, such as the square feet of a slab to be finished. The next step is to assign a productivity rate to determine the duration and resources necessary for the activity. The rate is the quantity of work accomplished per work hour (e.g., 30 square feet of concrete finished per work hour, or 30 SF/WH), or quantity of work accomplished per crew hour (e.g., 120 square feet/crew hour based on a four-person crew, or 120 SF/CH). The rate per man-hour (work hour) has to be adjusted for the size of the crew in order to determine duration.

Sources of Productivity Information

Sometimes if no formal estimate is prepared, the scheduler may have to come up with the productivity rate. The following sources may be useful.

- Company records on producing the same type of work on previous projects
- Published information in reference texts
- Observation and measurement of performance of the work as it is being put in place on another project
- Qualified expert opinion (Ask someone who is knowledgeable about this type of work.)

FACTORS AFFECTING PRODUCTIVITY

Variability of Productivity Rate

The amount of time it takes to accomplish a unit of work can vary appreciably for the same type of work from project to project and is the reason for unpredictability of construction labor costs. This variability in performance results from differences in communications, supervision/proper preplanning, layout of the work, crew balance, skill/craftsmanship, mental attitude of workers, purchasing practices, working conditions, continuously scheduled overtime, safety practices, work rules, and availability of work.

Communications

Good communications—in writing—cannot be stressed enough when discussing productivity. It is common practice for the superintendent to communicate orally with the foreman about the work to be accomplished, referring to the plans and specs and shop drawings. The foreman in turn communicates orally with craftspersons the direction, methods, and layout for accomplishing the work. This chain of oral communication leads to a great deal of rework justified by the comment "I built what I thought you wanted."

Many contractors now put communications in writing, using 10-day look-ahead and job-assignment sheets for in-depth planning and scheduling. An emphasis on sketches and drawings further reduces dependency on oral communications. The improved documentation improves productivity.

Supervision and Preplanning

Good supervision and proper preplanning also improve productivity. Commitment from top management to cost and schedule control leads to improvement of training, safety, scheduling, estimating, purchasing, and the emphasis on quality on all projects.

Efficient Layout of the Work

Of critical importance in the proper execution of the work is efficient layout:

- Storage of materials so the materials can be located quickly and easily when necessary
- Minimal handling of materials (on many work items, more time is spent handling the materials than actually putting the materials in place)
- Use of the right equipment for the job
- Efficient access to tools, utilities, drinking water, and sanitary facilities
- Where possible, work at waist level to reduce unnecessary motion and fatigue

Proper Crew Balance

A construction crew has a proper balance of workers for various aspects of each activity. In a masonry crew for example, the ratio of workers mixing the mortar to laborers transporting the mortar and stacking the blocks for placement and to masons actually placing the blocks must be balanced properly to accomplish the work efficiently. The ratio of highly paid skilled workers to lower paid helpers is important for peak efficiency.

Skill/Craftsmanship

Training of construction craftspersons to improve their skills is necessary because of ever-changing technology. Improvements in construction equipment are changing the way work is accomplished and improving productivity. Since a construction company's primary asset is its employees, nurturing and training them to increase performance is a wise investment.

Mental Attitude of Workers

People are not machines. The mental attitude of workers seriously affects productivity. Management should strive to promote workers' feelings of pride in the company, faith in a secure future, confidence in good management, the company's respect for employees, the company's regard for employee opinions, company growth, and the potential for individual growth. When employees get satisfaction from working with a company and feel their individual needs are being met, they will be more productive at the job site.

Purchasing Practices

A construction company's purchasing practices can have a tremendous impact on job site productivity. The use of prefabricated assemblies, or preassembled units, where labor is taken off site to a manufacturing-type controlled environment, can have a large impact on worker hours spent at the job site. Lower average wage rate, craft skills required, availability of specialized tools and jigs, working under a roof in a controlled environment, and availability of local labor can all be great advantages. A scheduling advantage is that fabrication can be concurrent with work at the job site, reducing the overall time for the project to be completed. Prefabricating materials in an environment more controlled than the job site can also have a radical impact on cost.

Working Conditions

In construction, the worker is typically exposed to the elements. Extremes can slow productivity. Conditions that influence job site productivity include some that cannot be controlled—heat, cold, rain, humidity, dust, wind, odor, and acts of God—and some that can be controlled, such as noise, climbing up or down, bending low or reaching high, and number of people on site.

Continuously Scheduled Overtime

Many studies have shown that continuously scheduled overtime has a disastrous impact on productivity. The construction industry's Business Round Table study titled "Cost Effectiveness Study C-3," (November 1980) shows that in working 60-hour weeks for nine straight weeks, in the ninth week the same amount of work is accomplished as in a normal 40-hour workweek without the overtime, which means that paying for the extra 20 hours accomplished nothing. Compounding this false economy is the excessive cost of paying overtime in excess of 40 hours per week. Spot overtime can be effective, but continuously scheduled overtime is decidedly ineffective from an economics point of view.

Safety Practices

Safety practices have both direct and indirect influences on productivity. The direct influence is that when someone is hurt, work at the job site usually stops or is at least impacted by the disturbance. It is the topic of conversation. Everyone is concerned. The company has lost the services of an employee at least temporarily. The loss also affects crew balance and hence productivity of a work unit.

The indirect impact relates to the effect accidents have on workers' morale and feelings of personal safety and security. Accidents affect the way employees think of the quality of the company's management and ability to manage the job site. Accidents also affect the employees' pride in the company and desire to stay with the company.

Work Rules

Sometimes labor constraints have a negative impact on productivity. Union constraints that may limit management's ability to organize for maximum productivity include jurisdictional disputes, production guidelines, limits on time studies, limits on piece work, and limits on prefabrication.

Availability of Work

If times are good and there is plenty of construction work available, employees know that other jobs are readily available. There is not as much pressure to produce at the existing job. Conversely, if things are tight, there is much more pressure to keep the existing job.

ACCURACY OF ESTIMATING DURATION

Schedulers sometimes make subjective judgments of duration of activities rather than taking the time to refer back to the estimate and perform an actual calculation based on productivity rates and quantity of work. Since time is money, the accuracy of a contractor's schedule depends directly on the reliability of estimates of activity durations.

EXAMPLE PROBLEM: Rough Manual Logic Diagram

Table 2-1 is a list of twenty-eight activities (activity ID and description) for a house put together as an example for student use (see the wood frame house drawings in the Appendix). The rough manual logic diagram (Figure 2-1a and b) was constructed using the list provided in Table 2-1.

Act. ID	Act. Description	Act. ID	Act. Description
100	Clear Site	1500	Ext Siding
200	Building Layout	1600	Ext Finish Carpentry
300	Form/Pour Footings	1700	Hang Drywall
400	Pier Masonry	1800	Finish Drywall
500	Wood Floor System	1900	Cabinets
600	Rough Framing Walls	2000	Ext Paint
700	Rough Framing Roof	2100	Int Finish Carpentry
800	Doors & Windows	2200	Int Paint
900	Ext Wall Board	2300	Finish Plumbing
1000	Ext Wall Insulation	2400	Finish HVAC
1100	Rough Plumbing	2500	Finish Elect
1200	Rough HVAC	2600	Flooring
1300	Rough Elect	2700	Grading & Landscaping
1400	Shingles	2800	Punch List

Table 2-1   Activity List—Wood Frame House

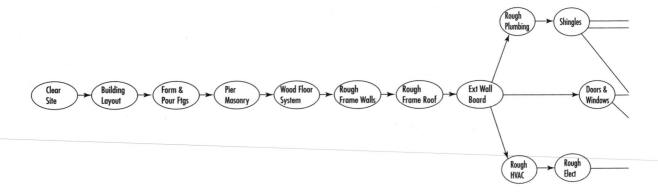

Figure 2-1a   Wood Frame House—Rough Logic Diagram, page 1

EXERCISES

1. Small Commercial Concrete Block Building

Prepare a rough manual logic diagram for the small commercial concrete block building located in the Appendix. Follow these steps:

- **a.** Prepare a list of activity descriptions (minimum of sixty activities).
- **b.** Establish activity relationships.
- **c.** Create the rough manual logic diagram.

2. Large Commercial Building

Prepare an on-screen bar chart for the large commercial building located in the Appendix. Follow these steps:

- **a.** Prepare a list of activity descriptions (minimum of 150 activities).
- **b.** Establish activity relationships.
- **c.** Create the rough manual logic diagram.

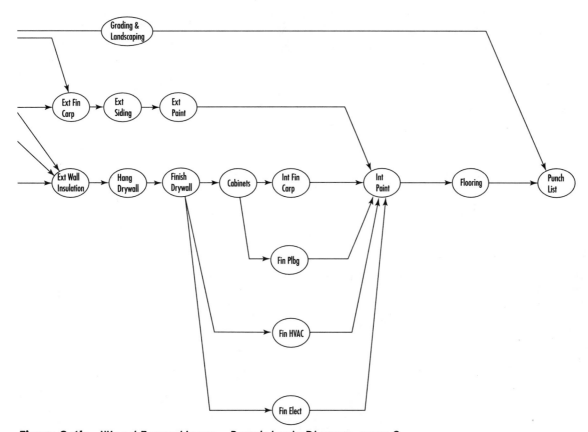

Figure 2-1b Wood Frame House—Rough Logic Diagram, page 2

Section 2

Scheduling

3

Schedule Calculations

Objectives

Upon completion of this chapter, the reader should be able to:

- Calculate forward pass
- Calculate backward pass
- Calculate total float
- Calculate free float
- Complete a data table
- Correlate ordinal and calendar days

DEFINITIONS

The mathematical calculations of a critical path method diagram follow a few simple rules. To make these calculations, the following definitions need to be understood:

Activity Construction projects are made up of a number of individual activities that must be accomplished in order to complete the project. Task activities have five specific characteristics. They consume time, consume resources, have a definable start and finish, are assignable, and are measurable.

Activity-on-Node Diagrams The nodes are the activities and are connected by arrows, which show relationships between activities. The nodes contain possible information about the activities. This information may contain any or all of the following: identification, description, original duration, remaining duration, early start, late start, early finish, late finish, total float. Since *P3* is written to support activity-on-node diagramming, this will be the diagramming method used in this book.

Node The node is the box (or other shape) at the intersection point of arrows containing activity information. Nodes contain varying amounts of information according to the requirements of the diagram. The nodes used in the diagrams in this chapter contain activity description, duration, early start, late finish, and float (see Figure 3-1).

Description	
Duration	Float
Early Start	Late Finish

Figure 3-1 Node Configuration

Critical Path The critical path is the continuous chain of activities with the longest duration; it determines the project duration.

Forward Pass The calculations for the forward pass start at the beginning of the diagram and proceed to the end; they determine early start, early finish, and project duration. All predecessor activities must be complete before an activity can start.

Early Start The earliest possible time that activity can start according to relationships assigned.

Early Finish The earliest time that an activity can finish and not prolong the project.

Backward Pass The backward pass calculations begin at the end of the diagram and go in a "backward" path to the beginning following logic constraints; they determine the late finish and late start of each activity.

Late Finish The latest time that an activity can finish and not prolong the project.

Late Start The latest time that an activity can start and not prolong the project.

Float The amount of "slack time," or time difference between the calculated duration of the activity chain and the critical path is called the float. It permits an activity to start later than its early start and not prolong the project. Float may be classified as total or free.

Total Float Total float is the measure of leeway in starting and completing an activity. It is the number of time units (hours, days, weeks, years) that an activity (or chain of activities) can be delayed without affecting the project end date.

Free Float Free float is also referred to as "activity float" because, unlike total float, free float is the property of an activity and not the network path an activity is part of. Free float is the amount of time the start of an activity may be delayed without delaying the start of a successor activity.

ACTIVITY RELATIONSHIP TYPES

The relationships between activities are defined as predecessor, successor, or concurrent.

Concurrent Activities are logically independent of one another and can be performed at the same time.

Predecessor Activity is one that must be completed before a given activity can be started.

Successor Activity is one that cannot start until a given activity is completed.

Relationships between the activity and its predecessor and successor activities can vary.

The options are

- Finish to Start (FS)
- Start to Start (SS)
- Finish to Finish (FF)
- Start to Finish (SF)

The FS relationship means the predecessor activity must finish before the successor activity can start. The SS relationship means the successor activity can start at the same time as the predecessor activity or later. The FF relationship means that the successor activity can finish at the same time as the predecessor or later. SF means that the predecessor activity must start before the successor activity can finish.

FORWARD PASS

Early Start

To calculate an activity's early start (forward pass), add to the early start of the preceding activity its duration. All predecessor activities to the activity must be complete. Activity R in Figure 3-2 is the first activity and therefore has no predecessors. It therefore has an early start of 1. Some schedulers prefer to start with day 0, rather than day 1, in order to get the correct number of days. Starting with day 1 produces the correct duration plus 1. This simply means the last day of the project is used up and the calculation proceeds to the beginning of the day after the project is finished. The early start of G is 5, or 1 (the early start of R, the preceding activity) plus 4 (the duration of the preceding activity).

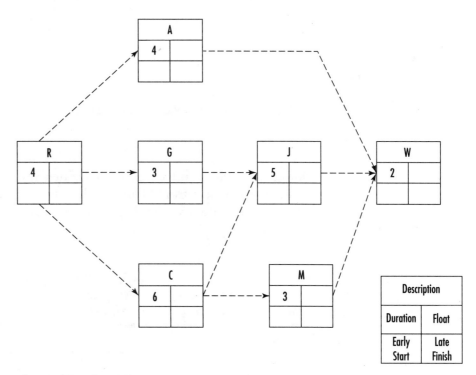

Figure 3-2   Precedence Diagram—Durations

Larger Value

Where two chains converge on a single activity, such as activity J in Figure 3-3, the larger early start value controls, since that path has the

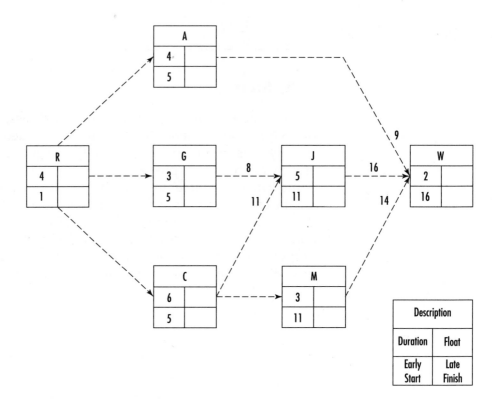

Figure 3-3 Forward Pass—Early Starts

longer duration and *all* prior activities must be finished. The two choices are 8 days from G and 11 days from C. The 11 controls since it is the larger value. See Table 3-1 for the early start calculations for Figure 3-3.

Activity	Preceding Activity	Early Start of Preceding Activity	Duration of Preceding Activity	Early Start of Activity (*Controls)
R				1
A	R	1	4	5
G	R	1	4	5
C	R	1	4	5
J	G	5	3	8
	C	5	6	11*
M	C	5	6	11
W	A	5	4	9
	J	11	5	16*
	M	11	3	14

Table 3-1 Early Start Calculations

BACKWARD PASS

Late Finish

The backward pass is used to determine activity late finishes. As the name implies, calculations are commenced at diagram end and pass in a backward direction to the beginning. The formula for the late finish of an activity is the late finish of the following activity minus the duration of the following activity. The late finish of activity W in Figure 3-4 is 18. Since W is the last activity, it has to be on the critical path. W has an early start of 16 and a duration of 2. W has a late finish of 18. The late finish of J is 16, or 18 (late finish of following activity) minus 2 (duration of following activity).

Smallest Value

When two paths converge on a single activity, the smallest value is used. The backward pass to C from the path through J has a late finish of 11 (16 − 5). The path to C through M has a late finish of 13 (16 − 3). Since 11 is the smaller value, it is used as the late finish of C. See Table 3-2 for the late finish calculations for Figure 3-4.

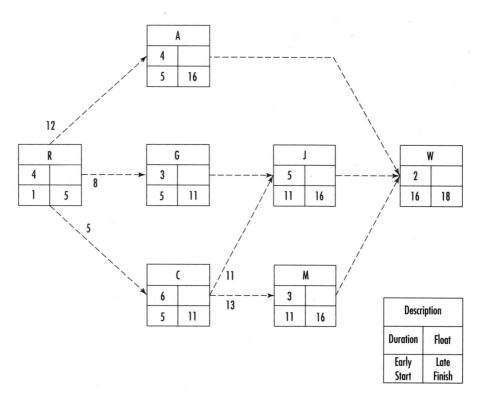

Figure 3-4   Backward Pass—Late Finishes

Activity	Following Activity	Late Finish of Following Activity	Duration of Following Activity	Late Finish (*Controls)
W				18
M	W	18	2	16
J	W	18	2	16
C	M	16	3	13
	J	16	5	11*
G	J	16	5	11
A	W	18	2	16
R	C	11	6	5*
	G	11	3	8
	A	16	4	12

Table 3-2 Late Finish Calculations

FLOAT

Flattened Requirements

Construction schedulers use the float to define which activities are critical and which activities have "slack" or "fluff." Float associated with noncritical activities can be used to "flatten" or level the requirements for resources, including personnel, materials, construction equipment, and cash. For example, if two activities had a scheduled early start of the same day and both required a carpentry crew, completing the critical activity first would reduce the total number of carpenters needed. Similarly, such sequencing can reduce material, equipment, and cash-flow requirements. Noncritical activities do not have the same priority as critical activities and do not determine the critical path or, therefore, the project duration; but if the float time is used up, the critical path changes and previously noncritical activities may become critical.

Formulas

The formulas for calculation of total float require late start and early finish calculations. The formulas are:

Late Start = Late Finish – Duration
(Determined from Backward Pass)

Early Finish = Early Start + Duration
(Determined from Forward Pass)

There are three formulas for the calculation of activity total float:

Total Float = Late Finish – (Early Start + Duration)

Total Float = Late Start – Early Start

Total Float = Late Finish – Early Finish

Given the information contained in the nodes in Figure 3-5, the first formula above is the most suitable. All three of these formulas must yield the same value, or a mistake has been made. See Table 3-3 for total float calculations for Figure 3-5. By examining Table 3-4 for the example problem, the formulas return the same value.

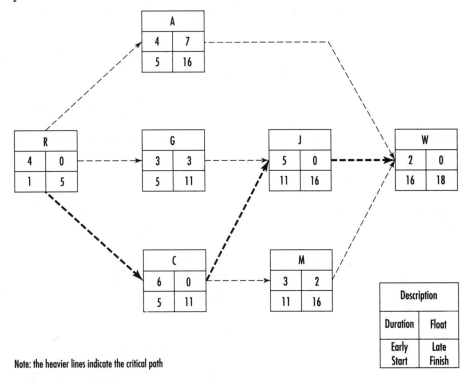

Note: the heavier lines indicate the critical path

Figure 3-5   Float

Activity	Late Finish	Early Start	Duration	Float
R	5	1	4	0
A	16	5	4	7
G	11	5	3	3
C	11	5	6	0
J	16	11	5	0
M	16	11	3	2
W	18	16	2	0

Table 3-3   Total Float Calculations

DATA TABLE

The following is a classic scheduling data table (tabular sort) sorted by early starts. The table includes late starts and early finishes. Again the formulas for late starts and early finishes are:

$$\text{Late Start} = \text{Late Finish} - \text{Duration}$$

$$\text{Early Finish} = \text{Early Start} + \text{Duration}$$

Activity	Duration	Early Start	Late Start	Early Finish	Late Finish	Float
R	4	1	1	5	5	0
A	4	5	12	9	16	7
G	3	5	8	8	11	3
C	6	5	5	11	11	0
J	5	11	11	16	16	0
M	3	11	13	14	16	2
W	2	16	16	18	18	0

Table 3-4   Data Table

CALENDARS

So far in this chapter, the scheduling units used have been workdays. To be more useful, these workdays may be converted to calendar days (Figure 3-6).

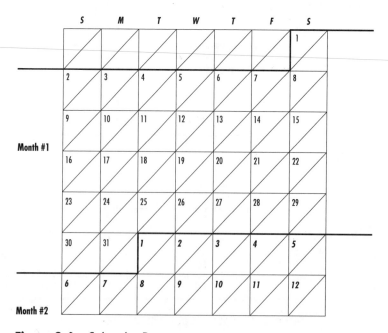

Figure 3-6   Calendar Days

Nonworkdays (Saturdays, Sundays, holidays, and a possible allowance for rain or other bad weather) must be decided upon. These calendar decisions must be made to convert from work to calendar days (Figure 3-7).

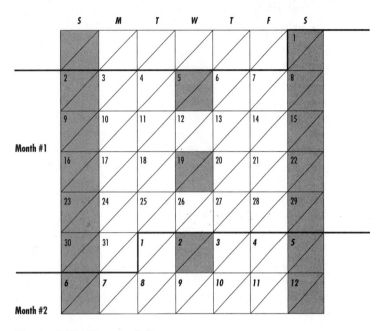

Figure 3-7 Nonwork Days

After allowing for nonworkdays, project workdays can now be correlated with calendar days. The project starts the tenth day of the first month (Figure 3-8).

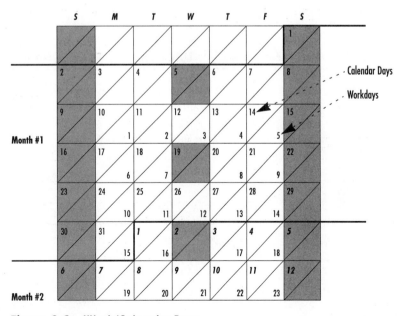

Figure 3-8 Work/Calendar Days

EXAMPLE PROBLEM: Calculations

Figure 3-9 is an example of the forward pass, backward pass, and float calculations prepared for readers' use. Table 3-5 is an example of the same information in tabular format.

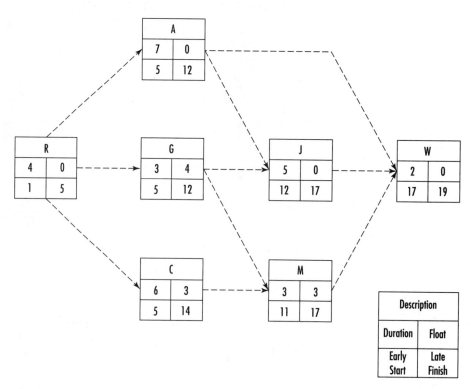

Figure 3-9   Example Problem—Precedence Diagram

Activity	Duration	Early Start	Late Start	Early Finish	Late Finish	Float
R	4	1	1	5	5	0
A	7	5	5	12	12	0
G	3	5	9	8	12	4
C	6	5	8	11	14	3
J	5	12	12	17	17	0
M	3	11	14	14	17	3
W	2	17	17	19	19	0

Table 3-5   Example Problem—Data Table

EXERCISES

1. Calculations

Complete the forward pass, backward pass, and float calculations for Figure 3-10. Also complete the tabular information for Table 3-6. Follow these steps:

1. Forward pass calculations (early starts)
2. Backward pass calculations (late finishes)
3. Total float calculations
4. Early finish calculations
5. Late start calculations

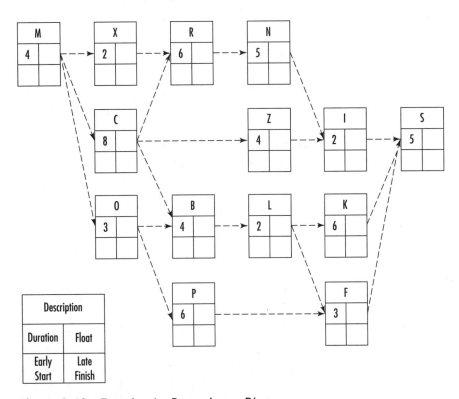

Figure 3-10   Exercise 1—Precedence Diagram

Activity	Duration	Early Start	Late Start	Early Finish	Late Finish	Float
M						
X						
C						
O						
R						
B						
P						
N						
Z						
L						
I						
K						
F						
S						

Table 3-6   Exercise 1—Data Table

2. Calculations

Complete the forward pass, backward pass, and float calculations for Figure 3-11. Also complete the tabular information for Table 3-7. Follow these steps:

1. Forward pass calculations (early starts)
2. Backward pass calculations (late finishes)

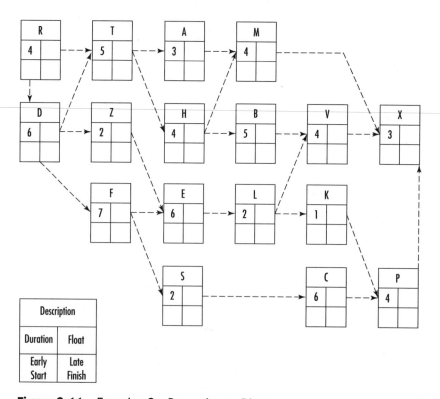

Figure 3-11   Exercise 2—Precedence Diagram

3. Total float calculations
4. Early finish calculations
5. Late start calculations

Activity	Duration	Early Start	Late Start	Early Finish	Late Finish	Float
R						
D						
T						
Z						
F						
A						
H						
E						
S						
M						
B						
L						
V						
K						
C						
X						
P						

Table 3-7   Exercise 2—Data Table

4

P3 Schedule Preparation

Objectives

Upon completion of this chapter, the reader should be able to:

- Add a project
- Add activities
- Define relationships
- Calculate the schedule
- Edit an activity

P3 SYSTEM REQUIREMENTS

Primavera Systems, Inc.'s most powerful project management/scheduling software package is named *Primavera Project Planner for Windows*, or *P3*. *P3 for Windows* runs comfortably on the following system:

- 486 (or better) computer
- 16 MB total memory
- 73 MB of free disk space for a complete installation; 32 MB of free disk space for a minimal installation
- DOS 5.0 or later (running Share or Vshare)
- *Windows 3.1* (enhanced mode), *Windows 95*, or *Windows NT*
- VGA or SuperVGA monitor
- Mouse (optional, but recommended)

This book is based on *P3 for Windows*, Version 2.0.

P3 START-UP

To start *P3*, double click the left mouse button on the *Primavera* application icon (Figure 4-1) at the Windows Program Manager screen. A blank or entrance *P3* screen (Figure 4-2) is the first screen to appear when *P3* is started (unless user name and password must be entered).

P3 Icon

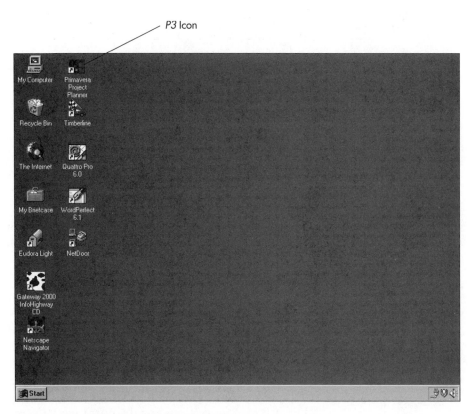

Figure 4-1　Windows 95 Desktop Screen

Menu Bar ——
Button Bar ——

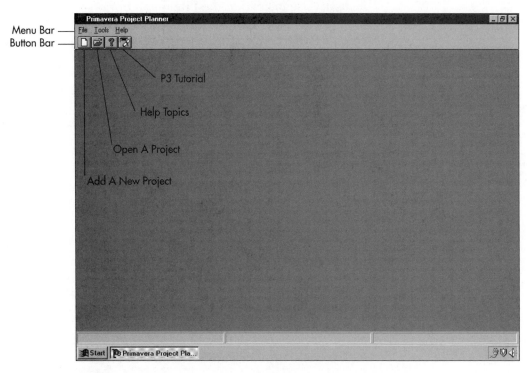

Figure 4-2　*Primavera Project Planner* Entrance Sceen

MENU OPTIONS

Pull-down menu options for the entrance *P3* screen are:

File   Click this option word to create a new project or to open an existing project.

Tools   Includes the **Draw**, **Look**, **Web Publishing Wizard**, **Project Utilities**, and **Options**. The **Draw**, **Look**, and **Project Utilities** functions will be discussed later in the book. **Draw** is used to create or modify clip art. **Look** is used to preview and save on-screen captures of hard copy prints. **Project Utilities** is used to **Copy...**, **Delete...**, **Merge...**, **Summarize...**, **Back Up...**, and **Restore...** project files.

Help   Click to see the help index.

BUTTON BAR OPTIONS

Default toolbar options for the entrance *P3* screen are as follows:

Add a New Project   Click on this option to add a new project to the system. (Use instead of the **File** menu option.)

Open a Project   Click to open an existing project. (Use instead of the **File** menu option.)

Help Topics   Click to see the help index. The on-line help in *P3* is extensive. The "?" button appears on both the entrance screen and the bar chart screen. Click the help button to produce the Help Topics menu. The menu files are:

- **Contents**. Displays *P3* help contents by category (books).
- **Index**. Displays the **Search** dialog box that enables you to find help information using keywords.
- **Find**. Enables you to search for specific words and phrases in help topics instead of searching by category.

Another convenience of *P3* is that you can pull up the Help Topics screen for almost any field within *P3* by depressing the F1 (first function) key.

P3 Tutorial   Provides access to *P3*'s internal tutorial with a click of the mouse.

ADD A NEW PROJECT DIALOG BOX

From the button bar, click on the Create New Project button. The **Add a New Project** dialog box (Figure 4-3) will appear.

Fields of the Add a New Project Dialog Box

The **Add a New Project** dialog box information fields are as follows:

Current directory: The current directory (Figure 4-3) is usually *c:\p3win\projects,* but it can be any other as well. Having a separate file to store project files (as opposed to program files) is convenient. To change the current directory, click on **Dir....**

Project name: The four-character alphanumeric field will appear at the top of the project screen. It is also the primary sort field in project file listings (when opening an existing schedule). Also, when you are looking in the *c:\p3win\projects* directory, you can start the project files for the created schedule with this project name.

Number/Version: For updating and control purposes, you may find it necessary to copy a schedule. Click on this option to compare the changes in the updated schedule to the original (or baseline) schedule.

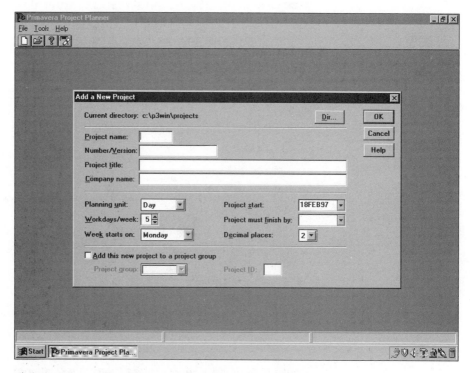

Figure 4-3   Add a New Project Dialog Box

The **Number/Version:** field is simply a way to keep track of the relationships of the multiple schedules on a project.

Project title: The **Project title:** field is a larger field than the **Project name:** field for naming a project. By default, the project title is printed on hard copy (the project title on both tabular reports and plots comes from this field).

Company name: By default the **Company name:** is printed on hard copy (both tabular reports and plots come from this field). If you don't want to specify a company name, leave this field blank.

Planning unit: Click on the down arrow (or pull-down menu) next to the **Planning unit:**, and a menu of options appears (see Figure 4-4). The choices are hour, day, week, or month. The default is day. On most construction projects the day is the ideal unit, although on some industrial projects hourly planning may be necessary, and for some broad, conceptual, and long-lasting projects the weekly or monthly units may be more appropriate.

Workdays/week: Click on the up or down arrow to designate the number of workdays per week. The default is 5 days per week, but the 4-day workweek has become popular in recent years. From a scheduling point of view, the 4-day workweek is efficient, since the fifth day can be used to make up lost workdays.

Week starts on: Some tabular and graphical reports require the start of the workweek be defined. The default is Monday, but the workweek can begin on any day of the week. Click on the pull-down menu arrow to define the week start day.

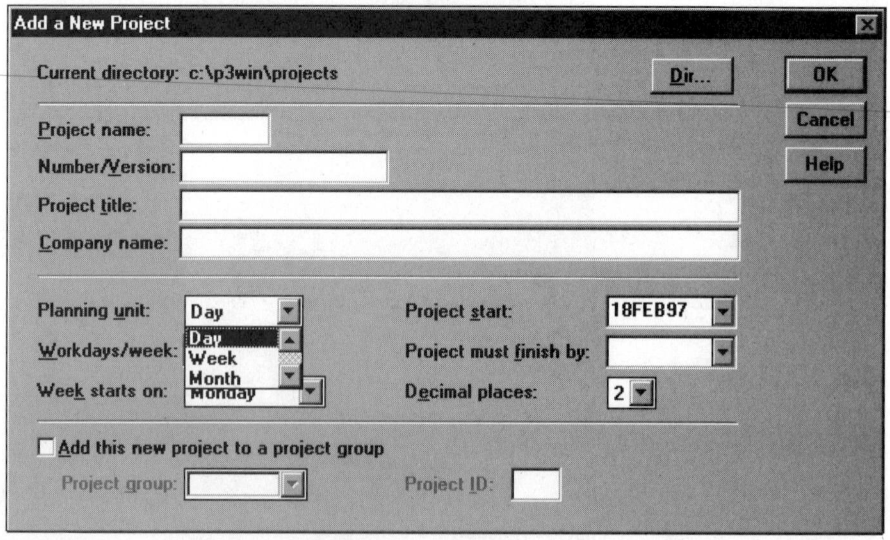

Figure 4-4   Add a New Project Dialog Box—Planning Unit

Project start: The **Project start:** (see Figure 4-5) is a required entry that initiates the starting point of the bar chart graphics. The motion bar to the right of the pull-down calendar is used to select the month. Click the mouse below the slide button on the motion bar to advance the calendar one month at a time. Click the mouse above the slide button to back up the calendar a month at a time. Click on the appropriate date and then click on the **OK** button to select the project start date.

Project must finish by: The **Project must finish by:** is an optional entry. If there is a superimposed finish date, the date can be input using the pull-down calendar the same way the project start date was input. If there is no required finish date, this field can be left blank and *P3* will calculate the finish date based on activities, durations, and logic restraints.

Decimal places: Select the number of decimal places to use for resource and cost information.

Add this new project to a project group Click on this check box to designate a project as part of a project group. This function breaks down a large project into more manageable units. If a large project is to include five buildings, each of the buildings can be a member project under a project group. Information such as the Resource Dictionary can be put in at the project group level and filter down to the member project level. Information input at the member project level can be merged into reports at the project group level.

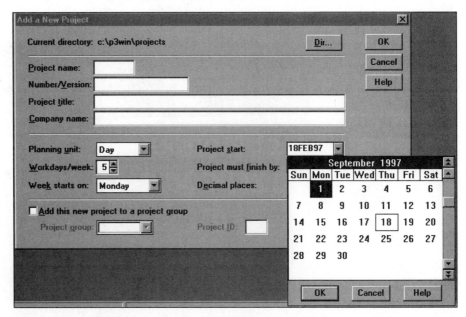

Figure 4-5 Add a New Project Dialog Box—Project Start

Project group: If you add a project that is part of a group, click on the down arrow to select a project group to assign to the new project. The new project must be created in the same directory as the project group.

Project ID: Identifies projects that are part of a group. *P3* automatically assigns a two-character alphanumeric ID to appear in the first two positions of the activity ID for each project activity. Accept the default, or type another prefix.

BUTTONS ON THE ADD A NEW PROJECT DIALOG BOX

The **Add a New Project** dialog box button options are:

OK Click on the **OK** button to add the new project to the projects directory and create the necessary files (Figure 4-6).

Cancel Click on the **Cancel** button to cancel the **Add a New Project** dialog box without creating a new project.

Help Takes you to the help menu relating to the **Add a New Project** dialog box.

Dir... Click on this button to change the default directory and to create the project on a drive or directory other than what was defined when *P3* was installed (usually *c:\p3win\projects*).

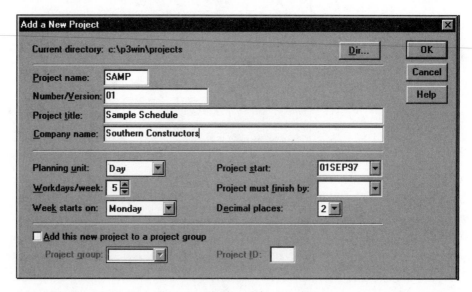

Figure 4-6 Add a New Project Dialog Box—Completed

ADDING ACTIVITIES

Click on the **OK** button in the **Add a New Project** dialog box to create the new project, with the blank bar chart screen (Figure 4-7) ready for input of schedule information. Notice that when the bar chart screen comes up, the insert activity mode is already initiated.

The first step in creating the schedule is to create activities and define relationships and durations. Steps in adding an activity are as follows:

Activity ID   When in the insert activity mode, the cursor is blinking in the **Activity ID** field on the edit bar. Activity 100 was input in Figure 4-8. To input the rest of the information relating to Activity 100, click on the **<u>V</u>iew** pull-down menu and select **<u>A</u>ctivity Form**. The **Activity** form will appear at the bottom of the screen (Figure 4-8). Giving the activity an alphanumeric name (up to ten characters) establishes a sort field. Each activity will have a unique name, so *P3* can use the relationships established to calculate project duration and float.

When *P3* assigns a new ID, it checks whether the ID already exists in the project. If it exists, *P3* searches in increments of 10 until it finds an

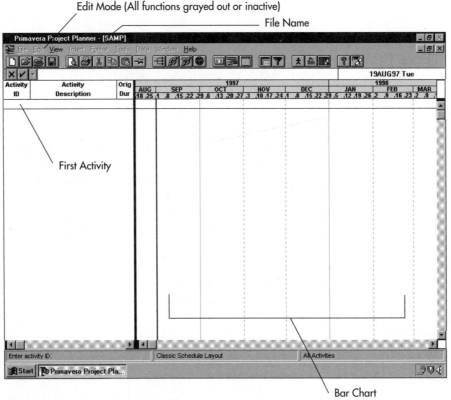

Figure 4-7   On-Screen Bar Chart

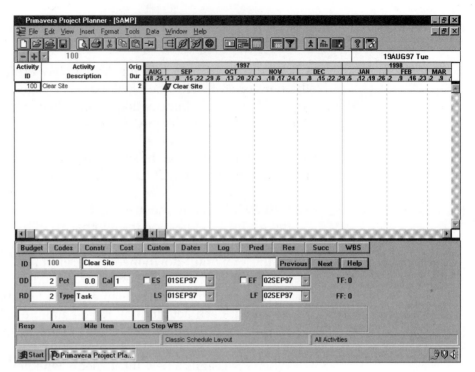

Figure 4-8 On-Screen Bar Chart—Activity Form

unused ID. If you delete an activity, that ID becomes available for a new activity. If you want *P3* to increment IDs by a value other than 10, choose the **Tools** main pull-down menu, **Options**, and then the **Activity Inserting...** option. Type a value in the **Increment:** field or turn off **Automatically number activities**, so that you can choose your own activity IDs.

The **ID** field can be used to sort activities into activity ID categories within a project. To use this option, select the **Data** main pull-down menu and then **Activity Codes....** The **Activity Codes Dictionary** dialog box will appear. Click on the **Activity ID** option to identify the ID codes. For example, you may want to add an "A" prefix to all activities in the subproject for "Area A." Similarly a "B" for "Area B," and so on.

After entering the ID, press the Tab key to move to the **Activity Description** field.

Activity Description   Use this alphanumeric field to name or describe the activity. It is usually not a sort field within *P3*. Use short, concise descriptions—usually two or three words will suffice in a noun + verb format. After entering the activity description, press Tab to move to the **OD** field.

OD OD stands for original duration, the basis of the original schedule. If the day is the scheduling unit, then input number of days for the activity duration.

RD RD stands for remaining duration. Do not use this field when inputting activities. Instead, use it when updating an activity to claim partial completion (see later chapters).

To accept the activity and place it on the bar chart, click on the **Previous** or **Next** buttons. If you need to edit the activity, simply click the activity on the bar chart, and the **Activity** form screen will fill with the activity information. Any information other than the activity ID can readily be changed. To edit activity ID in this screen requires exclusive access. To obtain exclusive access to a project, close the project, then click the **File** main pull-down menu from the *P3* entrance window (Figure 4-2, p. 71). The **Open a Project** dialog box will appear. Select the project to open with exclusive access. Then click on the **Exclusive** check box. When the schedule is brought up, you can edit the **ID** field.

Create next activity To create the next activity, click in the open field under the activity just added on the bar chart. A blank **Activity** form and a new highlighted activity will appear. The new activity information can be input using the same procedure as above. Figure 4-9 shows all the new activities added for the sample schedule.

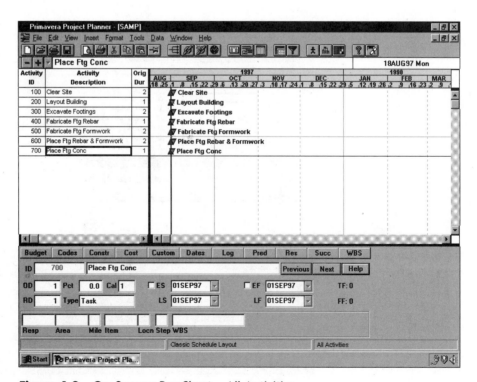

Figure 4-9 On-Screen Bar Chart—All Activities

RELATIONSHIPS

All the activities in Figure 4-9 start on the same date, the first day of the project. The reason is that no relationships between the activities have yet been established. After establishing the activity ID, giving the activity description, and a duration, the next step is to establish relationships among the activities. Steps in establishing relationships are as follows:

Successor/Predecessor An activity relationship can be either a predecessor or a successor relationship. If Activity A must occur before B, A is a predecessor to B and B is a successor to A. Either of these relationships can be identified to *P3*. If one is identified, the program automatically assumes the other. The **Successors** or **Predecessors** dialog box can be accessed in one of two ways. Click on the activity to which relationships are to be established. In Figure 4-10, Activity 100, Clear Site, is selected. Then click on the **Pred** or **Succ** buttons on the **Activity** form button bar. Alternately, click on the **View** main pull-down menu, then click on **Activity Detail** or press the right mouse button to access the same choices from the **Activity** form. The **Succ** button on the **Activity** form button bar was selected in Figure 4-10. Click on the + button to add a relationship.

Activity ID To identify the activity with which a relationship will be established, click on the down arrow button (Figure 4-11). A pop-up

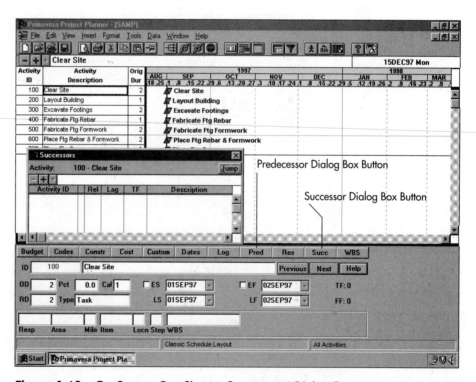

Figure 4-10 On-Screen Bar Chart—Successors Dialog Box

screen will appear that lists activity IDs and descriptions. To establish the relationship, click on the activity (Figures 4-12 and 4-13).

Rel: When Activity 200 is picked as a successor to activity 100, the default relationship (**Rel**) is **FS - Finish to start** (Figures 4-14 and 4-15). Activity 100 must be finished before Activity 200 can start. Click in

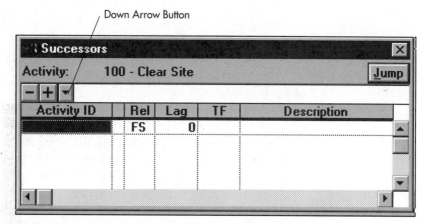

Figure 4-11   Successors Dialog Box—Adding Relationships

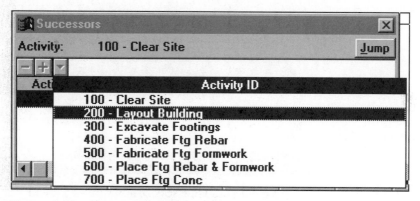

Figure 4-12   Successors Dialog Box—Choosing Successor Activity

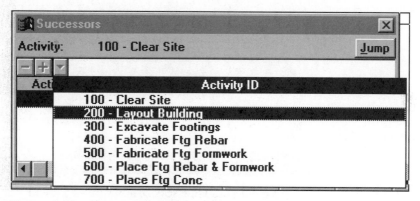

Figure 4-13   Successors Dialog Box—Successor Activity Chosen

the **Rel** field and then click the down arrow on the button bar to see a pull-down menu of other relationship types. There are four relationship types. The other three are **SS, FF,** and **SF.** An **SS - Start to start** relationship means the successor activity can start at the same time as or later than the predecessor activity. An **FF - Finish to finish** means that the successor activity can finish at the same time as or later than the predecessor. An **SF - Start to finish** means that the predecessor activity must start before the successor activity can finish. Clicking on one of the other options changes the default FS relationship. Multiple successors to an activity can be seen in Figure 4-16. This can be done by simply repeating the preceding steps.

Lag   Lag staggers or delays the relationship of one activity to another (Figure 4-17). In Figure 4-18, Activity 100 and Activity 400 did not have a pure finish to start relationship. Activity 400 will start 2 days after the finish of Activity 100. A lag of 2 days is imposed on the relationship. Click in the **Lag** field of the activity that should lag, then an increase/decrease number field button appears on the button bar. Click on the increase button twice (or type "2") to get 2; then press

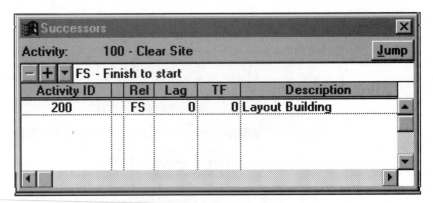

Figure 4-14   Successors Dialog Box—FS Relationship

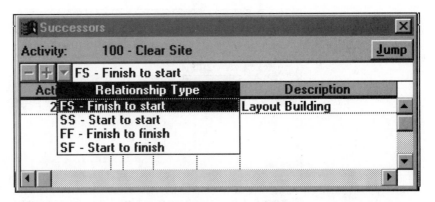

Figure 4-15   Successors Dialog Box—Relationship Type Options

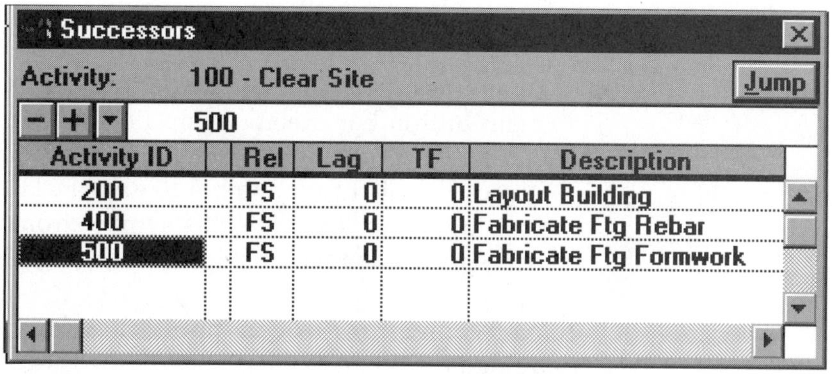

Figure 4-16   Successors Dialog Box—Adding Other Relationships

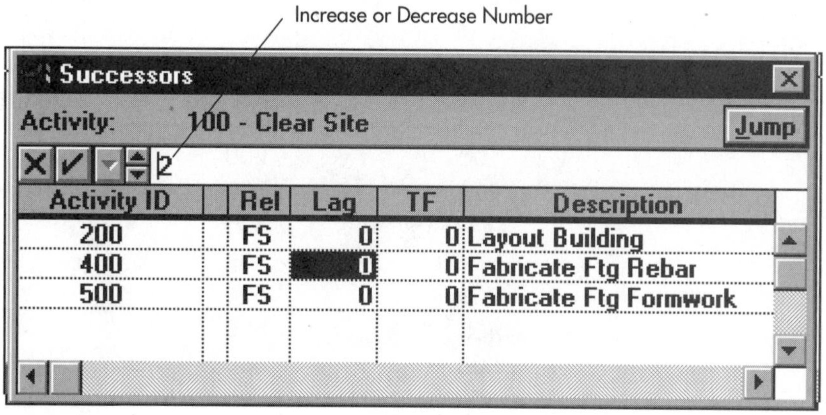

Figure 4-17   Successors Dialog Box—Lag

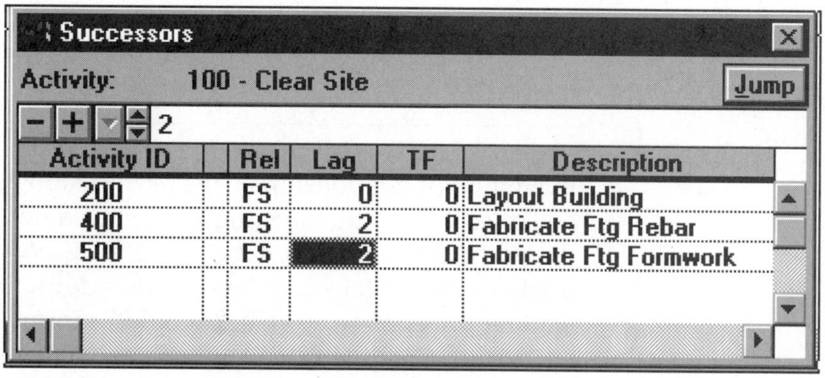

Figure 4-18   Successors Dialog Box—Lag

the Enter key on the keyboard or click on the check button to enter the 2 days of lag for Activity 100. An example of lag is time given to allow elevated concrete to cure to a certain level before the formwork wrecking reshoring activity can begin.

Relationships   Another way to establish relationships is to draw them on the screen. Click on the **View** pull-down menu, then select **Relationships**. *Note:* The pitchfork icon can also be selected from the main button bar. Relationship lines and arrows will appear on the screen. Placing the cursor at the beginning or end of an activity causes the pitchfork symbol to appear. Holding the left mouse button down and dragging this symbol from activity to activity establishes the same relationships as using the above methods.

To correct an error made in inputting activity relationships, see the section titled Editing Activities.

Table 4-1 lists the activities, the successor activity IDs, type relationships, and the lags as input for the sample schedule as calculated in the following section.

Activity	Successor Activity ID	Relationship	Lag
100	200	FS	0
200	300	FS	0
	400	FS	0
	500	FS	0
300	600	FS	0
400	600	FS	0
500	600	FS	0
600	700	FS	0
700			

Table 4-1   Sample Schedule Activity Relationships

CALCULATING THE SCHEDULE

Even though relationships have been established between the activities, all the activities still have the same start date, September 1 (01SEP97). If the schedule were automatically recalculated every time a change was made in the *P3* schedule being worked on, there would be substantial waiting time before more data could be input. Instead, the user controls when the schedule is calculated (or recalculated). *P3* provides three ways to begin the schedule process.

The Schedule Dialog Box

Method 1: Click the **Tools** pull-down menu selection (Figure 4-19). Then click on **Schedule....**

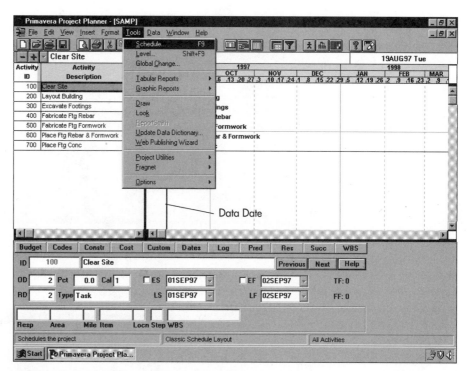

Figure 4-19 Tools Pull-Down Menu

Method 2: Press the F9 function key on the keyboard. *Note:* Using the F9 option does not permit the user to choose options for the calculation, but does allow the schedule to be recalculated without generating an error report.

Method 3: Click the toolbar button that looks like a round red clock.

The **Schedule** dialog box is displayed (Figure 4-20). The data date that was entered when the project was added can now be modified. The data date, marked by a heavy blue vertical line (Figure 4-19), establishes the date that will be the basis of the schedule. The data date can change with each update of the schedule. This concept will be important when updating is discussed in Chapter 10.

Options...   Click on the **Options...** button to reveal the **Schedule/Level Calculation Options** dialog box (Figure 4-21). This dialog box is critically important when inputting progress into the schedule. Use the **When scheduling activities apply** field to determine whether the original activity relationship logic controls the mathematical forward/ backward pass calculations or the actual progress durations input during schedule update controls. The two check box options are:

- **Retained logic**
- **Progress override**

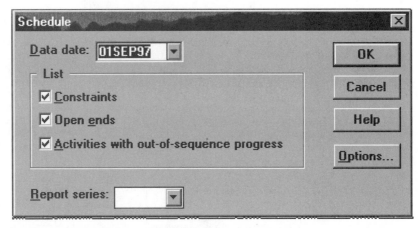

Figure 4-20   Schedule Dialog Box

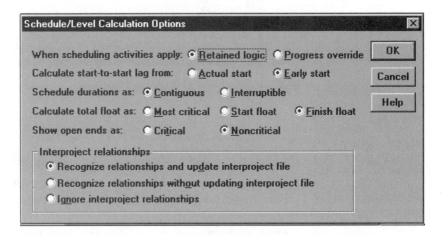

Figure 4-21   Schedule/Level Calculation Options Dialog Box

When you choose **Retained logic,** *P3* does not schedule the remaining completion of an activity until all its predecessors are complete. **Progress override** is used when updating the schedule. Its use will be discussed in Chapter 10.

Next, click on the **OK** button. The **Output Options** dialog box will appear (Figure 4-22). The options for viewing the calculated schedule are: **View on screen, Print immediately,** or **Save to the following Look file.**

View   Choose **View on screen** to generate a **Primavera Look** screen (Figure 4-23). **Primavera Look** shows what a report or graphic will look like before you print it. Since the schedule report is used for

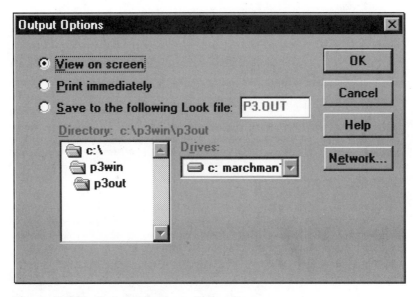

Figure 4-22   Output Options Dialog Box

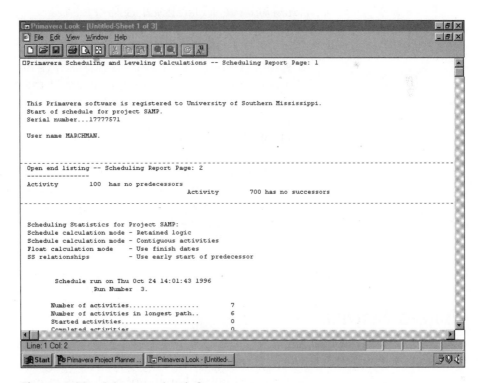

Figure 4-23   Primavera Look Screen

Primavera Look control menu

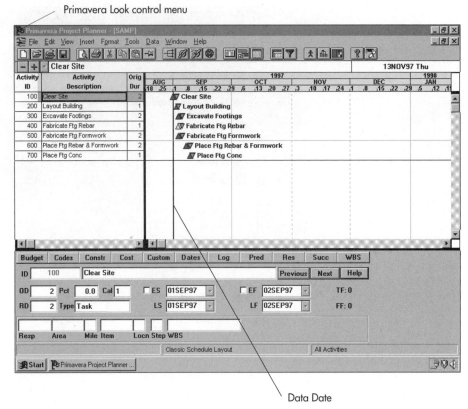

Data Date

Figure 4-24 Calculated Schedule

finding errors, a hard copy is not always needed. The **Print immediately** option provides a paper copy of the schedule report. The last option, **Save to the following Look file:**, is useful when the errors have been solved and the schedule is being recalculated. The report is simply saved to a specified file.

To turn off the **Primavera Look** screen, click on the **Primavera Look** control menu, then click on **Close.** This will return you to the bar chart screen of the newly scheduled project (Figure 4-24). Now all of the activities are not lined up on the data date but are staggered according to the logic defined by the relationships.

EDITING ACTIVITIES

Activity descriptions, original durations, relationships, or other input data can easily be changed or added to an existing activity. Click on the activity to be modified. Turn the **Activity** form back on, if it has been

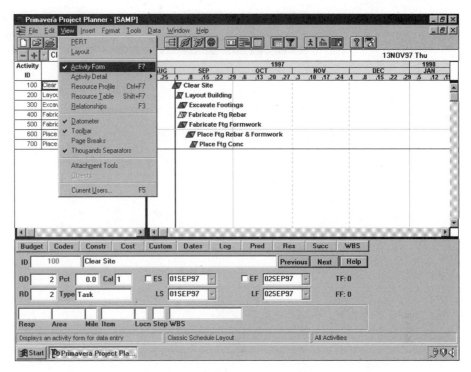

Figure 4-25   View Pull-Down Menu

turned off, by clicking on **View** from the pull-down menu. Then click on **Activity Form** (Figure 4-25).

Any of the fields for the highlighted activity (in this case, Activity 100, Clear Site) may be modified. Click on the button bar options on the **Activity** form to add new information (budget, cost, and resources) about the activity or change information already input, such as successors and predecessor. Another simple way to edit activity information is in Columns.

With the **Activity** form turned on, you can edit other activities by clicking the particular activity to be edited.

PERT VIEW

The on-screen bar chart is a convenient way to view the schedule. There is another option that simplifies the viewing of logic relationships. Figure 4-26 is a view of the sample schedule in PERT format. This format is

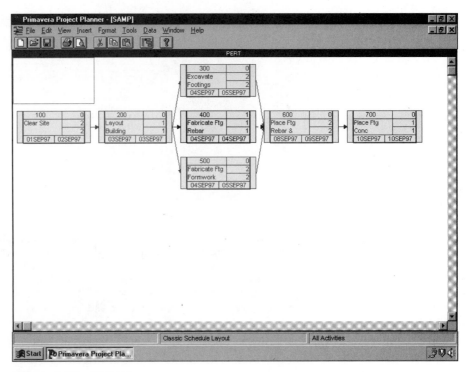

Figure 4-26 PERT View

obtained by selecting **PERT** at the **View** main pull-down menu. The information contained within and the size of the nodes can be modified. Different shapes, colors, and ends can be selected for designated activities. Select the nodes to be changed by holding down the Shift key and clicking on the activities. Then select **Format** and **Activity Box Ends and Colors...**.

OPENING A PROJECT

To open an existing project, rather than creating a new project as produced in this chapter, click on the **File** main pull-down menu and select **Open**, or click on the Open Existing Project button. The **Open a Project** dialog box that lists existing projects appears (Figure 4-27). You can change directories and drives when selecting a project. Click the **Overview...** button to view the **Project Overview** dialog box (Figure 4-28). Compare Figure 4-28, project information of the existing project, to Figure 4-6 (p. 76), the same project information input as the project was added. Sometimes it is necessary to change project information after a project has been created. This is the dialog box for making such

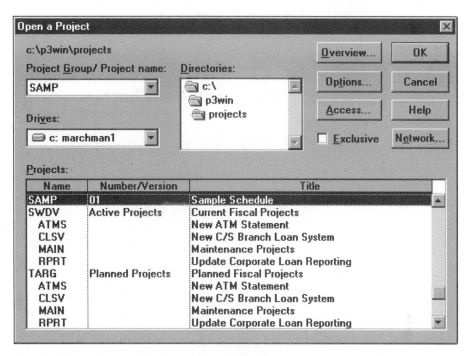

Figure 4-27 Open a Project Dialog Box

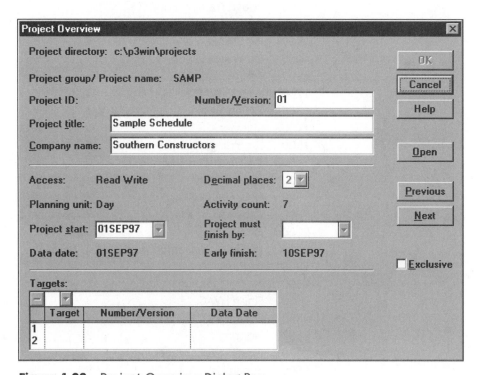

Figure 4-28 Project Overview Dialog Box

changes. You can also go directly to the **Project Overview** dialog box while already in an open project from the **File** main pull-down menu.

EXAMPLE PROBLEM: On-Screen Bar Chart

Table 4-2 is a tabular list of twenty-eight activities (activity ID, description, duration, and successor relationships) for a house put together as an example for student use (see the wood frame house drawings in the Appendix). Figure 4-29 is the on-screen bar chart of the house based on the list provided in Table 4-2.

Act. ID	Act. Description	Duration (Days)	Successors
100	Clear Site	2	200
200	Building Layout	1	300
300	Form/Pour Footings	3	400
400	Pier Masonry	2	500
500	Wood Floor System	4	600
600	Rough Framing Walls	6	700
700	Rough Framing Roof	4	900
800	Doors & Windows	4	1000, 1600
900	Ext Wall Board	2	800, 1100, 1200
1000	Ext Wall Insulation	1	1700
1100	Rough Plumbing	4	1400
1200	Rough HVAC	3	1300
1300	Rough Elect	3	1000
1400	Shingles	3	1000, 1600, 2700
1500	Ext Siding	3	2000
1600	Ext Finish Carpentry	2	1500
1700	Hang Drywall	4	1800
1800	Finish Drywall	4	1900, 2400, 2500
1900	Cabinets	2	2100, 2300
2000	Ext Paint	3	2200
2100	Int Finish Carpentry	4	2200
2200	Int Paint	3	2600
2300	Finish Plumbing	2	2200
2400	Finish HVAC	3	2200
2500	Finish Elect	2	2200
2600	Flooring	3	2800
2700	Grading & Landscaping	4	2800
2800	Punch List	2	

Table 4-2   Activity List with Durations and Relationships—Wood Frame House

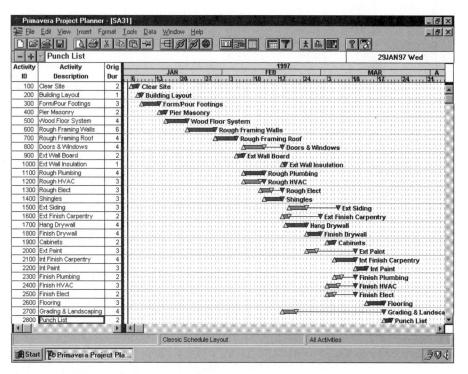

Figure 4-29   *P3* On-Screen Bar Chart—Wood Frame House

EXERCISES

1. Small Commercial Concrete Block Building— On-Screen Bar Chart

Prepare an on-screen bar chart for the small commercial concrete block building located in the Appendix. This exercise should include the following steps:

1. Prepare a list of activity descriptions (minimum of sixty activities).
2. Establish activity durations.
3. Establish activity relationships.
4. Create the *P3* on-screen bar chart.

2. Large Commercial Building—On-Screen Bar Chart

Prepare an on-screen bar chart for the large commercial building located in the Appendix. This exercise should include the following steps:

1. Prepare a list of activity descriptions (minimum of 150 activities).
2. Establish activity durations.
3. Establish activity relationships.
4. Create the *P3* on-screen bar chart.

5 Resources Using *P3*

Objectives

Upon completion of this chapter, the reader should be able to:

- Modify on-screen bar chart timescale
- Modify resource dictionary
- Assign activity resources
- Use resource profiles
- Use resource tables
- Define driving resources
- Set resource limits
- Level resources

NECESSITY FOR CONTROLLING RESOURCES

The contractor must be able to control resources—labor hours, bulk materials, construction equipment, and permanent equipment. This ability to get the biggest bang for the buck in putting the resources in place usually determines a contractor's success. Control of resources involves:

- maximizing resources
- attention to cost and time
- control of waste
- attention to detail
- preplanning
- attention to efficiencies

"Resource loading" the schedule (defining the people, materials, equipment you plan to use) is an effective way to control resources. Loading the labor resources identifies weeks in advance the exact activities to be performed on a particular day and the number of workers per craft that are required. Knowing the subcontractors' labor plans is also an effective way to measure subcontractor performance.

Remember that time is money. A contractor who uses only the cost system to control resources is not getting the complete picture in the effective use of the resources. The entire scheduling process should be looked at from the point of view of efficient use of time, money, and resources.

MODIFYING THE TIMESCALE

Figure 5-1 is the on-screen bar chart of the small project developed in Chapter 4. Since the sample project is of such short duration, the timescale of the bar chart can be modified to make the information easier to read.

Timescale Definition Dialog Box

The **Timescale Definition** dialog box (Figure 5-2) can be accessed in one of two ways. The first way is clicking on the **Format** pull-down menu. Then click on the **Timescale...** option. The other method is to double-click anywhere on the shaded timescale bar itself.

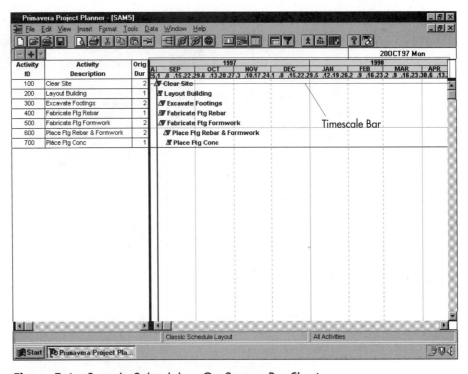

Figure 5-1   Sample Schedule—On-Screen Bar Chart

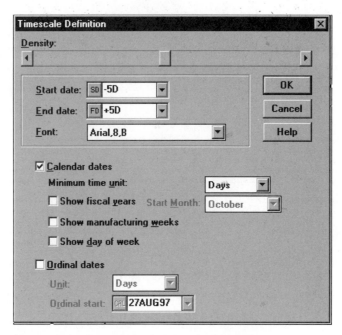

Figure 5-2 Timescale Definition Dialog Box

Density: The slide bar under **Density:** controls the screen timescale format. Click on the arrows, or click and hold the slide button and slide it to either side to modify the timescale of the on-screen bar chart. The scale here was modified to show only one month (Figure 5-3) instead

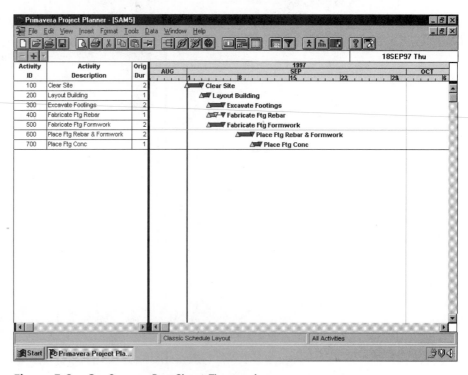

Figure 5-3 On-Screen Bar Chart Timescale

of eight. When the timescale is modified and the **OK** button is clicked on the **Timescale Definition** dialog box, *P3* modifies the appearance of the activities to relate to the new timescale.

The other modifications that may be made using the **Timescale Definition** dialog box are:

Start date: The default is a –5D. This indicates that the on-screen bar chart starts at five days before the original data date given when the schedule was created. To increase or decrease the number, click on the down arrow button for this option. For the **Start date:**, there are four possible options. They are **Calendar date**, **Start Date**, **Data date**, and **Finish date**.

End date: The default is a +5D. This indicates that the on-screen bar chart will have a minimum length of at least the longest scheduled duration for the project plus five days. For the **End date:**, the same four options are available as for the **Start date:**.

Font: Modifying this option changes only the font used for the letters on the timescale bar. To change the type, size, and bold options of the font, click on the down arrow button for the available choices.

Minimum time unit: The timescale option shown in Figure 5-3 (p. 96) is weeks. September 1, 8, 15, 22, and 29 are shown on the timescale bar. This represents the Mondays (beginning of the week) that are available. The timescale can be changed to months, quarters, or years as the on-screen bar chart timescale. The time frame can be changed by clicking the down arrow button.

Ordinal dates Usually it is easier to think of the construction schedule in terms of calendar days. But sometimes it is simpler to see the ordinal (or numerical) workdays. An example is the first ordinal day of this project, which is August 27, a calendar day. Click on the option button on this selection to change the timescale bar from calendar to ordinal days.

RESOURCE DICTIONARY

Determining Requirements

The scheduler has a great deal of flexibility in determining the resource requirements of a project. Resources are typically thought of as labor, material, or construction equipment. For the purposes of this chapter, only direct labor will be analyzed. The labor requirements of each activity will be defined, then the requirements of the entire project can be evaluated and refined if necessary.

Defining Resources

The resources must be defined in the resource dictionary before the requirements of each activity can be addressed. Click on the **Data** main pull-down menu, then click on the **Resources...** option. The **Resource Dictionary** dialog box will appear (Figure 5-4).

Transferring a Resource Dictionary

There are two ways to define the resources in the dictionary for our particular project. The first and easiest is to transfer the resource dictionary from another project. By clicking on **Transfer...**, the **Resource Dictionary will be updated and appended** caution box will appear. Click the **OK** button and the **Transfer** dialog box will appear. It includes a projects directory for scanning *P3* files. Select a project in the **Project:** field, then click on the **OK** button. The resource dictionary of the selected project is the basis of the dictionary for the new project. Obviously, once you have developed a dictionary that meets the needs of the type project you intend to build, there is no need to reinvent the wheel and redefine the resource dictionary for every new project. Also, using a previously defined dictionary speeds up the process.

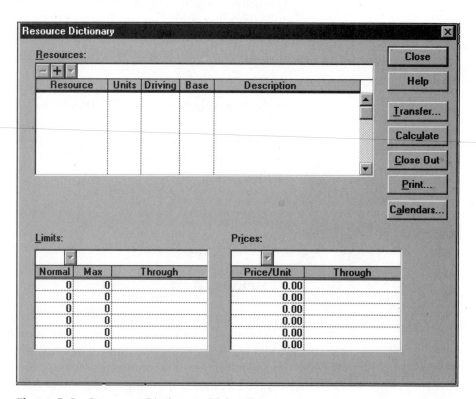

Figure 5-4   Resource Dictionary Dialog Box

Creating a New Resource Dictionary

The second method of defining resources is to create a new resource dictionary. For the purposes of this chapter, a new resource dictionary will be created, instead of using **Transfer...** to copy an existing dictionary from an earlier project. Click on the + button bar option of the **Resource Dictionary** dialog box (Figure 5-5). The top portion of the dialog box is used to define the resources for the project.

Resource The **Resource** column shows the eight-character name for the resource that will be assigned to the activity. Type the resource abbreviation (see CARP 1 in Figure 5-5). Press the Enter key to add the resource to the dictionary. Further information fields are available to define the resource.

Units This is used to describe how *P3* should measure the resource (such as days or weeks). It is important to realize that this is just a name for the unit. *P3* does not use this information to calculate usage.

Driving A driving resource is one that will be the controlling resource if more than one resource is used for an activity. A driving resource can be defined in the **Resource Dictionary** dialog box or the activity

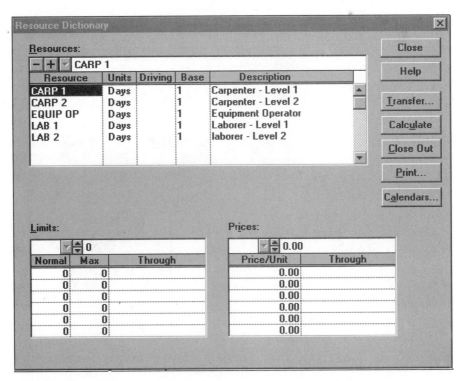

Figure 5-5 Resource Dictionary Dialog Box—Resource Field

Resources form. Since a resource may not be the controlling one for all activities for which it is used, care should be taken in defining it in the resource dictionary.

Description   This is used to enter the full description of the resource.

ASSIGNING RESOURCES

Activity Requirements

When the crafts (resources) have been defined in the resource dictionary, the requirements of each activity can be specified. The **Resources** form can be accessed in one of three ways. One way is to click on the **View** main pull-down menu, click on **Activity Form,** then click on the **Res** (resources) button on the button bar. The second method is to click the right mouse button and select **Activity Detail**. The third method is to click the **View** main pull-down menu, select **Activity Detail** (Figure 5-6), and then click on **Resources.** See Table 5-1 for a listing of the resources and the units per day as input into the sample schedule.

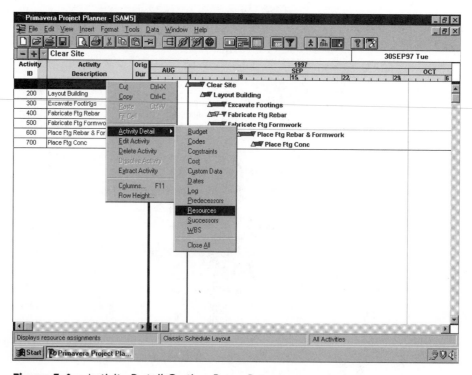

Figure 5-6   Activity Detail Option Box—Resources

Activity	Resource	Units per Day
100	CARP 1	1
200	CARP 1	1
	CARP 2	1
	LAB 1	1
300	LAB 1	1
	EQUIP OP	1
400	LAB 1	1
	LAB 2	1
500	CARP 1	1
	CARP 2	1
	LAB 2	1
600	CARP 1	1
	LAB 2	1
	EQUIP OP	1
700	CARP 1	1
	LAB 1	1
	EQUIP OP	1

Table 5-1   Sample Schedule Resource List

Resources Dialog Box

The **Resources** dialog box (Figure 5-7) is the method for specifying activity resource requirements within *P3*. To add a resource requirement to an activity, click on the + button, then click the down arrow, and the resource listing as defined in the resource dictionary will appear. Click on the desired resource. In Figure 5-8, CARP 1 (Carpenter - Level 1) was selected.

Units per day   To complete the next field, the **Units per day** (Figure 5-9), use the up/down arrow to specify the number of units, then press the Enter key on the keyboard. The time period in the **Units per day** field may be hour, day, week, etc. The time period for the particular resource is defined in the resource dictionary.

As can be seen from Figure 5-10, more than one resource per activity can be specified. This figure shows the requirements for Activity 300, Excavate Footings. LAB 1 and EQUIP OP are specified. Notice that each time the Units per day is input and accepted, *P3* automatically calculates the quantities for the **Budgeted quantity,** the **Actual to date,** and the **At completion** fields.

Budgeted quantity   The budgeted quantity is calculated by multiplying the units per day by the duration. The LAB 1 requirement for Activity 300 is 1 unit per day times a duration of 2 days equals a budgeted quantity of 2. The budgeted quantity, the to complete, and

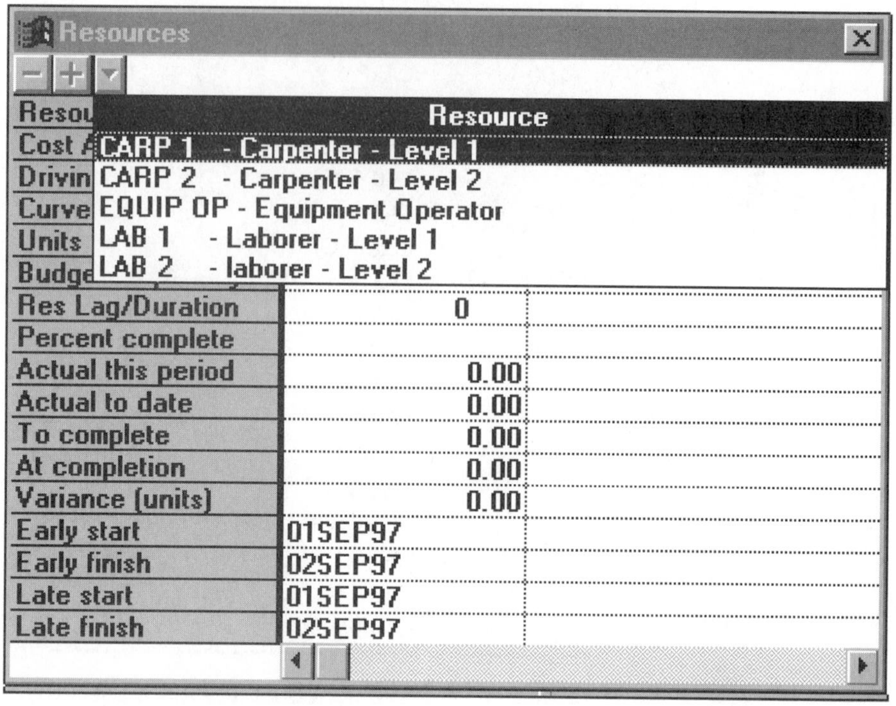

Figure 5-7 Resources Dialog Box

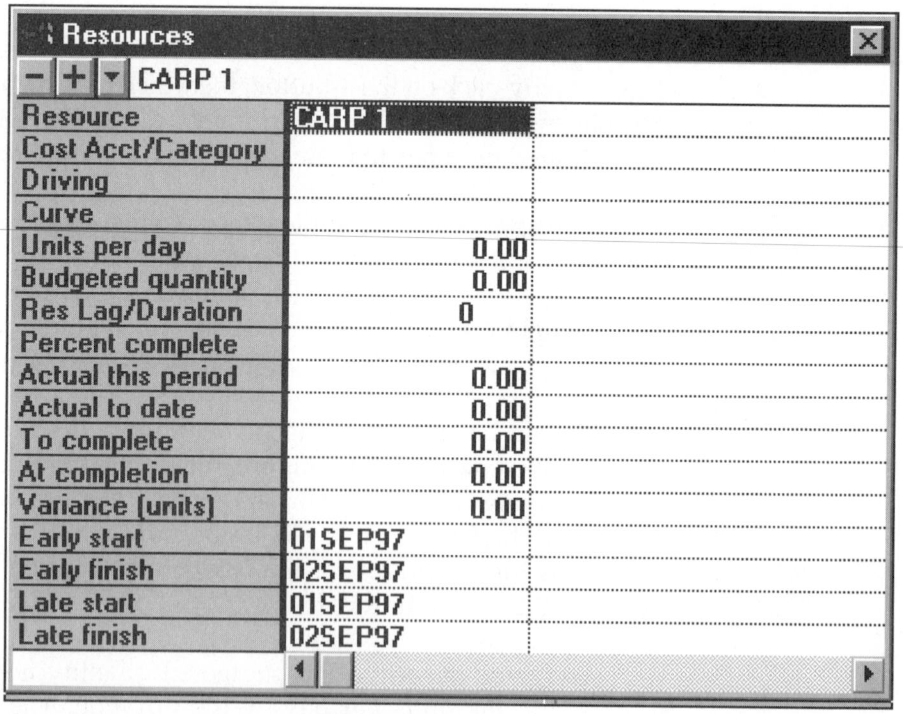

Figure 5-8 Resources Dialog Box—Resource Field

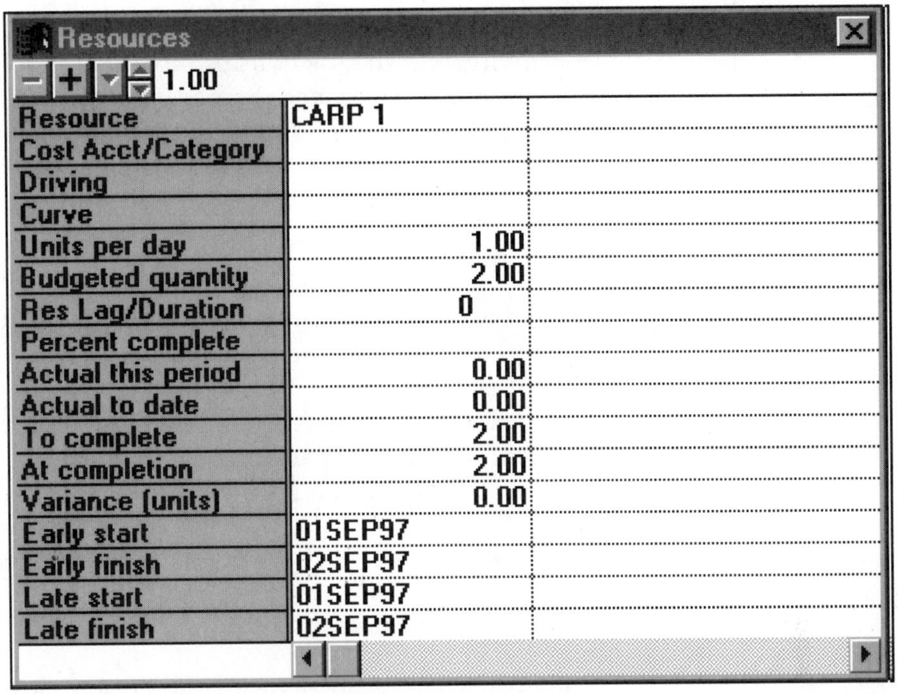

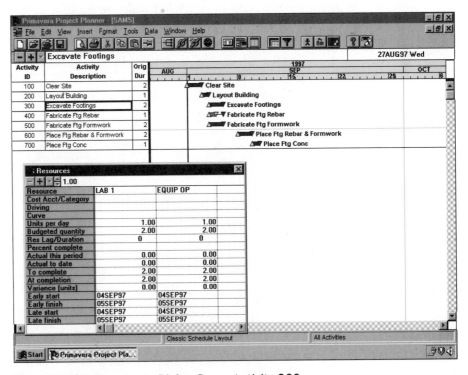

Figure 5-9   Resources Dialog Box—Units per Day Field

Figure 5-10   Resources Dialog Box—Activity 300

at completion are always the same until progress is made on the activity, and then actual resource usage is input using these fields. This is true unless the **Autocost Rules...** (select **Options** menu from the **Tools** main pull-down menu) is used to break the tie between the budgeted quantity and the at completion (click on rule #7).

Next Activity

Instead of closing the **Resources** dialog box, after inputting the resource requirements of each activity, simply click on the next activity and a blank **Resources** dialog box will appear for that activity. The **Resources** dialog box does not have to be closed until all resource requirements are defined. *Note:* If the **Activity** form is active, the **Previous** and **Next** buttons can be used to scroll through the activities.

RESOURCE PROFILES

The purpose of resource profiles and tables is to create graphical representations of the resource requirements of the project. The graphical representations make analysis simpler. In the previous sections, the resource dictionary was completed, and the individual activity requirements were defined on the **Resources** dialog box. Now that all the information is in the system, you can use the information. To do this, look at either resource profiles or tables.

To access resource profiles, click on **View** pull-down menu (Figure 5-11). Then click on the **Resource Profile**. The resource profile will appear at the bottom of the screen (Figure 5-12). The resource profile can also be accessed by clicking on the resource profile button on the button bar.

A profile is a side view or cross section of something sliced through. Notice in the **Resource Profile/Table** dialog box in Figure 5-12 that total or all resources appear in the profile. All carpenters, laborers, and equipment operators, as defined by the resource dictionary and on the **Resources** dialog box of the individual activities, are shown.

The profile shows:

Sep 1 to Sep 3—2 days—1 craftsperson

Sep 3 to Sep 4—1 day—3 craftspersons

Sep 4 to Sep 5—1 day—8 craftspersons

Sep 5 to Sep 6—1 day—6 craftspersons

Sep 6 to Sep 8—2 days—0 craftspersons (weekend)

Sep 8 to Sep 11—3 days—3 craftspersons

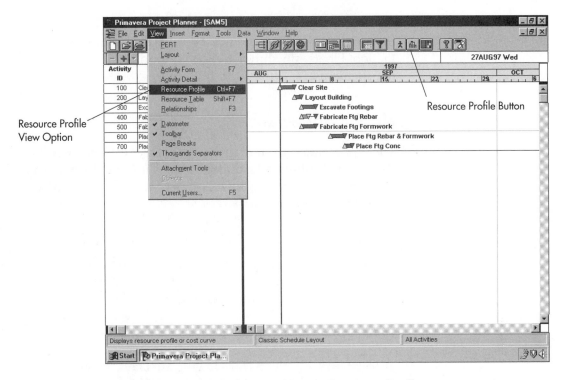

Resource Profile
View Option

Figure 5-11   View Pull-Down Menu—Resource Profile

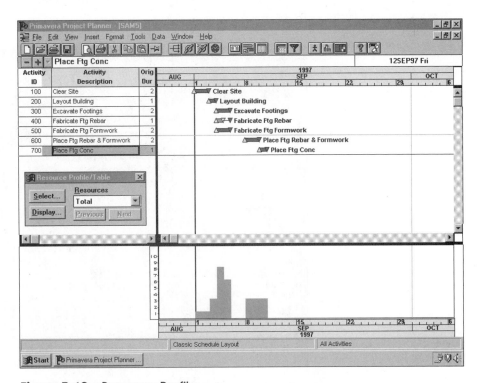

Figure 5-12   Resource Profile

Obviously, this resource profile is not very flat. The wild swing in needs from one to eight people in a week may create a problem in manning the job. Floats can shift activities around to "flatten" the peak resource requirements.

Modify Appearance

The appearance of the resource profile can be modified using the **Resource Profile/Table** dialog box. Click on the **Display...** button to cause the **Resource Profile Display Options** dialog box to appear (Figure 5-13). The options are:

Display

Units   For a resource profile, the units option is typically selected. This option produces the profile in Figure 5-12 (p. 105) with time on the horizontal axis and number of units on the vertical axis.

Costs   Choose this option for cash-flow analysis.

Time interval: Options for viewing are **Days, Weeks, Months, Quarters,** and **Years** (or on an hourly project, **Hours**). The default is whatever units the on-screen bar chart is configured to. To change the default, click on the down arrow and then click on the option preferred.

Histogram   Figure 5-12 (p. 105) is an example of a histogram showing an area for each day. If the **Bar** option is chosen, each day will have a

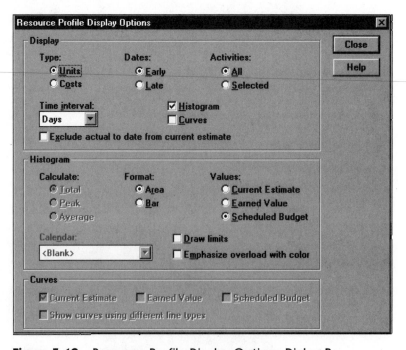

Figure 5-13   Resource Profile Display Options Dialog Box

stand-alone bar. Sometimes this makes the graphic easier to understand.

_C_urves   Check this box to superimpose a cumulative resource curve on the profile histogram.

Histogram

Total   The profile used in Figure 5-12 (p. 105) was calculated using **_T_otal**. Use **_P_eak** and **A_v_erage** when using a time interval other than a day.

If you wish to view a particular craft or resource on screen, rather than total requirements, click on the **Resources** down arrow (Figure 5-14). Notice the difference in the profile when only the CARP 1 is selected rather than the total requirements. The maximum requirements for CARP 1 (2) happens on September 4, the same day as the maximum for all crafts (8).

RESOURCE TABLES

To access resource tables, click on the **_V_iew** main pull-down menu (Figure 5-15). Then click on the **Resource _T_able**. The resource table will appear at the bottom of the screen (Figure 5-16). The resource table can also be accessed by clicking on the resource table button on the button bar.

Craft Numerical Totals

The resource table provides a numerical total for each day by craft or resource. Selecting any individual resource in the **Resource Profile/Table** dialog box (for example, CARP 1 in Figure 5-16) causes all used resources to appear (all resources below the one selected will appear alphabetically; for example, if CARP 2 is selected, CARP 1 would disappear off the

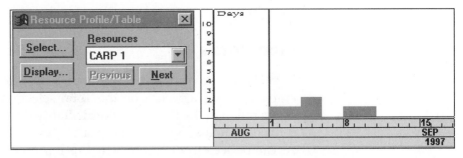

Figure 5-14   Resource Profile/Table Dialog Box—CARP 1 Resource Profile

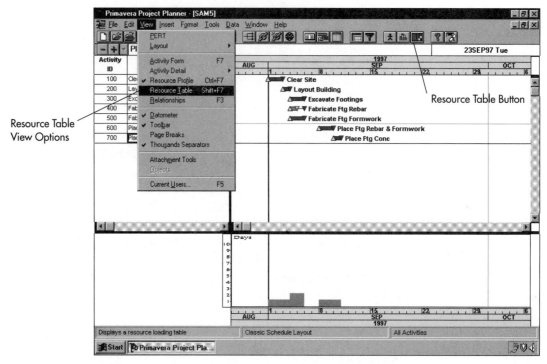

Figure 5-15 View Pull-Down Menu—Resource Table

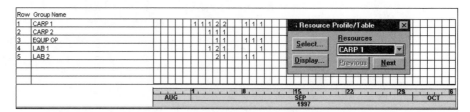

Figure 5-16 Resource Profile/Table—Detailed Resource Table

top). The numbers of persons hired as carpenters, laborers, and equipment operators as defined by the resource dictionary and on the **Resources** form of the individual activities are shown on the resource table.

Labor Totals

If totals per day are wanted, rather than the breakdown of numbers employed by craft, click on the **Total** selection on the **Resources** pull-down menu in the **Resource Profile/Table** dialog box. Figure 5-17 represents only totals. It shows the peak labor requirement on September 4. As can be seen from Figure 5-16, this represents: 2 CARP 1, 1 CARP 2, 1 EQUIP OP, 2 LAB 1, and 2 LAB 2.

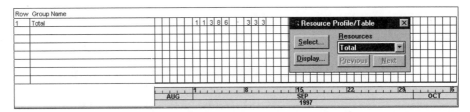

Figure 5-17   Resource Profile/Table—Total Resources

DRIVING RESOURCES

A driving resource is a resource that determines the duration of the activity to which it is assigned. *P3* automatically calculates the activity duration based on the quantity to complete (amount of work) and the units per time period (productivity rate) of the driving resource. If an activity has more than one resource, the driving resource may need to be established. The driving resource can be established by double-clicking the **Driving** check box from the **Resources** dialog box (Figure 5-18).

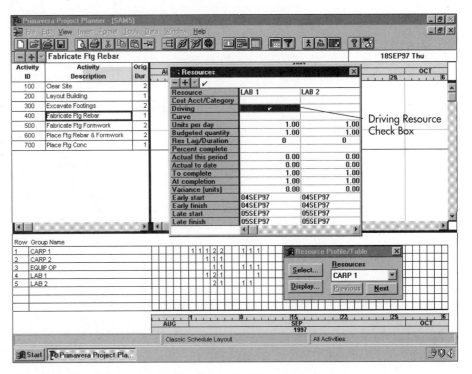

Figure 5-18   Resources Dialog Box—Driving Resource

SETTING RESOURCE LIMITS

Resource Dictionary Dialog Box

The bottom portion of the **Resource dictionary** dialog box is used to set limits on the availability of a resource (see Figure 5-19). There are three columns under **Limits:**

Normal   The typical availability of the resource.

Max   The maximum or highest availability of the resource at one time.

Through   Specify time intervals for entering cutoff dates.

Limit per Unit of Measure

Limits relate to the units used to define the resource. Use a meaningful code of up to four characters, such as HRS (hours), EA (each), or CY (cubic yards). The units for LAB 1 are days. Since the project being scheduled is daily, a better choice would have been EA (each). Entering 1 for the normal means that only one LAB 1 is normally available to the project per day. Any combination of activity requirements requiring

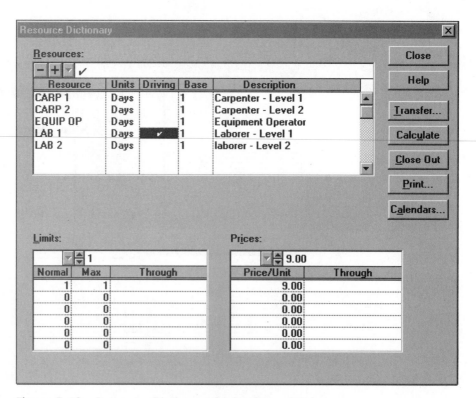

Figure 5-19   Resource Dictionary Dialog Box—Driving

more than one LAB 1 per day would be an overload and a possible problem. Also, specifying 1 as the maximum limit leaves no flexibility to exceed the normal limits of the resource for spot periods within the project.

Overload

Resources that go over the limit are easily identified in *P3*. When the resource table is engaged (click on the **Resource Table** from the **View** pull-down menu), the resource that is over a maximum limit will appear in a different color (red) from the other resources (black). A resource that is over a normal limit will appear in yellow (see Figure 5-20). This color change only happens when the resource table time interval (here it is days) matches the project planning unit (also days and set when the project was created). Activity 300, Excavate Footings, and Activity 400, Fabricate Ftg Rebar, both require a LAB 1. Both activities are scheduled for September 4. This would require the services of two LAB 1s. Since we have established a limit of 1, this would be an overload.

Another View

Another way to view the requirements for a resource is to view the resource profile. Click on **Resource Profile** from the **View** main pull-down menu (Figure 5-21). To view the LAB 1 resource, click the down arrow under **Resources** and click on LAB 1. To reconfigure the profile, click on **Display...**. The **Resource Profile Display Options** dialog box will appear (Figure 5-22). Check on the **Emphasize overload with color** and **Draw limits** check boxes. The **Normal resource usage:**, **Above normal usage:**, and **Above maximum usage:** colors may be modified by clicking the **Format** main pull-down menu, then selecting **Screen Colors:**. Click on the colored box next to the limit to be changed and a color menu will appear. As can be seen in Figure 5-21, the limits are shown at 1, and where the limits are exceeded, a different color is used to show the problem.

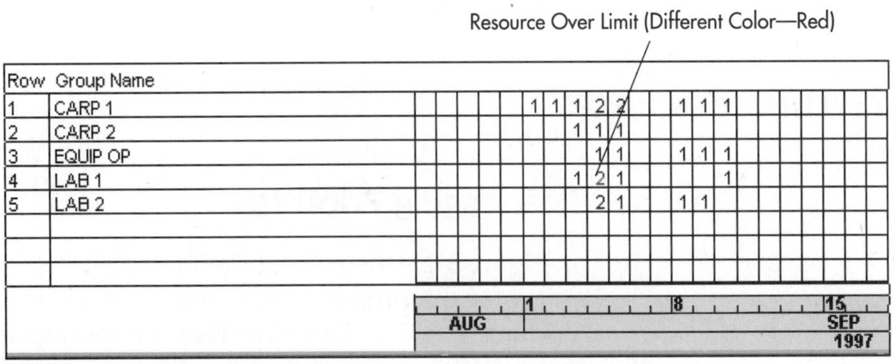

Figure 5-20 Resource Profile/Table—Over Limit Resource

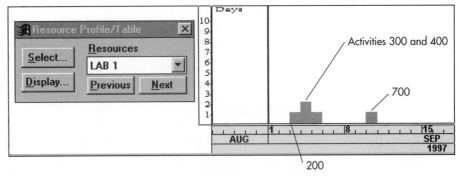

Figure 5-21 Resource Profile/Table—Resources—Activities

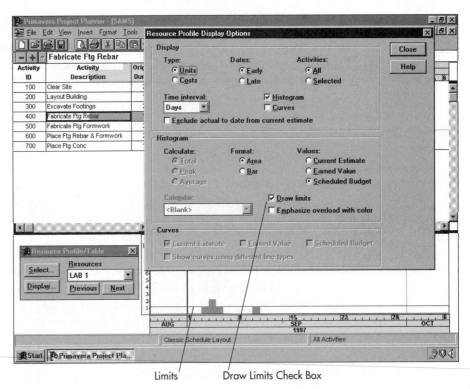

Figure 5-22 Resource Profile/Table—Resource Limits

LEVELING RESOURCES

Rearranging Priorities

In Figure 5-22, the normal limit for LAB 1 is exceeded. Rearranging priorities of the project so that the resource limit will not be exceeded is resource leveling. Resource leveling means taking the peaks out of the profile to lower the overall need for a resource, thus lessening the problems associated with mobilizing and demobilizing personnel. Of course, the first option in trying to level resources is to use positive floats to rearrange activities without prolonging project duration.

Leveling Resources   To begin *P3*'s leveling process, click on the **Level...** option on **Tools** main pull-down menu (Figure 5-23), and the **Resource Leveling** dialog box will appear.

Dialog Box for Resource Leveling   This dialog box's options are as follows:

Forward   Rearrange the activities to level resource requirements using positive floats and early dates. The object is to start the leveled activity as soon as possible.

Backward   Rearrange the activities to level resource requirements using late finishes. The object is to start the leveled activity as late as possible.

Data Date: The default data date for resource leveling is the same as the one used when scheduling the project.

Smoothing   Resource smoothing uses positive floats to minimize sharp changes in resource usage. This goes beyond just a resource exceeding the limits to smoothing or flattening the entire shape of the resource profile.

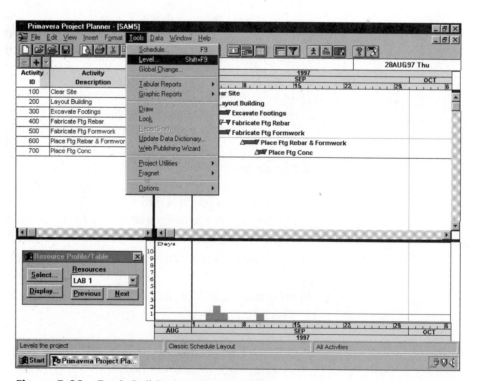

Figure 5-23   Tools Pull-Down Menu—Level

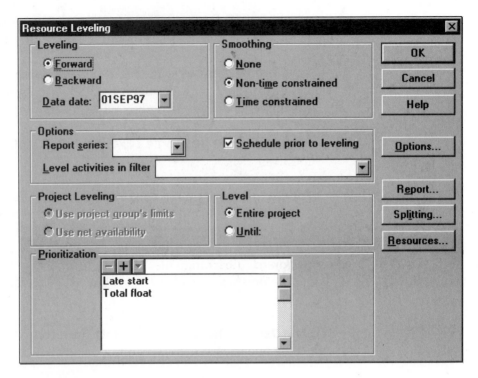

Figure 5-24 Resource Leveling Dialog Box

None   Click on this box if no smoothing is wanted.

Non-time constrained   Click on this option to override a time constraint defined in the resource dictionary.

Time constrained   If this resource is time constrained as defined in the resource dictionary, this option will not override the time constraint. Activities will not be delayed beyond their late dates.

Schedule prior to leveling   Click on this option if you want the project to use the latest version of the leveling process. Since changes may have been made since the last time the project was rescheduled (calculated), it is always a good idea to have this box checked.

Level Entire project   This is the default.

Level Until: Click on this option to impose start and finish dates on the leveling process.

Prioritization   Identifies priorities in the leveling process. The default is late start and total float, but you can change these priorities.

Note: Hierarchy (i.e., relationship) takes precedence over this prioritization.

Resources...   Clicking on this button causes the **Resource Selection** dialog box to appear (Figure 5-25). If only certain resources are to be leveled, select those resources. In this figure, LAB 1 was selected, since that was the only resource over the limit. Click on the **OK** button after completing resource selection.

OK   Click on the **OK** button to initiate the leveling process by *P3*.

Results

Figure 5-26 displays the results of the leveling process by *P3*. Compare the profile in Figure 5-26 with the one in Figure 5-21 (p. 112) to see the results of the resource-leveling exercise. Notice that Activity 400 was the only activity moved and that the project duration was extended by one day. Activity 400 had only one day of float (given our original logic), so there was no way to prevent it from going on simultaneously with Activity 300 except by extending the project duration by one day. On a larger project with more activities, relationships, and floats, a lot of manipulation can take place without extending project duration.

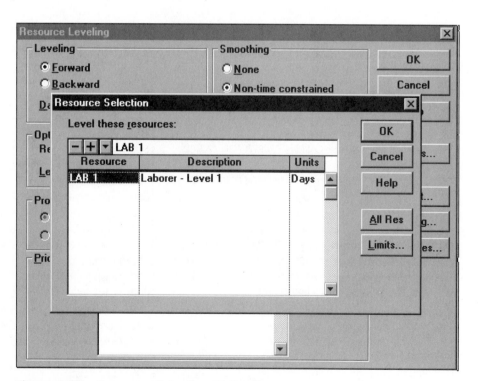

Figure 5-25   Resource Selection Dialog Box

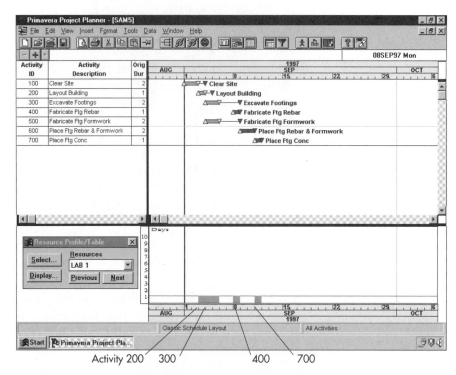

Figure 5-26   Schedule with Leveled Resources

Example Problem: Resources

Table 5-2 is an Activity List with labor resources for a house put together as an example for student use (see the wood frame house drawings in the

Act. ID	Act. Description	Labor Classification	Act. ID	Act. Description	Labor Classification
100	Clear Site	Sub	1400	Shingles	Sub
200	Building Layout	CARPENTR, CRPN FOR, LAB CL 1, LAB CL 2	1500	Ext Siding	Sub
			1600	Ext Finish Carpentry	CARPENTR, CRPN FOR
300	Form/Pour Footings	CARPENTR, CRPN FOR, LAB CL 1, LAB CL 2	1700	Hang Drywall	Sub
			1800	Finish Drywall	Sub
400	Pier Masonry	CARPENTR, CRPN FOR, CRPN HLP, MASON, LAB CL1, LAB CL 2	1900	Cabinets	Sub
			2000	Ext Paint	Sub
500	Wood Floor System	CARPENTR, CRPN FOR, CRPN HLP	2100	Int Finish Carpentry	CARPENTR, CRPN HLP
600	Rough Framing Walls	CARPENTR, CRPN FOR, CRPN HLP	2200	Int Paint	Sub
700	Rough Framing Roof	CARPENTR, CRPN FOR, CRPN HLP	2300	Finish Plumbing	Sub
800	Doors & Windows	CARPENTR, CRPN HLP	2400	Finish HVAC	Sub
900	Ext Wall Board	CARPENTR, CRPN FOR, CRPN HLP	2500	Finish Elect	Sub
1000	Ext Wall Insulation	Sub	2600	Flooring	Sub
1100	Rough Plumbing	Sub	2700	Grading & Landscaping	Sub
1200	Rough HVAC	Sub			
1300	Rough Elect	Sub	2800	Punch List	

Table 5-2   Activity List with Labor Resources—Wood Frame House

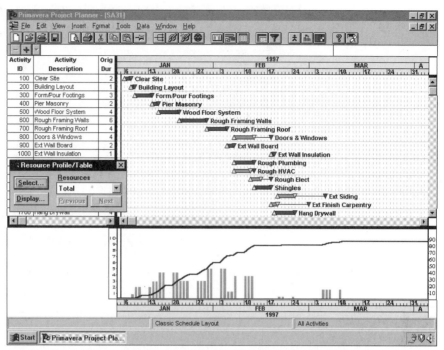

Figure 5-27   Total Resources Profile—Wood Frame House

Appendix). Figures 5-27 and 5-30 contain additional resource information. The on-screen total resources profile (Figure 5-27) was constructed using the tabular list of twenty-eight activities (activity ID, description, and labor classification). Figures 5-28 and 5-29 are a resource profile and a resource table for the CARPENTR craft. Figure 5-30 is a total resource table.

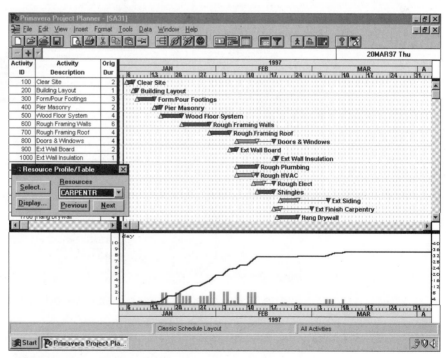

Figure 5-28   CARPENTR Profile—Wood Frame House

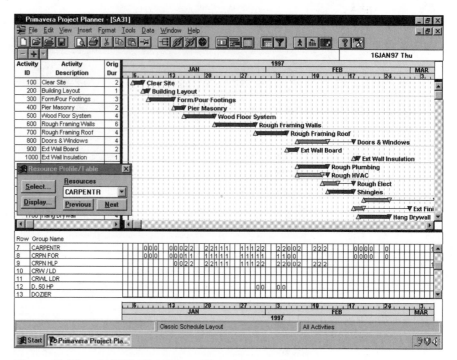

Figure 5-29   CARPENTR Table—Wood Frame House

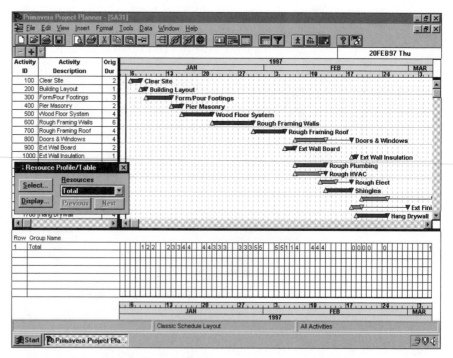

Figure 5-30   Total Resources Table—Wood Frame House

EXERCISES

1. Small Commercial Concrete Block Building—Resources

Prepare an on-screen bar chart for the small commercial concrete block building located in the Appendix. This exercise should include the following steps:

1. Prepare a list of labor resources required by activity (minimum of sixty activities).
2. Create the on-screen resource profiles.
3. Create the on-screen resource tables.

2. Large Commercial Building—Resources

Prepare an on-screen bar chart for the large commercial building located in the Appendix. This exercise should include the following steps:

1. Prepare a list of labor resources required by activity (minimum of 150 activities).
2. Create the on-screen resource profiles.
3. Create the on-screen resource tables.

6

Using *P3* to Calculate Costs

Objectives

Upon completion of this chapter, the reader should be able to:

- Modify cost accounts dictionary
- Assign activity costs
- Use cost profiles
- Use cost tables
- Use cumulative costs

NECESSITY FOR CONTROLLING COSTS

Financing

Money   The most important resource needed for building is money. Scheduling and controlling the expenditure of funds is critical to the building process. The contractor must finance the project. Having the amount necessary and at the right time is a tricky business for the contractor. Ultimately, the owner pays for a project through a contractor. But the contractor must make sure that money is available to finance interim periods until the monthly payment from the owner is available. With a good estimate, accurate cash projections, and intelligent distribution of funds, the contractor can finance the project without having to borrow funds. If the contractor has to borrow funds, a potential source of profit is lost through interest payments to a banker.

Conflict   Assume a contractor's bid to an owner for a new project is $10 million. The contractor never has that much money in the bank, ready to completely finance the project, to be reimbursed when the project is complete. The contractor depends on progress payments to recoup costs as the project is being put in place. Every contractor wants to finance the project completely with the owner's money, and every owner wants the contractor to have some money at financial risk in the project.

Cash Needs   The contractor has to be able to finance not only the particular project, but other projects he or she is building and have funds to finance the home office. Monies have to be available to meet payroll costs each week for the craftspersons who work directly for the contractor. Monies have to be available each month to pay for materials and supplies used at the job site and for project subcontractors. The contractor also must finance the construction equipment used at the job site, whether owned by the company or rented.

Timing of the Expenditures   The contractor needs to maintain cash flow in order to finance projects without having to borrow funds or take cash from accounts that are earning interest. The goal is to have each project stand on its own. This requires:

- Distribution of cost by activity
- Tracking of cost by activity
- Control of payment request to owner
- Control of labor productivity and costs
- Control of material and supplies cost
- Control of subcontractors
- Control of overhead costs
- Control of payment of funds

Owner's Requirements

Cost Control   The owner is as concerned about controlling the expenditure of funds as controlling time. Most project contracts call for the contractor to be paid monthly as the project progresses. The owner needs a way to make sure that not only is the project progressing toward a satisfactory completion, but funds are being properly spent and the contractor is not being overpaid according to progress accomplished on the project.

Progress Measures   The primary reason for the owner requiring a schedule is to have a document to evaluate the physical progress of the contractor as the project progresses. Usually, within a short period of time after the contract is signed, the contractor must turn over to the owner two documents. The first is the schedule, and the second is the schedule of values. The schedule of values is a breakdown of cost by category or phase of work. When the contractor turns in a pay request at the end of the month, a judgment is made as to the percent completion of each category. Since this judgment is usually subjective, there is room for argument. Many owners now require a cost-loaded logic diagram. The primary advantage is a greater breakdown of cost. The process of cost loading the activities and coming to an agreement at the beginning as to their value reduces potential conflict. There is less argument at each pay period as to percent complete, since the completion of a particular activity is much easier to judge than progress in broad phases of cost.

The purpose of this chapter is to address the planning of cash needs, expenditures, and the timing of the expenditures so that the contractor is placed in a constant positive cash position.

COST ACCOUNTS DICTIONARY

Existing Cost Account Structure

Cost Accounts Dictionary Dialog Box   Like the resources in Chapter 5, a cost structure must be implemented in the *P3* dictionary before activity costs can be input. This dictionary is called cost accounts. Click the **Data** main pull-down menu and select **Cos̱t Accounts...** (Figure 6-1). The **Cost Accounts Dictionary** dialog box will appear (Figure 6-2).

Consistency   The cost account structure can be reinput for each new project. But, this is usually not the case. Most construction companies have a cost account structure already defined for estimating and accounting purposes. The account structure will be consistent from project to project. If the construction company is primarily a building

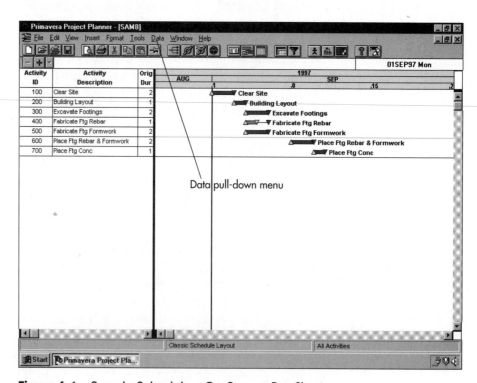

Figure 6-1   Sample Schedule—On-Screen Bar Chart

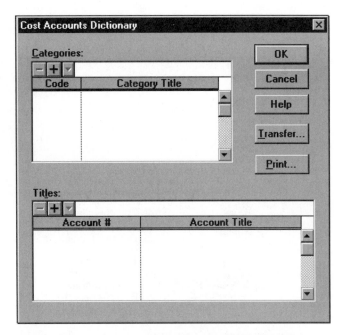

Figure 6-2   Cost Accounts Dictionary Dialog Box

contractor, the structure will probably be Construction Specifications Institute format. Consistency from project to project is critical for personnel to understand and communicate the cost code system. Consistency is necessary to develop historical information to use as a basis for the estimating and scheduling systems. This historical information is a history of the performance of a construction company.

Copying a Previous Cost Code Structure   It is possible to copy a cost account structure from another project to be used on a new schedule, as follows.

Transfer...   Click on the **Transfer...** button in the **Cost Accounts Dictionary** dialog box. The warning **Transfer will overwrite your dictionary** caution box will appear on screen (Figure 6-3). Since the schedule is new, there is nothing in the dictionary. If cost account information is input and then a structure is copied from another project, the original information would be overwritten when the

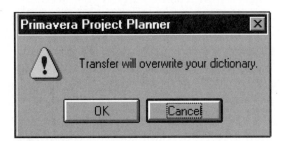

Figure 6-3   Transfer Error Message

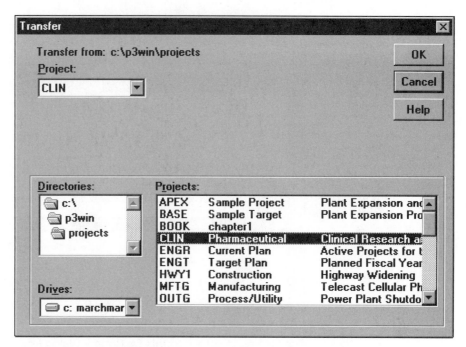

Figure 6-4   Transfer Dictionary Dialog Box

account code structure is transferred. As a rule, it is best to copy the cost account structure from another schedule when setting up the new schedule.

Projects: The **Transfer** dialog box appears when you click the **Transfer...** button and the **OK** button is selected from the **Transfer will overwrite your dictionary** caution box. Scroll the **Projects:** listing in the lower right to find the project to be used as a basis for the cost account structure. Then double-click on the project, and the project name will appear in the **Project:** field in the upper left (Figure 6-4). Click on the **OK** button to transfer the account codes from the chosen project to the new project.

New Cost Code Structure

Dialog Box for Cost Accounts Dictionary   If copying from an existing project is not feasible, a new cost account structure must be added. Click on the + button of the **Cost Accounts Dictionary** dialog box (Figure 6-5). The cost accounts dictionary has two components: **Categories:** and **Titles:**.

Categories: A cost category could be anything needed, but the usual categories are: L for labor costs; M for material costs; E for equipment costs; S for subcontractor costs; and O for other costs. When an activity is identified by account code for the work being done on the activity, the category of work must also be identified. This is convenient for

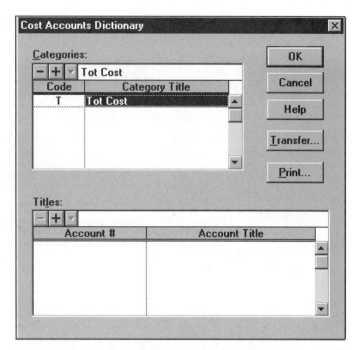

Figure 6-5   Cost Accounts Dictionary Dialog Box—Categories

sorting work and entering actual cost from the project as the work is erected. Loading the schedule with costs from the original estimate gives it the same amount of detail as the original estimate by category. For the purposes of this book and to simplify the discussion, only one category, T for total costs, will be used.

Category Title   After entering T in the code field, either strike the right arrow key or click in the **Category Title** field, and enter an eight-character description of the field. To enter more categories, either click the + button again or click under the category just entered.

Titles: Entering new cost accounts works the same way as adding categories. Go to **Titles:** in the **Cost Accounts Dictionary** dialog box. Click the + button to add new accounts (Figure 6-6). Here cost account 2104, Clear and Grub, was added. Keep adding account codes until the entire account code structure is added (Figure 6-7). Use the slide bar to the right of the **Account Title** box to scroll through the accounts field to find the wanted account. The number of account codes added can be as large as needed. Remember to simplify the work by using as few account codes as necessary to properly define the work.

If it is necessary to add an account code number between two accounts already entered, simply click on the **Account #** field under the account where you want to add the new account. This highlights the field. Depress the Insert key on the keyboard, or click on the + button on the **Titles:** button bar. Since *P3* automatically alphabetizes the new accounts, they can be input in any sequence.

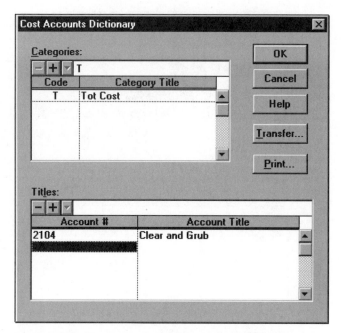

Figure 6-6 Cost Accounts Dictionary Dialog Box—Titles

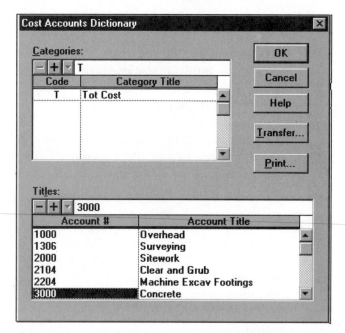

Figure 6-7 Cost Accounts Dictionary Dialog Box—Titles Account Number

ASSIGN COSTS

Cost Dialog Box

Now that the cost code structure has been defined in the dictionary, it is time to enter costs per activity. For the purposes of this chapter, total costs per activity is used rather than breaking costs down into the categories of labor, materials, subcontractors, indirect, and possibly other costs. To enter costs per activity, select (click) on the activity for which costs are to be entered. (See Activity 100, Clear Site in Figure 6-8.) The selected activity will become highlighted. Now, click the right mouse button to bring up a pull-down menu. Click **Activity Detail** and then **Cost** and the **Cost** dialog box appears (Figure 6-9).

Cost Acct/Category   To add costs to Activity 100, click the + button. Next, click on the **Cost Acct/Category** field (Figure 6-10). The **Cost Acct/Category** field will become highlighted, meaning it has been selected. Click on the down arrow on the **Cost** dialog box to bring up

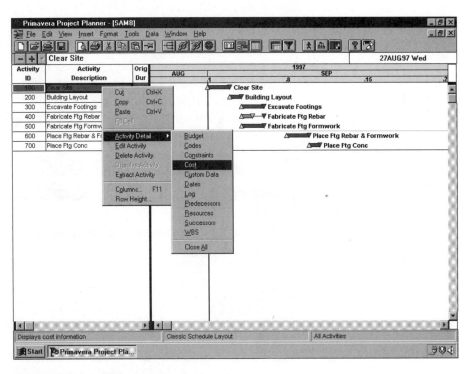

Figure 6-8   Activity Detail—Cost

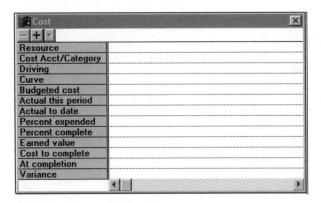

Figure 6-9   Cost Dialog Box

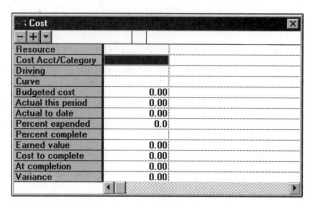

Figure 6-10   Cost Dialog Box—Cost Acct/Category

a listing of the **Cost Account Code** dictionary listing (Figure 6-11). If the dictionary has more than one page of entries, a slide bar will appear at the right of the listing to enable you to scroll to find the necessary account code. For Activity 100, select account 2104, Clear and Grub. Click on the selected account, and it will appear in the **Cost Acct/Category** field in the **Cost** dialog box (Figure 6-12).

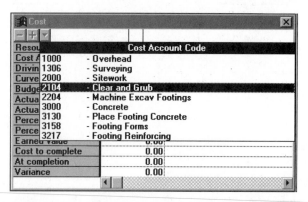

Figure 6-11   Cost Dialog Box—Cost Account Code

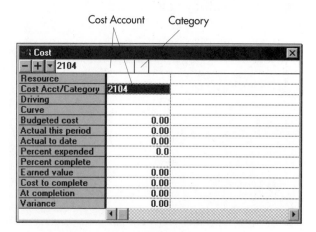

Figure 6-12   Cost Dialog Box—Cost Account

Category Next, establish the category to which to allocate the costs. Notice in Figure 6-12 (p. 128) that when the cost account was selected, a field to the right of the account entry on the button bar line appeared. The only category established earlier in the chapter was T for total costs. Manually type T in the field and depress the Enter key on the keyboard (Figure 6-13). Notice from Figure 6-14, that the **Cost Acct/Category** is actually two fields: the cost account field (2104) and the category field (T).

Budgeted cost Next, click on the **Budgeted cost** field to enter the monies allocated to Activity 100, Clear Site. Notice when this field is selected, a numeric field is opened on the button bar, with up and down arrow buttons. Either click on the numeric field and type in the budgeted cost number, or use the up and down arrow buttons to identify the number. For Activity 100, enter 1000.00. The budgeted cost is $1,000. Note the conservative assumption: "All project costs are paid for the day used." The actual cash flow will lag this projected cash flow.

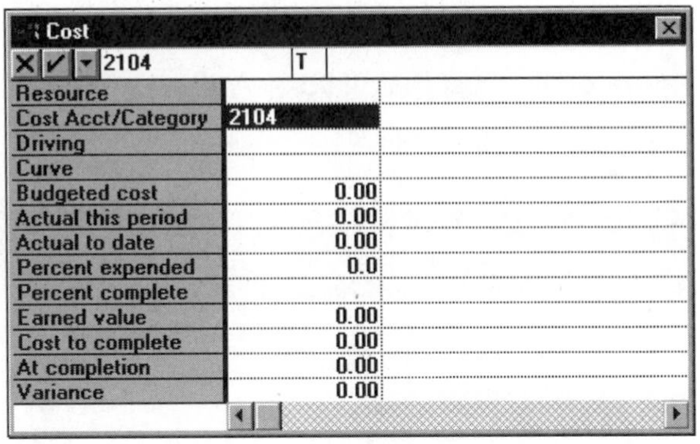

Figure 6-13 Cost Dialog Box—Cost Category

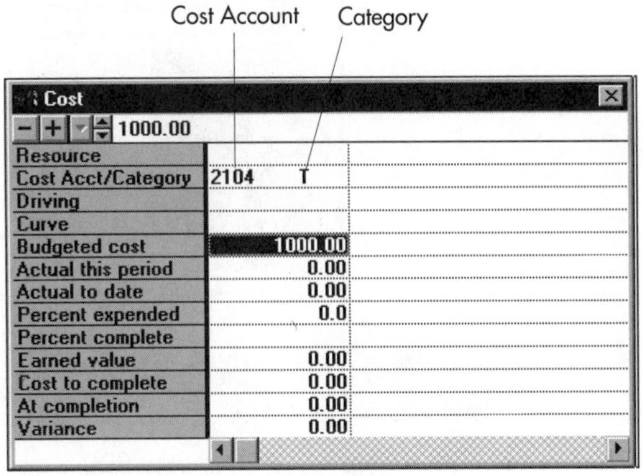

Figure 6-14 Cost Dialog Box—Budgeted Cost

Notice in Figure 6-15 that the $1,000 is also automatically entered into the **Cost to complete** and the **At completion** fields. When actual costs are entered during the construction process, these two fields will change.

Next Activity   To enter costs for the next activity (Activity 200, Building Layout), click on that activity (Figure 6-16). It is not necessary to close the **Cost** dialog box to continue entering activity cost informa-

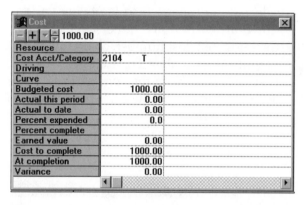

Figure 6-15   Cost Dialog Box—At Completion

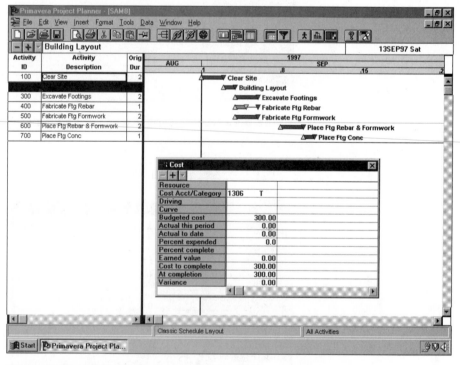

Figure 6-16   Cost Dialog Box—Activity 200

tion. When finished entering the cost information for an activity, simply click on the next wanted activity. The information input for Activity 200, Building Layout is: cost account code 1306; category T; and budgeted cost of $300.00. The $300.00 cost at completion and the $300.00 at completion were established by *P3*.

COST PROFILES

Resource Profile/Table Dialog Box

To view the information in the cost account dictionary and in the cost form at the activity level click on the **View** main pull-down menu (Figure 6-17). Then click on the **Resource Profile** option. The **Resource Profile/Table** dialog box will appear at the bottom of the screen (Figure 6-18).

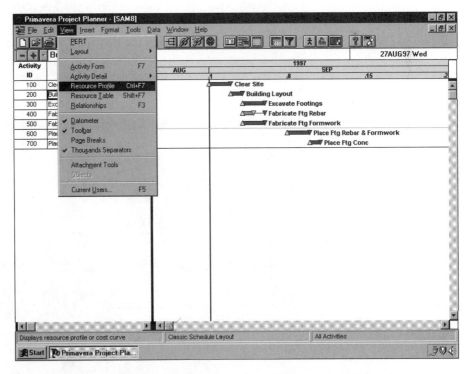

Figure 6-17 View Main Pull-Down Menu—Resource Profile Option

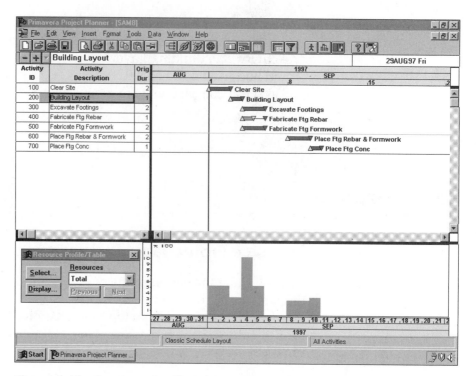

Figure 6-18 Resource Profile Table—Cost Profile

Resource Profile Display Options

Configuration   If, in viewing the resource profile, no usable information is included, the table graphic must be configured. Click on the **Display...** button of the **Resource Profile/Table** dialog box, and the **Resource Profile Display Options** dialog box will appear (Figure 6-19). Under **Display**, click on the **Costs** option. In Chapter 5 when you wanted to

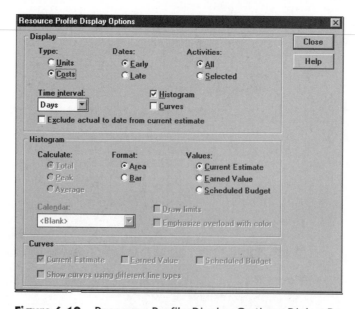

Figure 6-19   Resource Profile Display Options Dialog Box

view resources you clicked on **Units** using this same dialog box. Also, make sure to check the **Histogram** option box. Choosing the **Area** or **Bar** option under **Histogram** depends on personal preference in the appearance on the cost profile. When finished, click the **Close** button.

Resources   If the desired information still does not appear on the cost profile, the **Resources** selection (in the **Resource Profile/Table** dialog box) must be changed. Note that choosing an individual resource (such as CARP 1) causes a problem. When configuring the resource dictionary, costs at the craft level were not included. If the price per unit for CARP 1 had been included in the resource dictionary, there would be monies appearing at the detail level.

Category   When setting up the cost accounts dictionary, only one cost category was input, total cost (T) per activity. Click on the down arrow key under **Resources** and the menu of the resource dictionary will appear. Click **Total**, and a histogram appears in the cost profile window.

Axes   The horizontal axis on the cost profile is calendar days. The vertical axis is dollars. The shaded area within the histogram represents the cost per day that is planned to be expended on a project according to the predecessor/successor logic and the estimated cost per activity. The peak cost expended per day appears to be $1,000 ($10 \times 100$) from Figure 6-18. The cost profile is useful, but a table with the exact dollar values per day to be expended may be more useful.

Cost Table

Resource Profile/Table Dialog Box

To change from the cost profile to the cost table, click on the **View** main pull-down menu (Figure 6-20). Then click on the **Resource Table** option, and the cost table will appear (Figure 6-21). Obviously the cost table in Figure 6-21 must be configured so that the information is readable. Steps need to be taken to make the information usable. First, if the **Resource Profile/Table** dialog box is in the way of viewing, click and hold the mouse on the title bar and drag it out of the way.

Timescale Definition   Next, the timescale needs to be adjusted to make the numbers in the table large enough to read. Double-click on the timescale bar. The **Timescale Definition** dialog box will appear (Figure 6-22). The slide scale **Density:** bar controls the timescale spacing of days (time units) of the on-screen bar chart and cost table. Click and hold the button on the slide bar and move it to the right. The timescale or on-screen space allotted to each time unit will be increased. Between Figures 6-21 and 6-23, the button was moved about $\frac{1}{2}$ inch. To try the new spacing, click on the **OK** button. Now the numbers appearing in the

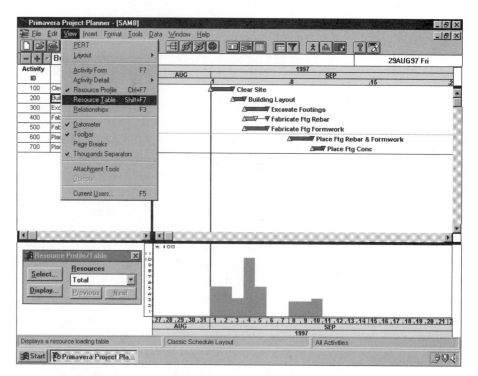

Figure 6-20   View Main Pull-Down Menu—Resource Table Option

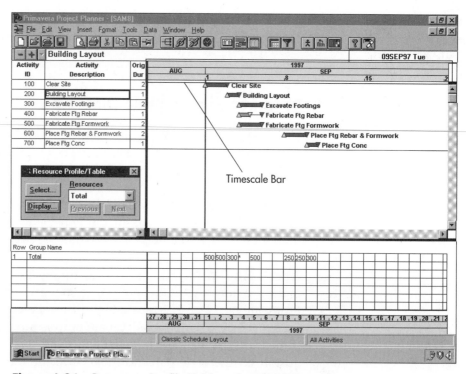

Figure 6-21   Resource Profile/Table—Cost Table

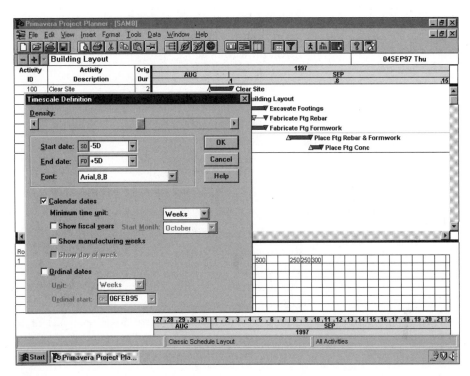

Figure 6-22   Timescale Definition Dialog Box

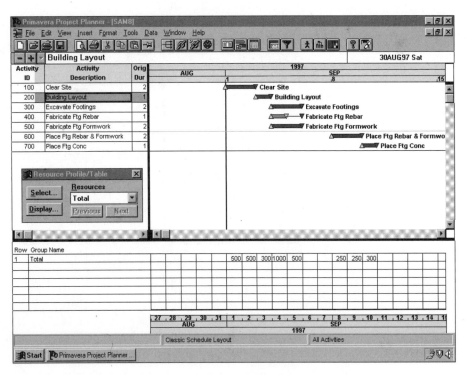

Figure 6-23   Resource Profile/Table—Expanded-View Cost Table

cost table are clearly legible. The cost per day information in the cost profile and the cost table is helpful, but cumulative costs are also useful information.

CUMULATIVE COSTS

Resource Profile/Table Dialog Box

Cumulative Curve   From the **View** main pull-down menu, click on the **Resource Profile**. The cost profile, displaying cost per day, appears at the bottom of the screen (Figure 6-24). Click the **Display...** button on the **Resource Profile/Table** dialog box. The **Resource Profile Display Options** box appears (Figure 6-25). The only change from the last time this box was used (Figure 6-19, p. 132) is the check box for **Curves**. Notice that as soon as the box is clicked, the cumulative curve appears in the cost profile screen in Figure 6-26.

Area Under the Curve   The area under the curve represents the cumulative amount of money spent on the project to date. The vertical

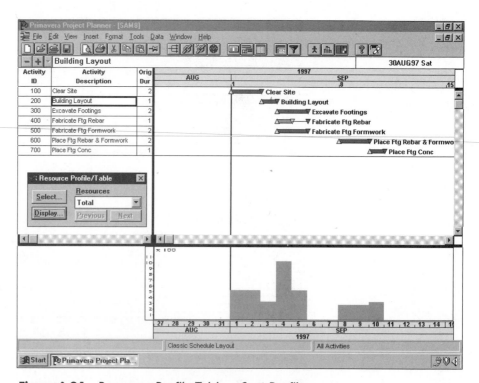

Figure 6-24   Resource Profile/Table—Cost Profile

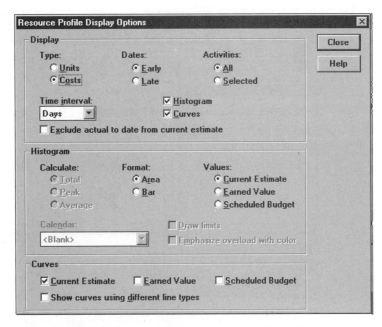

Figure 6-25 Resource Profile Display Options Dialog Box

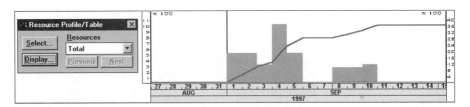

Figure 6-26 Resource Profile/Table—Cost Profile with Curves

scale on the right of the cost profile represents total cumulative cost for the project. Here, the total monies anticipated to be spent (cash flow), for the project is $3,600 (36 × 100). No money is spent until the beginning of the day on September 1. The last activity for which money is expended takes place on September 10. The $3,600 that is anticipated to be spent on the project will be spent between September 1 and 10. The height of the curve shows the monies to be spent as of any particular day.

View The example project has few activities and a short duration. Obviously, the typical construction project would have many more activities and a longer duration, but the entire cumulative cost curve could still be viewed on a single screen. Again, double-click on the on-screen bar chart timescale. The **Timescale Definition** dialog box will appear. Observe what happens when you move the slide bar button to

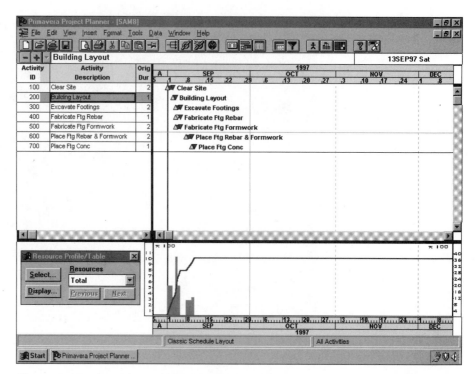

Figure 6-27   Resource Profile/Table—Cost Profile with Curves, Condensed View

the left an inch or so and click on the **OK** button. As can be seen from Figure 6-27, not only does this condense the on-screen bar chart timescale, but it also modifies the cost profile. In Figure 6-26 (p. 137), less than a month was being observed, while in Figure 6-27, an approximate four-month time frame can be evaluated. If the project is of a longer duration, say a year, adjusting the timescale will show the cash requirements for the entire project (by day and cumulative) on a single screen.

EXAMPLE PROBLEM: Cost

Table 6-1 is an activity list with costs for a house put together as an example for student use (see the wood frame house drawings in the Appendix). Figures 6-28 to 6-32 provide additional resource information. The on-screen resource CARPENTR requirement for Activity 300 (Figure 6-28) was prepared as an example of resource cost input. The cost profiles (detailed, Figure 6-29, and total, Figure 6-30) were prepared using the tabular list of twenty-eight activities (activity ID, description, and total activity costs). The cost tables (detailed, Figure 6-31, and total, Figure 6-32) are also provided.

Act. ID	Act. Description	Costs
100	Clear Site	$1,280.00
200	Building Layout	$ 130.49
300	Form/Pour Footings	$1,174.32
400	Pier Masonry	$ 967.29
500	Wood Floor System	$4,181.04
600	Rough Framing Walls	$3,323.99
700	Rough Framing Roof	$3,468.40
800	Doors & Windows	$3,995.40
900	Ext Wall Board	$ 409.97
1000	Ext Wall Insulation	$ 385.32
1100	Rough Plumbing	$ 750.00
1200	Rough HVAC	$1,168.75
1300	Rough Elect	$ 940.00
1400	Shingles	$1,091.29

Act. ID	Act. Description	Costs
1500	Ext Siding	$1,710.54
1600	Ext Finish Carpentry	$ 190.06
1700	Hang Drywall	$1,844.60
1800	Finish Drywall	$ 790.54
1900	Cabinets	$1,618.00
2000	Ext Paint	$ 525.00
2100	Int Finish Carpentry	$1,203.15
2200	Int Paint	$4,725.00
2300	Finish Plumbing	$3,000.00
2400	Finish HVAC	$3,506.25
2500	Finish Elect	$1,410.00
2600	Flooring	$1,583.34
2700	Grading & Landscaping	$ 600.00
2800	Punch List	

Table 6-1   Activity List with Costs—Wood Frame House

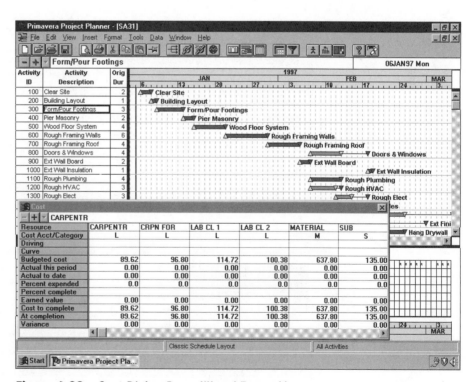

Figure 6-28   Cost Dialog Box—Wood Frame House

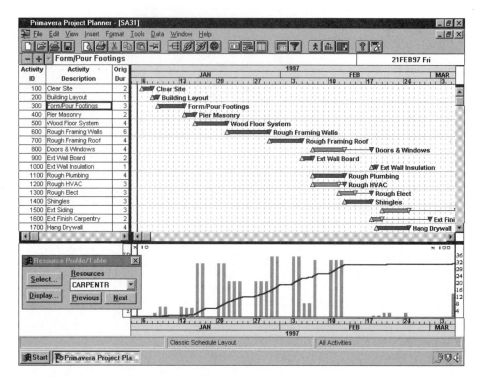

Figure 6-29   CARPENTR Cost Profile—Wood Frame House

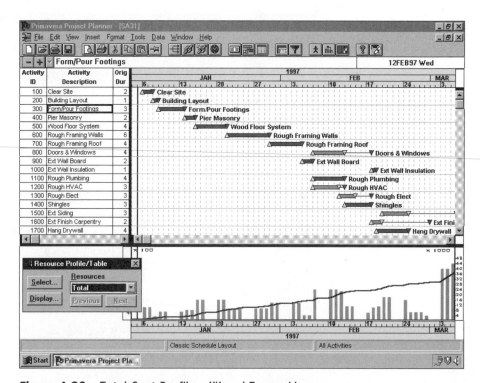

Figure 6-30   Total Cost Profile—Wood Frame House

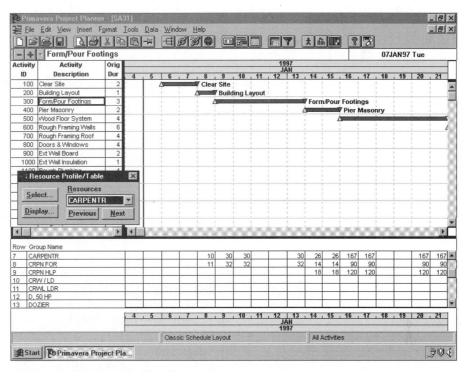

Figure 6-31   CARPENTR Cost Table—Wood Frame House

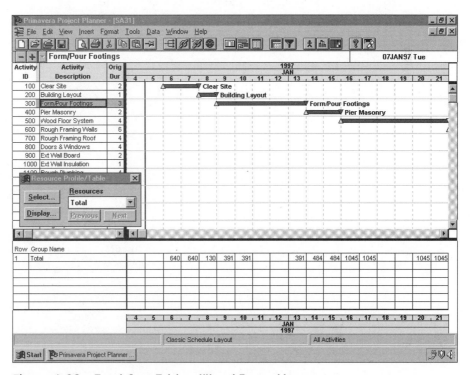

Figure 6-32   Total Cost Table—Wood Frame House

EXERCISES

1. Small Commercial Concrete Block Building—Costs

Prepare an on-screen cost profile and cost table for the small commercial concrete block building located in the Appendix. This exercise should include the following steps:

1. Prepare a list of labor resources required by activity.
2. Assign total costs per activity.
3. Create the on-screen cost profile.
4. Create the on-screen cost table.

2. Large Commercial Building—Costs

Prepare an on-screen cost profile and cost table bar chart for the large commercial building located in the Appendix. This exercise should include the following steps:

1. Prepare a list of labor resources required by activity.
2. Assign total costs per activity.
3. Create the on-screen cost profile.
4. Create the on-screen cost table.

Bar Chart Presentation Using *P3*

Objectives

Upon completion of this chapter, the reader should be able to:

- Modify on-screen bar chart printout
- Print on-screen bar chart
- Create new tools, bar chart printout
- Modify tools, bar chart printout
- Print tools, bar chart

NECESSITY FOR GOOD PRESENTATION

Obviously putting a schedule together at the beginning of a project is valuable. The process forces you to plan, organize, sequence, and show the interrelationships among different activities or functions of the project. This information is useless, however, unless it is communicated to all relevant parties on the project. Good scheduling involves disseminating and getting feedback from all key parties regarding the original plan, updates, and progress toward completion of the project. At present, the easiest and most efficient way to disseminate the computerized *P3* schedule is through hard copy (paper) graphic or tabular reports.

GRAPHIC REPORTS

A graphic report is a printout or plot depicting information about the schedule. A plotter produces large drawings (Table 7-1). The plotter offers multiple colors, line types, line widths, and shading capabilities for producing professional presentations. Graphic printers are usually of the laser jet or bubble jet variety. Multiple colors are an option with these printers and the paper is usually letter size (8.5″ × 11″) or legal

U.S.	Letter	8 $\frac{1}{2}$ x 11 in.
	Legal	8 $\frac{1}{2}$ x 14 in.
ANSI	B	11 x 17 in.
	C	17 x 22 in.
	D	22 x 34 in.
	E	34 x 44 in.
Architectural	A	9 x 12 in.
	B	12 x 18 in.
	C	18 x 24 in.
	D	24 x 36 in.
	E	36 x 48 in.
ISO	A4	210 x 297 mm
	A3	297 x 420 mm
	A2	420 x 594 mm
	A1	594 x 841 mm
	A0	849 x 1189 mm

Table 7-1   Plotter Print Sizes

size (8.5" × 14"). Whether you choose a printer or a plotter depends primarily on the size of the hard copy you wish to produce.

Primavera calls the graphic reports of *P3* bar charts, timescaled logic diagrams, pure logic diagrams, resource/cost profiles, and curves. The bar chart graphic reports will be covered in this chapter.

ON-SCREEN BAR CHART

Simplicity

As mentioned in earlier chapters, the primary advantage of the bar chart is its overall simplicity. It is easy to read and interpret, and therefore can be an effective communications tool. The primary disadvantage is that it does not show the interrelationships among project activities. If a certain activity is delayed, it is difficult to determine the impact on the rest of the schedule. The bar chart is undoubtedly the most common print form of *P3* used.

A hard copy of the *P3* bar chart can be produced in one of two ways. It can be accessed either from the **File** or the **Tools** pull-down menus.

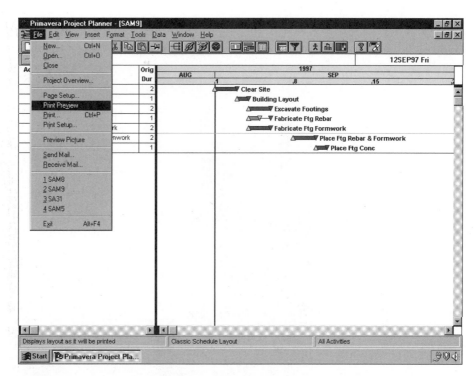

Figure 7-1   File Main Pull-Down Menu—Print Preview Option

Each method offers different options and capabilities, so both will be covered in detail.

File Pull-Down Menu

Accessing the bar chart print from the **File** pull-down menu (Figure 7-1) has the advantage of simplicity of usage over accessing it from the **Tools** pull-down menu. The disadvantage is that not as many options are available to modify and customize the report. The **File** option gives essentially a copy of the on-screen bar chart. There are four bar chart print functions available from this menu: **Print Preview**, **Page Setup...**, **Print...**, and **Print Setup...**.

Print Preview

Page tabs   First, click on the **Print Preview** option, and the project bar chart will appear on the screen (Figure 7-2). It is an on-screen image of the hard copy print of the *P3* bar chart. The on-screen image can be manipulated using the button bar at the top of the screen. The first four buttons, at the top left of the screen, are arrows representing page tabs. The schedule print can be composed of multiple pages. Use the horizontal arrows (right and left) to scroll the pages horizontally and the vertical arrows (up and down) to scroll pages vertically. In Figure 7-2, three of the four arrows are "grayed out," meaning the hard copy print of the sample project is composed of a single page.

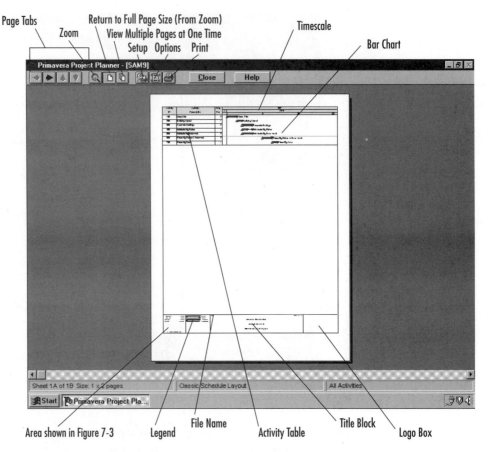

Figure 7-2 Print Preview

Zoom The next button to the right is a picture of a magnifying glass. This initiates the *zoom function*. When the cursor is outside the boundaries of the reproduction of the hard copy page it appears as an arrow, its normal appearance. When it is inside the boundaries of the page, it appears as a magnifying glass. Steps to using the zoom function are:

Step 1. Move the cursor (magnifying glass) to the location to be zoomed (or magnified).
Step 2. Click on the area to be magnified. Figure 7-3 shows the blowup of the lower left of the schedule selected from Figure 7-2.

Return to full page size To return to a full-page view, click on the button to the right of the zoom button. The button has a picture of a page with a bent right corner.

View multiple pages The next button to the right of the **Return to full page size** button shows a picture of a stack of pages. If the hard copy print of the schedule is more than one page, click on this button to view multiple pages at the same time.

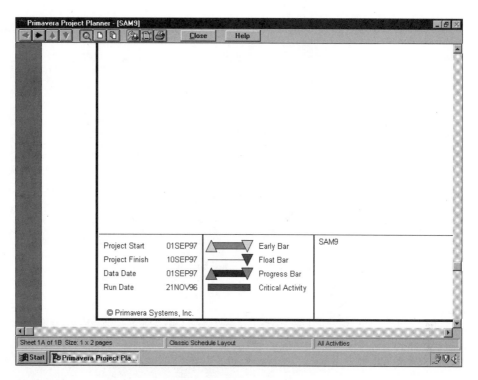

Figure 7-3   Print Preview—Zoomed

Print Options   The next three button bar buttons from the on-screen **Print Preview** option are **Setup, Options**, and **Print**. Although these are the same options available from the **File** pull-down menu, the advantage here is that the print setup and options can be modified and a hard copy print executed without leaving print preview.

Close   If, after viewing the copy of the hard copy print of the bar chart in the print preview mode, you decide no other action is needed, click the **Close** button. The user is returned to the on-screen bar chart mode of *P3* and out of print preview.

Page Setup   From the **File** pull-down menu, click on **Page Setup....** The **Page Setup** dialog box will appear (Figure 7-4). The two primary areas are **Print** and **Page settings**.

Start date: The options in the **Print** area can be used to control the length of the bar chart. Click on the down arrow next to **Start date:** and a calendar pop-up screen will appear (Figure 7-5). Move the button on the slide bar or click on the arrow keys to find the year and month of the start date. Click on the desired day to select and highlight the date (for example, August 31, 1997). Click on the **OK** button to accept the start date. The start date of the bar chart does not have to agree with the start date assigned when the schedule was calculated. The start date selected in the **Page Setup** dialog box simply

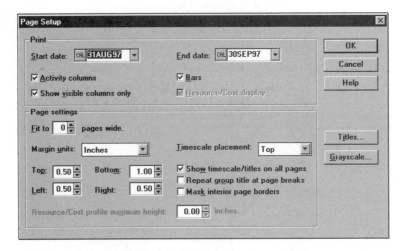

Figure 7-4 Page Setup Dialog Box

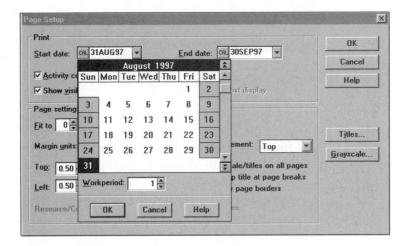

Figure 7-5 Page Setup Dialog Box—Start Date

determines the start date of the timescale for the hard copy print of the bar chart. Being able to choose the start date and end date of the printout gives you some control in printing only the information that is needed.

End date: The **End date:** calendar pop-up screen operates the same way as the **Start date:** option. Again, the end date assigned here has nothing to do with the end date as calculated within the *P3* schedule. To prevent an error, make sure the **End date:** is later than the **Start date:. End date:** is used to determine the length of the printed bar chart.

Activity columns Also appearing in the **Print** group of options is the **Activity columns** check box. The **Activity columns** refers to the activity ID, activity description, and original duration printed in tabular form on the bar chart hard copy. If the check is removed from this box, the activity table will not be printed, and only the bar chart will appear.

Show visible columns only   The **Show <u>v</u>isible columns only** check box refers to the activity table columns that appear on the hard copy print. In Figure 7-2 (p. 146), only the activity table columns of activity ID, activity description, and original duration appear on the hard copy print. If this check box is unchecked, all columns of the full activity table will appear in the hard copy print. The default columns available are: activity ID, activity description, original duration, remaining duration, percent, early start, early finish, total float, resources, and budgeted cost. If you don't want the original duration column to appear, go to the on-screen bar chart, hide the column, and it will not appear in the hard copy.

<u>B</u>ars   If the **<u>B</u>ars** check box is unchecked, the bars of the bar chart will disappear and only the activity table information will appear on the hard copy print. Printing the entire activity table in this way is a handy method to get a tabular report.

Resource/Cost display   The last check box in the **Print** area of the **Page Setup** dialog box is the **<u>R</u>esource/Cost display**. In Figure 7-4 (p. 148), this field is grayed-out, or nonfunctional. The reason is that the resource profile was not activated when the **Page Setup** dialog box was pulled up. To correct this problem, close the **Page Setup** dialog box, click on the **<u>V</u>iew** main pull-down menu, select the **Resource Pro<u>f</u>ile** option. Now go back to the **Page Setup** dialog box, and the **<u>R</u>esource/Cost display** field is darkened, or operational. Click on the field, and the hard copy will appear as in Figure 7-6. The activity table, the bar chart, and cost information (here cost and not resource information was requested) will appear on the hard copy print.

Timescale Placement: **<u>T</u>imescale placement:** is one of the options under **Page settings**. As can be seen from Figure 7-2 (p. 146), the timescale placement was at the top of the hard copy. Click the down arrow for this option box (Figure 7-7) to place the timescale at the top, bottom, or on both top and bottom.

<u>F</u>it to _ pages wide   The **<u>F</u>it to _ pages wide** is a print-to-fit option. For example, if the hard copy print were a little over two pages, choosing this option and setting it at two pages would resize all the components of the hard copy to fit a maximum of two pages.

Sho<u>w</u> timescale/titles on all pages   The **Sho<u>w</u> timescale/titles on all pages**, **Repeat gr<u>o</u>up title at page breaks**, and **Mas<u>k</u> interior page borders** check boxes control whether the title boxes, timescales, and borders will be printed on all pages. Sometimes it is necessary to

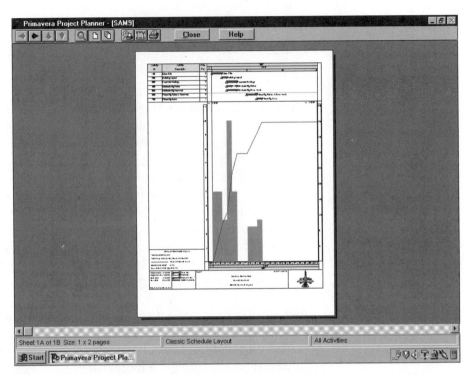

Figure 7-6   Print Preview—Resource Profile Option

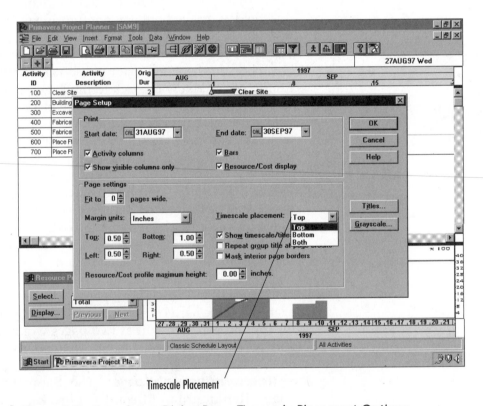

Timescale Placement

Figure 7-7   Page Setup Dialog Box—Timescale Placement Options

place all the hard copy pages together for a combined large drawing of the schedule for presentation purposes. If these boxes are left unchecked, these items will not print.

Margin units: The last five option boxes in the **Page settings** group of options control the page margin units and sizes. The **Margin units:** options are inches, centimeters, and points. The inch is a U.S. unit of measure. The centimeter is metric. The point is a measurement of type size, equal to approximately $\frac{1}{72}$ of an inch. The arrow option boxes **Top:**, **Bottom:**, **Left:**, and **Right:** control the size of the margins of the bar chart hard copy print.

Titles...   Select the **Titles...** button and the **Page Titles** dialog box will appear (Figure 7-8). Use this dialog box to input the text for the title block. Four centered lines are available. **Center title 1:**, **Center title 2:**, and **Center title 3:** can be used for the schedule title or name. **Revision title:** is used to identify the schedule revision number or name. Click on the down arrow under **Placement:** to define which pages of the hard copy the title block will appear on. The options are **First, Last, All**, and **None**.

Font: Click the down arrow under **Font:** (Figure 7-9) to change the type and size of the print font for the title block. Use this feature to customize the print and make it look more professional. Using special fonts to match the company's or client's logo or letterhead can make the schedule presentation very appealing.

Suppress revision box   The **Suppress revision box** is checked, so on the hard copy *P3* masks or hides the box that identifies the schedule revision. If you want the box to appear, simply uncheck this field and the text entered in the **Revision title:** field will appear in the title box on the hard copy.

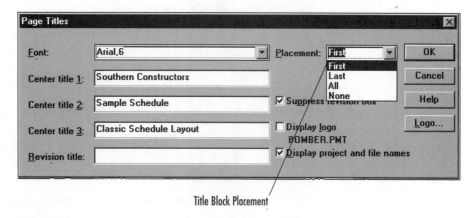

Title Block Placement

Figure 7-8   Page Titles Dialog Box—Placement Options

Display logo The next check box, **Display logo**, shows that the logo BOMBER.PMT has been selected but not checked to be displayed. Check this box to display the logo in the logo box (see Figure 7-10).

Display project and file names The last check box in the **Page Titles** dialog box is **Display project and file names**. As you can see from the upper left corner of the title block in Figure 7-3 (p. 147), the title SAM9 appears. This is the name of the *P3* file that contains the schedule being printed. If this information is to be suppressed, simply unclick this check box.

Logo... Use the **Logo...** button to assign a graphic symbol (clip art or logo) to the hard copy print. The logo will appear in the logo box on the hard copy print. Click on the **Logo...** button and the **Primavera**

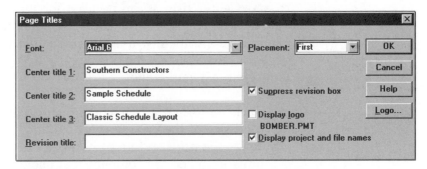

Figure 7-9 Page Titles Dialog Box—Font

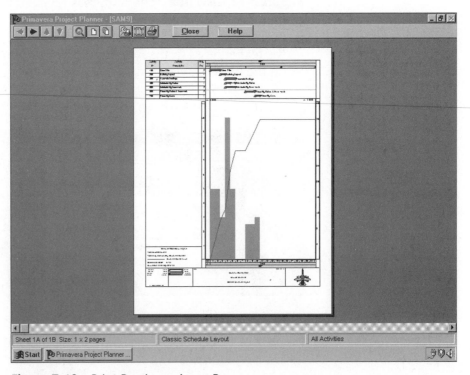

Figure 7-10 Print Preview—Logo Box

Symbol Selection dialog box will appear (Figure 7-11). *P3 for Windows* comes with many clip art files. These files may be found in *c:\p3win\p3progs\clipart*. They have a .pmt extension. Other images can be purchased, or company or project logo image files can be created by using scanners.

Preview   In Figure 7-12, the bomber symbol was selected by clicking on *bomber.pmt*. To preview the image, click on **Preview**. Click on **OK**, and the bomber will be accepted as the default logo for the hard copy of the schedule you are presently working on.

Grayscale...   The remaining option in the **Page Setup** dialog box for the bar chart hard copy is **Grayscale...**. The **Grayscale...** button is used to match screen colors with printer colors (Figure 7-13). Particularly with black-and-white printers, this scale can be used to achieve contrasts in the shades of gray to make the hard copy more readable.

Print Setup   Using the **File** pull-down menu (see Figure 7-14), click on **Print Setup...** and the **Print Setup** dialog box will appear (Figure 7-15). This dialog box can be used to change the default printer (as defined within *Windows*) to another printer.

Orientation   The **Orientation** of the hard copy print can be changed from portrait (Figure 7-10, p. 152) to landscape (Figure 7-16).

Figure 7-11   Primavera Symbol Selection Dialog Box

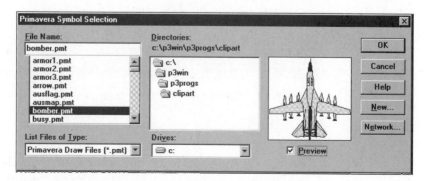

Figure 7-12   Primavera Symbol Selection Dialog Box—Preview

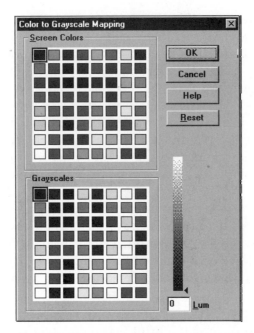

Figure 7-13　Printer Color to Grayscale Mapping Dialog Box

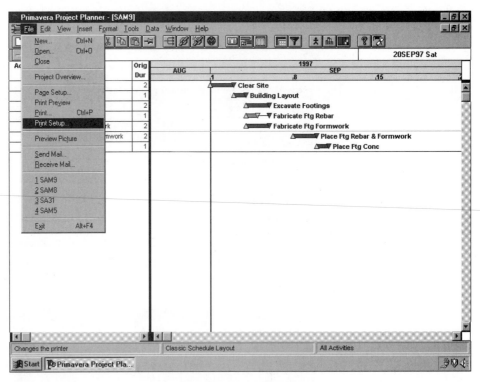

Figure 7-14　File Main Pull-Down Menu—Print Setup

Print　Of the four print options under the **File** pull-down menu, the one not yet discussed is the actual **Print...** option. Click **Print...** and the **Print** dialog box will appear (Figure 7-17).

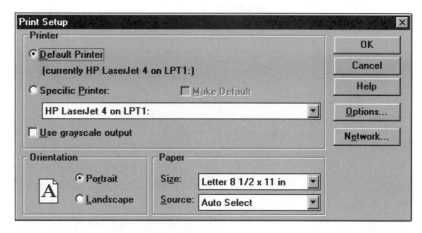

Figure 7-15   Print Setup Dialog Box

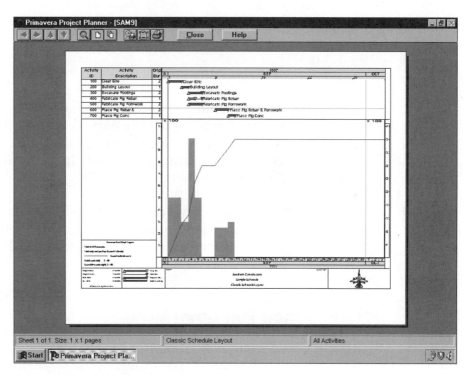

Figure 7-16   Print Preview—Landscape

Pages   A handy feature of this dialog box is the **Pages** function. For a schedule whose hard copy is six pages long, three pages horizontally (alpha characters—columns A, B, and C) and two pages vertically (numbers—rows 1 and 2), the *P3* numbering scheme would be:

1A	1B	1C
2A	2B	2C

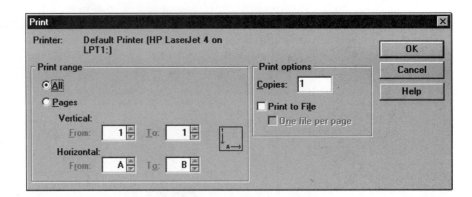

Figure 7-17   Print Dialog Box

This numbering scheme enables you to select only those pages that need to be printed.

Print to File   Another option of the **Print** dialog box is the **Print to File** function. This is useful when the printer or plotter is not connected to the computer where the schedule was created. Printing to a print file that can be copied to a portable disk greatly expands the optional use of other printers. The print file when saved will have a .prn extension. To print this file with *DOS* from the C: prompt, use *C:\COPY FILENAME.PRN LPT1*.

OK   Once you have configured the print setup, options, orientation, logo, and default printer of the bar chart, it is time to print. Click on the **OK** button on the **Print** dialog box to execute the hard copy of the bar chart (Figure 7-18).

Tools Pull-Down Menu

The second way to print a bar chart in *P3* is through the **Tools** main pull-down menu. The primary advantage of using this option, rather than **Print** from the **File** main pull-down menu, is that it offers greater customization in producing the bar chart, including greater flexibility in sorting, sizing, color selection, and formatting of the print.

Bar Charts Title Selection Box   Click on the **Graphic Reports** option of the **Tools** main pull-down menu (Figure 7-19). Select **Bar....** The **Bar Charts** title selection box will appear (Figure 7-20). All the customized bar chart configurations that have been saved are listed by bar chart ID. To create a new bar chart configuration to meet specific needs, click on the **Add...** button. The **Add a New Report** dialog box will appear (Figure 7-21). The default identification number will be the next sequential

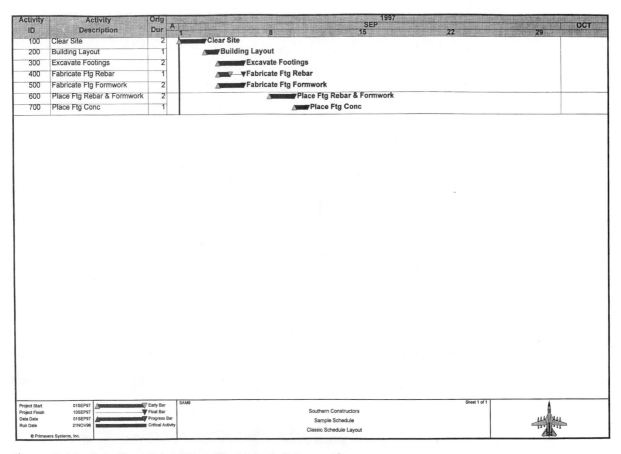

Activity ID	Activity Description	Orig Dur
100	Clear Site	2
200	Building Layout	1
300	Excavate Footings	2
400	Fabricate Ftg Rebar	1
500	Fabricate Ftg Formwork	2
600	Place Ftg Rebar & Formwork	2
700	Place Ftg Conc	1

Figure 7-18   Bar Chart Print Using File Main Pull-Down Menu

unused number, in this case 24. Click **OK** to accept the new report identification number.

Activity Data Screen   The next screen to appear is the **Bar Charts** dialog box, for bar chart configuration BC-24 (see Figure 7-22) for project SAM9. This options box has eight associated screens. The first screen to appear when a new bar chart configuration or identification is added (or modified) is the **Activity Data** screen (Figure 7-22).

Title: The **Title:** field is used to name the report configuration and will appear along with the bar chart ID number in the **Bar Charts** title selection box. In this example, the report is titled "Sample Schedule Bar Chart."

Specifications   Under **Specifications:** (Figure 7-23), the activity table information that will appear in the hard copy print is defined. In Figure 7-24, the **Specifications** field has been configured to provide activity ID, activity description, and original duration.

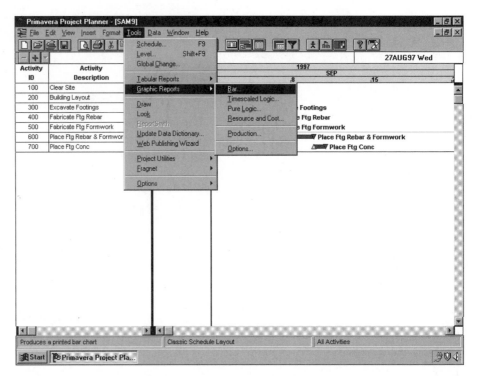

Figure 7-19 Tools Main Pull-Down Menu—Graphic Reports and Bar Options

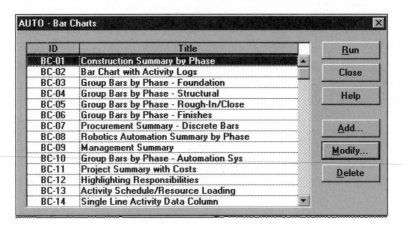

Figure 7-20 Bar Charts Title Selection Box

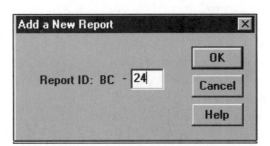

Figure 7-21 Add a New Report Dialog Box

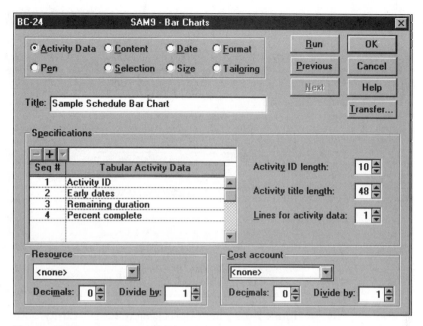

Figure 7-22   Bar Charts Dialog Box—Activity Data Screen

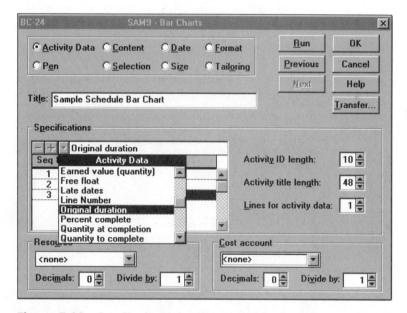

Figure 7-23   Bar Charts Dialog Box—Activity Data Screen Specifications Options

Also located in the **Specifications** section are the **Activity ID length:**, **Activity title length:**, and **Lines for activity data:** fields. These fields are used to limit the default size of printed fields. The **Activity title length:** was changed from the default of 48 in Figure 7-25 to 20 in Figure 7-26 to produce the printout shown in Figure 7-27.

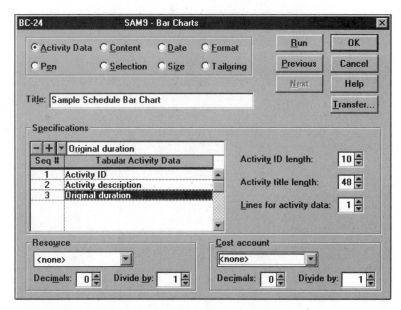

Figure 7-24   Bar Charts Dialog Box—Activity Data Screen Specifications

Resource and/or Cost account   To limit a bar chart to a certain resource or cost account, click on the down arrows for **Resource** and/or **Cost account** to identify the items from the resource and cost accounts library (Figures 7-25 and 7-26).

Transfer...   Before leaving the **Activity Data** screen, click on **Transfer...**. The **Transfer** dialog box will appear (Figure 7-28). This can be used

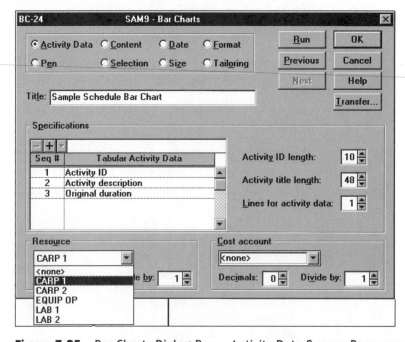

Figure 7-25   Bar Charts Dialog Box—Activity Data Screen Resource Options

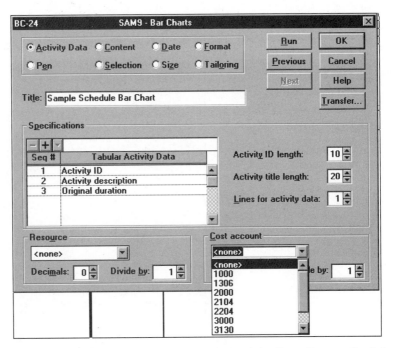

Figure 7-26   Bar Charts Dialog Box—Activity Data Screen Cost Account Options

to transfer a bar chart configuration from another project, possibly on a different directory or drive.

Pen Screen   To view the second of the eight screens of the **Bar Charts** options dialog box, click on **Pen** (Figure 7-29). The options on this screen are used to change the colors of such items as critical path on the hard copy print.

Elements: By varying the pen number designation, the grays of a black-and-white printer or the colors of a color printer can be changed to customize the hard copy print. With a multipen plotter or color printer, it is possible to specify a different number for each item in the **Item** field. Place a 0 in the **Pen Number** field to mask an item. The following default pen numbers and colors are available for each item, and any of the pen numbers can be changed.

1. Black
2. Red
3. Blue
4. Green
5. Dark magenta
6. Cyan
7. Yellow
8. Orange
9. Brown

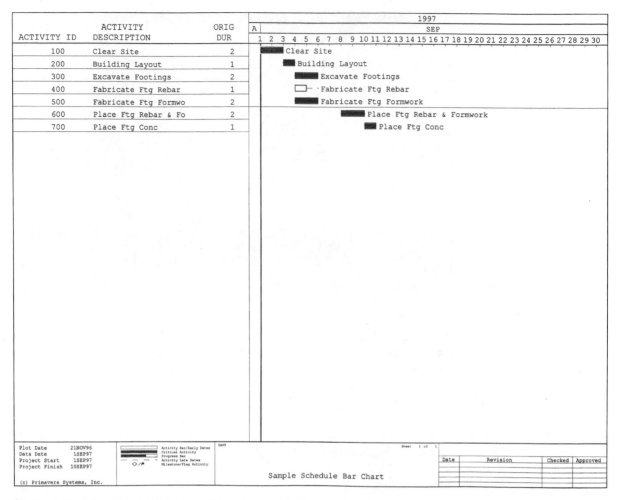

Figure 7-27   Bar Chart Print Using Tools Main Pull-Down Menu

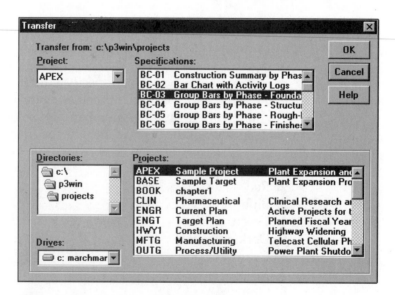

Figure 7-28   Transfer Dialog Box

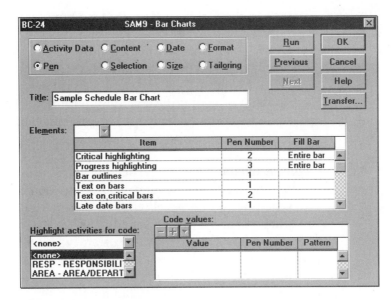

Figure 7-29   Bar Charts Dialog Box—Pen Screen

Highlight activities for code: Certain   activities   can   be   highlighted
   using the **Highlight activities for code:** and the **Code values:** fields.
   P3 assumes Pen 2 (red) for **Critical highlighting** and **Text on critical
   bars**, Pen 3 (blue) for **Progress highlighting**, and Pen 1 (black) for all
   other items.

Content Screen   The third screen of the **Bar Charts** dialog box is the
Content screen (Figure 7-30).

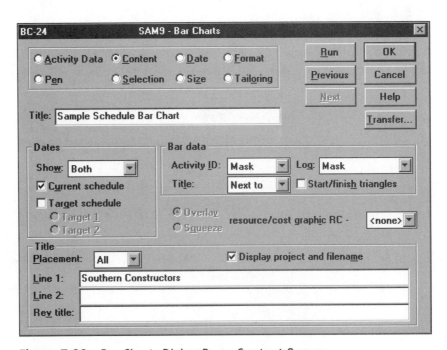

Figure 7-30   Bar Charts Dialog Box—Content Screen

Dates   In the **Dates** section, under **Show:**, you can define whether the hard copy print will be based on early starts, late finishes, or both (shows total floats). Click on the down arrow by the **Show:** options box to define which option will be used on the bar chart (see Figure 7-31). In Figure 7-32, the option **Both** was used. Notice that Activity 400 shows a noncritical activity with float associated. If either of the other two options were selected, the placement of the activity would have changed, and the float would not have been shown.

Current schedule and Target schedule   The next set of check boxes under the **Dates** section is used to determine which schedule or schedule comparisons will be printed. The check boxes used for this purpose are **Current schedule** and **Target schedule**. The current schedule means the most current version or the version to use for comparison purposes. The target version means the baseline version to use for updating or status reporting. These concepts will be covered in greater detail in Chapter 10.

Bar data   The next section under the **Content** screen is **Bar data**. The first option is the **Activity ID:** field (Figure 7-33). Other options would be to print activity ID above, below, or next to the bar, or to mask the activity. The next choice is the print options of the **Title:** field (Figure 7-34). Notice that the activity titles are masked in Figure 7-32. Figure 7-18 (p. 157) is an example of what the bar chart looks like with the title printed next to the bar on the bar chart side of the

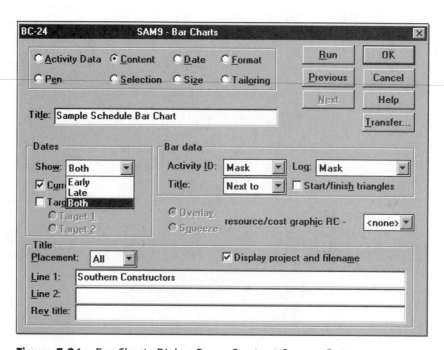

Figure 7-31   Bar Charts Dialog Box—Content Screen Dates

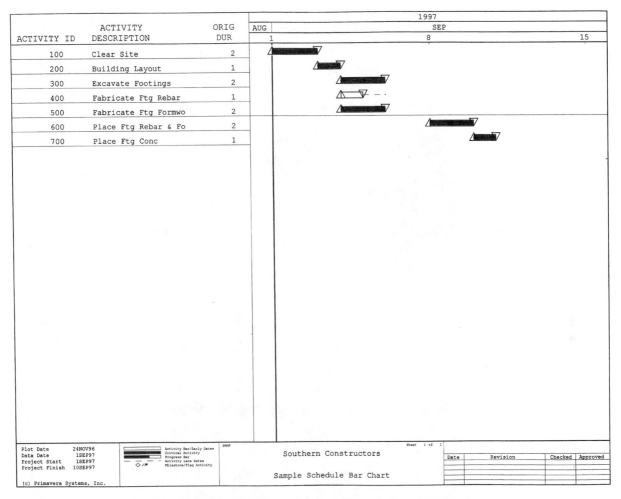

Figure 7-32   Bar Chart Print Using Tools Main Pull-Down Menu—Bars Modified

print. The **Log:** options are to mask, print in the activity area, or print in the graphic area (Figure 7-35). The log acts as an electronic note-pad for recording memos about a particular activity. You have the option of printing these logs (memos) on the hard copy bar chart. The last field under **Bar data** is **Start/finish triangles**. If you want triangles to appear at the beginning and ending of each activity as in Figure 7-32, click on this check box.

resource/cost graphic RC   The next field in the **Content** screen is the **resource/cost graphic RC**. To print a resource profile or cost histogram, configure the hard copy using this option box.

Title   The last group of fields on this screen deals with the title block placement and fields. Under the **Placement:** pull-down options, the title block can be specified to print on all, first, last, or no pages.

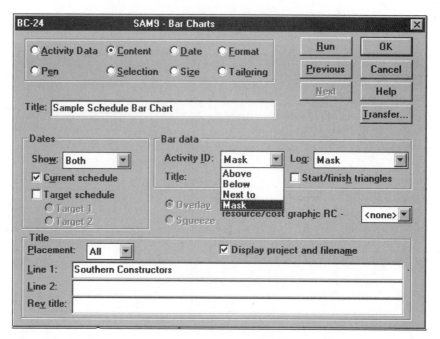

Figure 7-33 Bar Charts Dialog Box—Content Screen Activity ID Options

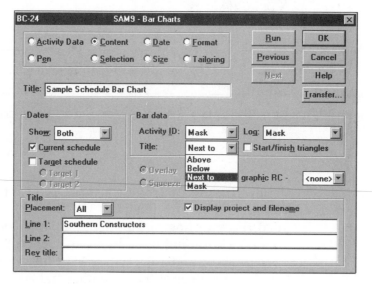

Figure 7-34 Bar Charts Dialog Box—Content Screen Bar Data Title

Check the **Display project and filename** check box to cause the assigned *P3* filename for the project to appear in the title box. For Figure 7-36, **Line 1:** of the title block legend was "Southern Constructors." **Line 2:** was left blank. In Figure 7-35, an optional third line to the title block, **Rev title:** was not used. In Figure 7-36, the "Sample Schedule Bar Chart" appearing in the fourth line of the title block is the name of the report being run.

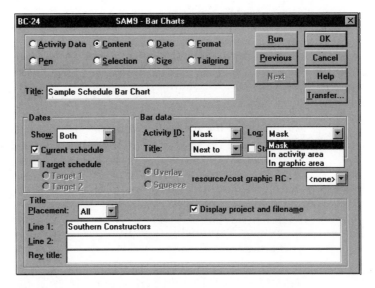

Figure 7-35   Bar Charts Dialog Box—Content Screen Log Options

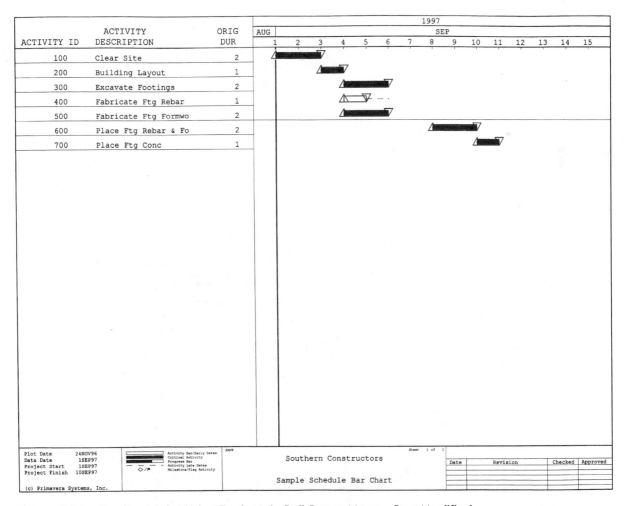

Figure 7-36   Bar Chart Print Using Tools Main Pull-Down Menu—Bars Modified

Selection Screen

Level   The fourth screen of the **Bar Charts** dialog box is the **Selection** screen (Figure 7-37). This option offers a choice of activities for print that meet certain criteria. *P3* allows up to four levels of selection criteria. Each level is defined by data in a separate column list box.

All or Any of the following criteria: The next criteria are the check boxes **All** or **Any of the following criteria:**. Click on **All** to print in the bar chart only those activities that meet all criteria. This ability to customize printouts is particularly beneficial for large projects with many activities and many parties using the schedule. You may want to show only activities relating to a certain subcontractor, a certain crew or responsibility requirement, or just a certain time frame (say a 30-day look-ahead).

Select if   *Note*: Leaving the **Select if** field blank automatically selects all activities. To add a selection criteria (or build a criteria statement), click on the plus button of the **Selection criteria:**. Click on the down arrow pull-down menu for **Selection criteria:**. A list of the selection criteria is given in Table 7-2.

Is   Activity ID was chosen under **Select if** in Figure 7-38 as the selection criteria for the criteria statement. Click on the **Is** field, and click on the down arrow pull-down menu. The following choices will appear:

EQ   Equal
NE   Not Equal

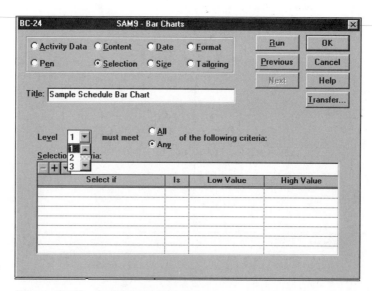

Figure 7-37   Bar Charts Dialog Box—Selection Screen Level Options

GT	Greater Than
LT	Less Than
WR	Within Range
NR	Not Within Range
CN	Contains
SN	Does Not Contain

In Figure 7-38, GT (greater than) was chosen for the **Is** criteria. Click under **Low Value** to highlight the field. Click on the down arrow pull-down menu to view activity choices. In this figure, the low value for Activity 100 was chosen. The criteria statement for the first line of entry would be, choose any activity "whose activity ID is greater than 100." The next line of criteria in Figure 7-38 limits choices to any activity "whose activity ID is Less than 400." In Figure 7-36 (p. 167), the only activities left that meet the above criteria are Activities 200 and 300. Figure 7-39 is a hard copy print with the above selection criteria chosen.

Activity description	Resume date
Activity ID	Suspend date
Activity type	Total float
Actual dates	Variance target 1 early start
Actual start	Variance target 1 early finish
Actual finish	Variance target 1 late start
Calendar ID	Variance target 1 late finish
Constraint	Variance target 2 early start
Cost account (11)	Variance target 2 early finish
Cost account (12)	Variance target 2 late start
Cost account category	Variance target 2 late finish
Early dates	WBS
Early start	RESPONSIBILITY
Early finish	AREA/DEPARTMENT
Free float	MILESTONE
Late dates	ITEM NAME
Late start	LOCATION
Late finish	STEP
Log record	Planned Start
Longest path	Planned Finish
Original duration	Specification
Percent complete	Submittal
Remaining duration	Orig Budget Cost
Resource	Orig Budget Qty
Resource curve	

Table 7-2   Selection Criteria

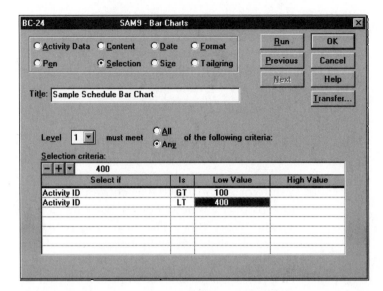

Figure 7-38 Bar Charts Dialog Box—Selection Screen Selection Criteria

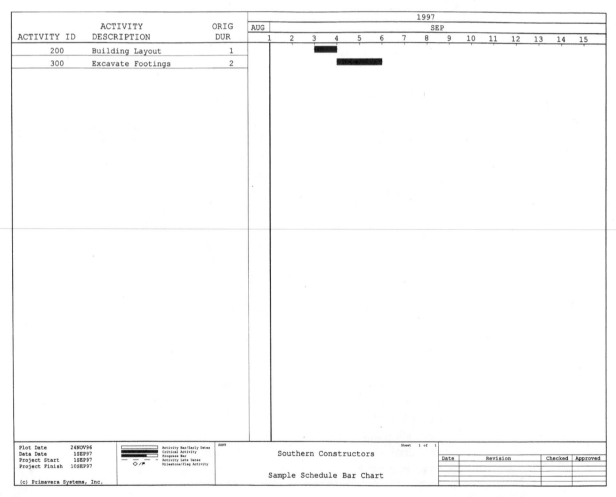

Figure 7-39 Bar Chart Print Using Tools Main Pull-Down Menu—Selection Criteria Modified

Date Screen

Start date: and End date: The fifth screen of the **Bar Charts** dialog box is the **Date** screen (Figure 7-40). The dates that are defined on this screen are the dates that determine the length of the bar chart hard copy print, not the dates of the schedule as determined by *P3*. The obvious advantages of being able to vary the hard copy print are that we may only want to look at a 30- or 60-day slice of the job and not the entire project duration. To set the start date of the hard copy print, click on the down arrow for the pull-down calendar next to **Start date:** (Figure 7-40). Use the slide bar to find the appropriate year and month, and then click on the desired start day for the hard copy print of the bar chart. Choose the end date of the hard copy bar chart print the same way using the **End date:** pull-down calendar.

Timescale options   Next under the **Date** screen are the **Timescale** options for modifying the timescale display of the hard copy print. Click on the **Display:** down arrow for options (Figure 7-41). The default is calendar dates. Figure 7-42 is an example of the hard copy bar chart print in ordinal (or numerical: 1, 2, 3, …) format. Sometimes, particularly during the planning stage and brainstorming sessions, the ordinal format is easier to use and communicate. If you choose the ordinal format under **Display:**, the two options at the bottom of the **Timescale** options section will no longer be grayed out. These two options are **Ordinal units:** and **Ordinal start date:**. The **Ordinal units:** options are day, week, month, and year. The **Ordinal start date:** is a pull-down calendar to determine what calendar date is ordinal day #1.

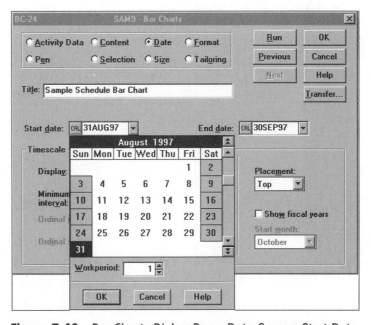

Figure 7-40   Bar Charts Dialog Box—Date Screen Start Date

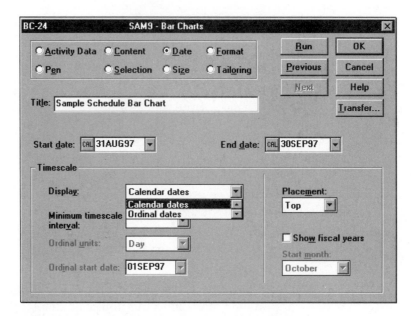

Figure 7-41 Bar Charts Dialog Box—Date Screen Timescale Display Options

ACTIVITY ID	ACTIVITY DESCRIPTION	ORIG DUR
100	Clear Site	2
200	Building Layout	1
300	Excavate Footings	2
400	Fabricate Ftg Rebar	1
500	Fabricate Ftg Formwo	2
600	Place Ftg Rebar & Fo	2
700	Place Ftg Conc	1

DAYS: -1 1 2 3 4 5 6 7 8 9 10 11 12 13 14 15

Clear Site
Building Layout
Excavate Footings
Fabricate Ftg Rebar
Fabricate Ftg Formwork
Place Ftg Rebar & Formwork
Place Ftg Conc

Plot Date 24NOV96
Data Date 1SEP97
Project Start 1SEP97
Project Finish 10SEP97

(c) Primavera Systems, Inc.

Activity Bar/Early Dates
Critical Activity
Progress Bar
Activity Late Dates
Milestone/Flag Activity

SAM9

Southern Constructors

Sample Schedule Bar Chart

Sheet 1 of 1

Date	Revision	Checked	Approved

Figure 7-42 Bar Chart Print Using Tools Main Pull-Down Menu—Ordinal Dates

Minimum timescale interval: Next, click on the **Minimum timescale interval:** (Figure 7-43). The choices are day, week, month, quarter, and year. Figure 7-44 is an example of the weekly timescale interval. Compare this figure to Figure 7-36 (p. 167). Compressing the timescale and using the weekly minimum unit displays more information on a page.

Placement: The **Placement:** pull-down menu places the timescale at the top, bottom, or on both the top and bottom of the hard copy bar chart print.

Size Screen

Point size The sixth screen of the **Bar Charts** dialog box is the **Size** screen (Figure 7-45). The **Size** screen is used to change the point size of type for parts of the hard copy print. Figure 7-46 is an example of changing the point size of the activity data from the default of 9 to 15. Besides the activity code titles, the other parts for possible modification are:

Activity bars	the actual bars themselves
Text on bars	the activity descriptions if printed on the bars
Activity data	activity data printed on the bars
Activity logs	printed memos associated with the activity
Activity code titles	the columnar portion of the bar chart (modified in Figure 7-46)
Title block	text within the title block
Timescale	text within the timescale

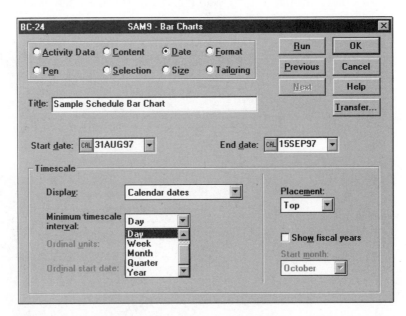

Figure 7-43 Bar Charts Dialog Box—Date Screen Minimum Timescale Interval

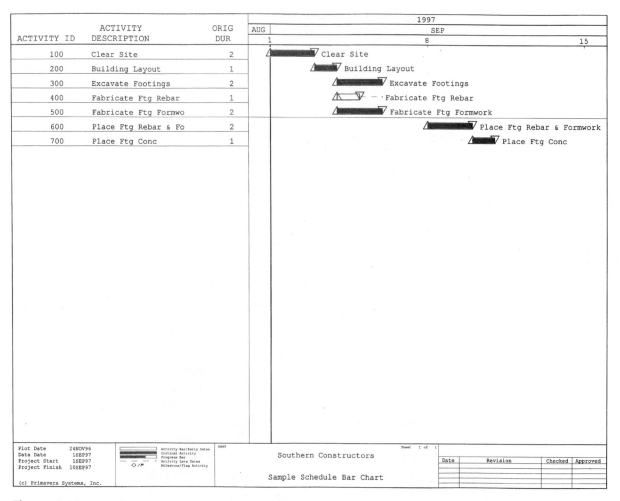

Figure 7-44 Bar Chart Print Using Tools Main Pull-Down Menu Weekly Timescale Interval

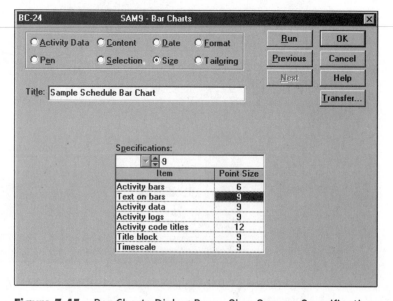

Figure 7-45 Bar Charts Dialog Box—Size Screen Specifications

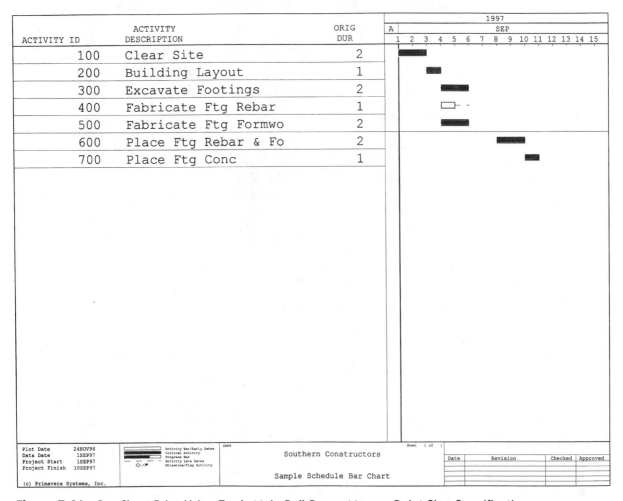

ACTIVITY ID	ACTIVITY DESCRIPTION	ORIG DUR	1997 SEP
100	Clear Site	2	
200	Building Layout	1	
300	Excavate Footings	2	
400	Fabricate Ftg Rebar	1	
500	Fabricate Ftg Formwo	2	
600	Place Ftg Rebar & Fo	2	
700	Place Ftg Conc	1	

Plot Date   24NOV96
Data Date   1SEP97
Project Start 1SEP97
Project Finish 10SEP97

(c) Primavera Systems, Inc.

Activity Bar/Early Dates
Critical Activity
Progress Bar
Activity Late Dates
Milestone/Flag Activity

SAM9

Southern Constructors

Sample Schedule Bar Chart

Sheet 1 of 1

Date	Revision	Checked	Approved

Figure 7-46   Bar Chart Print Using Tools Main Pull-Down Menu—Point Size Specifications

Experiment with different sizes of the different portions of the bar chart hard copy print until the finished copy has the exact look needed for final presentation.

Format Screen

Group by: The seventh screen of the **Bar Charts** dialog box is the **Format** screen (Figure 7-47). This screen is used to change the sequencing or grouping of activities on the hard copy print. Click on the **Group by:** down arrow for the possible grouping selections. The classifications are the same as the activity codes in the activity codes dictionary (Table 7-3). Access this dictionary through **Activity Codes...** from the **Data** main pull-down menu. Through the activity codes dictionary, new groups, or the options within groups, can be modified. Particularly for larger schedules, being able to sort by responsibility (crew, subcontractor) or area reduces the amount of paper each person has to wade through and therefore enhances communication.

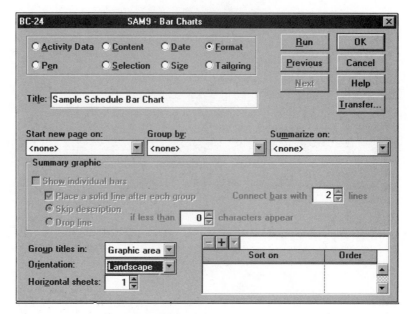

Figure 7-47 Bar Charts Dialog Box—Format Screen

Activity Codes:	RESP	RESPONSIBILITY
	AREA	AREA/DEPARTMENT
	MILE	MILESTONE
	ITEM	ITEM NAME
	LOCN	LOCATION
	STEP	STEP

Table 7-3 Activity Code Dictionary

Start new page on: You would probably want to **Start new page on:** with the selection of responsibility (Figure 7-48). This way each responsible party can have his or her own report without getting multiple copies or cutting up any pages.

Summarize on: With the **Summarize on:** pull-down menu options you can produce a summary bar chart based on the same groupings as the **Group by:** and the **Start new page on:** pull-down menus—the groupings defined in the activity codes dictionary. The ability to summarize a schedule, e.g., taking twenty activities associated with a particular responsibility and making it a single or summary activity with the overall characteristics of the group of activities, has advantages. From a communications point of view, not all people or positions need the same level or detail of information. Upper management, owners, and other parties may need only summary information; having the detailed schedule may confuse them. Providing

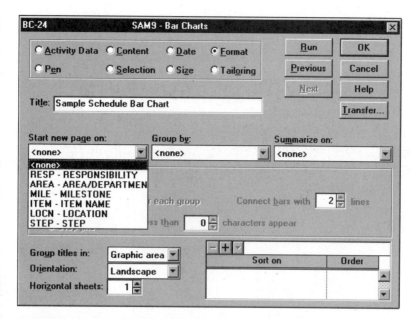

Figure 7-48   Bar Charts Dialog Box—Format—Start new page on

the right level of communication enhances the communication process; therefore the ability to produce hard copy prints based on summary information is beneficial.

Summary graphic   When a **Summarize on:** option is chosen, the **Summary graphic** options become operational. These options define the separation between groups.

Group titles in: There are four other options at the bottom of the **Format** screen. The **Group titles in:** pull-down menu gives the user the option of having the group titles print in the activity column area or the graphic bar area.

Orientation: The next option is the **Orientation:**. The choices are landscape and portrait. Figure 7-49 shows a portrait orientation.

Horizontal sheets: The **Horizontal sheets:** is a print-to-fit option. Use this option to force *P3* to reduce point size to make all the information fit into the specified number of horizontal pages.

Sort on   The last option is the **Sort on** field in the lower right-hand corner of the **Format** screen. This provides the ability to change the sort criteria. Clicking on the plus (+) button to add a sort criteria causes the **Schedule Parameters** menu to appear (Figure 7-50). By selecting the criteria, the sorting of activities can be defined. The options for the **Sort on** field are very similar to the **Selection criteria:** field in the **Selection** screen (see Table 7-2, p. 169).

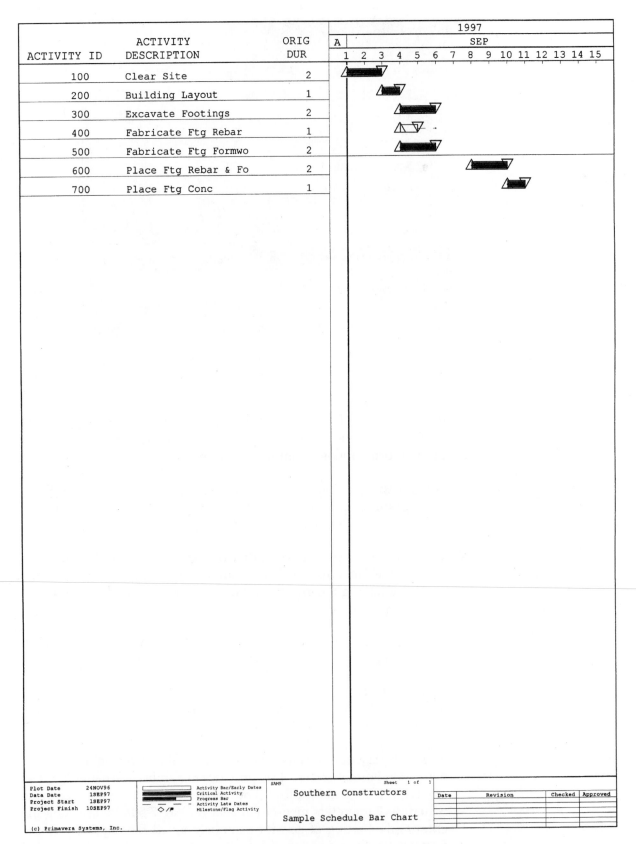

Figure 7-49 Bar Chart Print Using Tools Main Pull-Down Menu—Group By

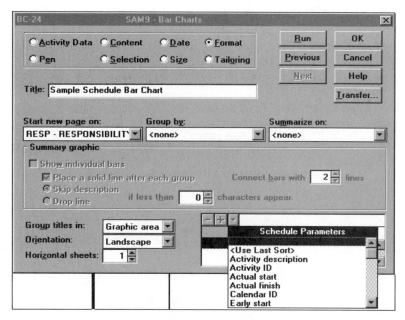

Figure 7-50   Bar Charts Dialog Box—Format Screen Sort On Options

Tailoring Screen

Vertical sight lines   The eighth and last screen of the **Bar Charts** dialog box is the **Tailoring** screen (Figure 7-51) for making visual changes to the bar chart before printing it. Figure 7-51 shows the normal *P3* defaults to this screen which produced the bar chart format in Figure 7-36 (p. 167). Changing the sight line, margin, and row-separation

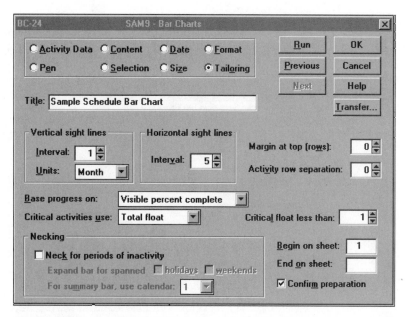

Figure 7-51   Bar Charts Dialog Box—Tailoring Screen

fields (Figure 7-52) will produce a bar chart that looks like Figure 7-53. Now there is a vertical sight line appearing for every day.

Horizontal sight lines   The **Horizontal sight lines** were changed from 5 to 2. As can be seen from Figure 7-53, the horizontal sight lines now appear after every other activity.

Margin at top (ro̲ws)   The **Margin at top (ro̲ws):** was changed from 0 to 2. Two blank lines, or a margin of two rows, is left at the top of the diagram (Figure 7-53). This space can be used for special text or for other reasons.

Activity row separation: The **Acti̲vity row separation:** was changed from 0 to 1. There now is a blank row left between activities.

Base progress on: The **B̲ase progress on:** field options are:

- Visible percent complete
- Entire percent complete
- Remaining duration
- Mask

P3 shows progress based on the fill pattern set in the **Pen** screen. These options on updating will be covered more thoroughly in Chapter 10.

Critical activities u̲se: The **Critical activities u̲se:** field defines how the critical path is to be determined. The options are **Total float** and **Longest path**. This determination comes into play on multicalendar projects, projects with constraints, or projects with lots of negative

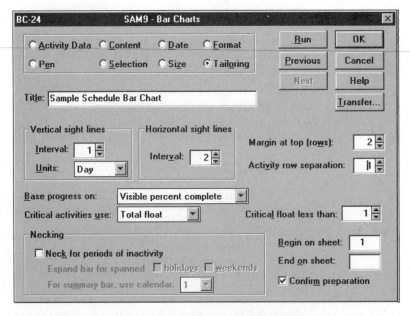

Figure 7-52 Bar Charts Dialog Box—Format—Tailoring—Sight Lines

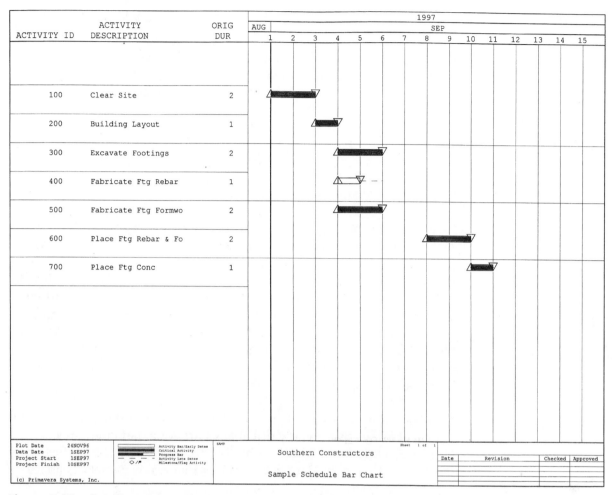

Figure 7-53   Bar Chart Print—Tools Main Pull-Down Menu—Sight Lines

float. The default and the usual choice for this field is **Total float**. An activity is normally considered critical when it has 0 float. The cutoff value for total float to determine critical-path highlighting using color can be defined. This value does not have to be 0.

Critical float less than: Use the **Critical float less than:** field to define the critical path for highlighting purposes.

Neck for periods of inactivity   Necking is the narrowing of the activity bars to show periods of inactivity for activities whose durations extend over weekends, holidays, or other nonwork periods. The necking can be turned on and off using the **Neck for periods of inactivity** check box.

Begin on sheet: and End on sheet: The **Begin on sheet:** and **End on sheet:** fields are handy if you do not want to print numerous pages of hard copy. These two fields enable you to print a range of pages. For example, in a 10-page bar chart, you could specify a range starting with page 4 and ending on page 6. The other pages simply would not be printed.

Confirm preparation The last field of the **Tailoring** screen is the **Confirm preparation** check box. Clear this check box if you do not want to confirm the number of hard copy sheets to be printed before printing.

REPORT

Bar Charts Title Selection Box

Now that you have gone through the eight screens of the **Bar Charts** dialog box, the next step is to use the buttons in the upper right corner (Figure 7-52, p. 181). Click on the **Cancel** button to erase the changes to the dialog box and return to the on-screen bar chart. Click on the **OK** button to save the current configuration for the next time the particular bar chart configuration (in this case BC-24, Sample Schedule Bar chart) is pulled up. If the hard copy of the bar chart is run without clicking the **OK** button first, the changes made to the bar chart hard copy configuration will be lost. Click on the **OK** button to return to the **Bar Chart** title selection box (Figure 7-54).

Modify... If for any reason you want to make changes to the configuration of BC-24, Sample Schedule Bar Chart, for a particular project (e.g., SAM9), click on the **Modify...** button from the **Bar Chart** title selection box to return to the **Bar Charts** options dialog box to make changes.

Graphic Report Options Dialog Box

Run **Run**, or the execution of the hard copy print, can be initiated using buttons on either the **Bar Chart** title selection box or the **Bar Charts** dialog box. When the **Run** button is clicked from either box, the **Graphic Report Options** dialog box appears (Figure 7-55). This screen provides the option of viewing the print file on screen, printing immediately, or saving the print to a file.

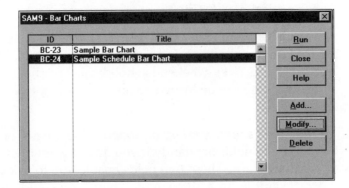

Figure 7-54 Bar Charts Title Selection Box

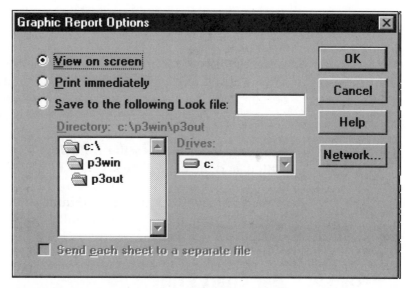

Figure 7-55 Graphic Report Options Dialog Box

View on screen Click on **View on screen** and click on the **OK** button to
initiate Primavera Look (see Figure 7-56). As can be seen, Primavera
Look does not provide the same amount of detail as the **Print Preview**
option under the **File** main pull-down menu for the bar chart print. It
is still a good way to check the print configuration before taking the
time to obtain a hard copy print. If the print has more than one page,
the slide button bars at the right and bottom of the graphic can be

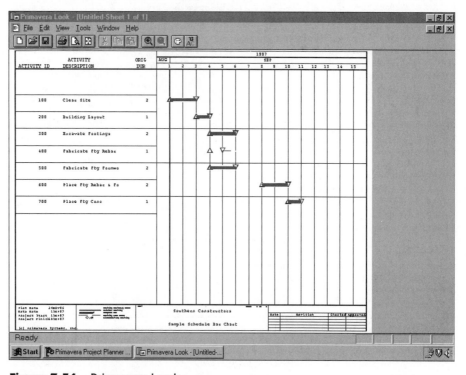

Figure 7-56 Primavera Look

used to view the different pages. Click on the picture of the printer on the button bar to initiate the hard copy print from Primavera Look.

Save to the following Look file: When you click on the **Save to the following Look file:** option, the grayed out **Directory:** and **Drives:** become black and operational. These options are used for file placement. The default placement of print files is usually the _P3OUT_ subdirectory, in the _P3Win_ directory. Type in the name of the print file to be saved under **Save to the following Look file:**, click on the **OK** button, and the file will be saved.

Print immediately Clicking on **Print immediately** initiates a printout of the report using the _Windows_ default options.

EXAMPLE PROBLEM: Bar Chart Print

Figures 7-57 and 7-58 are bar chart hard copy prints for a house put together as an example for student use (see the wood frame house

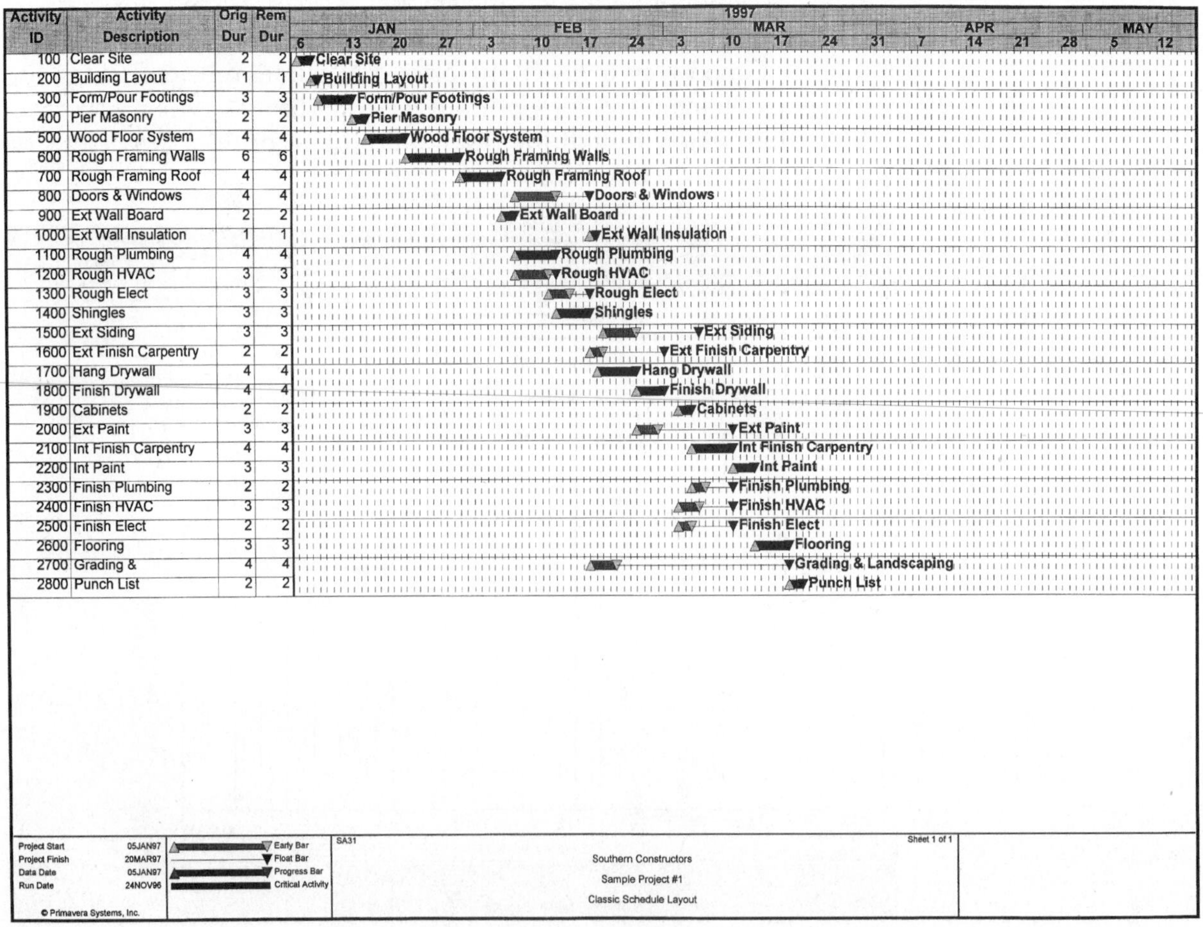

Figure 7-57 Bar Chart Print Using File Main Pull-Down Menu—Wood Frame House

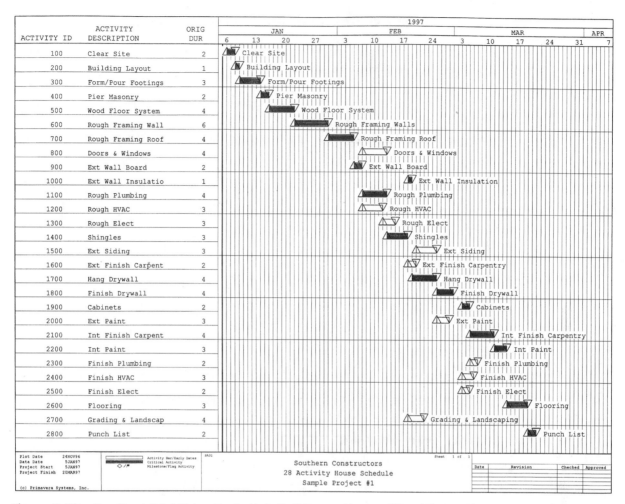

Figure 7-58 Bar Chart Print Using Tools Main Pull-Down Menu—Wood Frame House

drawings in the Appendix). Figure 7-57 is a bar chart hard copy print using the **File** main pull-down menu option. Figure 7-58 is a bar chart hard copy print using the **Tools** main pull-down menu option.

EXERCISES

1. Small Commercial Concrete Block Building—Bar Chart Print

Prepare bar chart prints for the small commercial concrete block building located in the Appendix. Follow these steps:

1. Prepare a hard copy print using the **File** main pull-down menu option.

2. Prepare a hard copy print using the **Tools** main pull-down menu option.

2. Large Commercial Building—Bar Chart Print

Prepare bar chart prints for the large commercial building located in the Appendix. Follow these steps:

1. Prepare a hard copy print using the **File** main pull-down menu option.
2. Prepare a hard copy print using the **Tools** main pull-down menu option.

Logic Diagram Presentation Using *P3*

Objectives

Upon completion of this chapter, the reader should be able to:

• Modify and print a timescaled logic diagram
• Modify and print a pure logic diagram

This chapter will discuss the hard copy presentation of the schedule in both timescaled logic and pure logic diagrams.

TIMESCALED LOGIC DIAGRAM

Advantage

The activity table (the left portion) of the bar chart is not an option in the timescaled logic format. All the information associated with the activity is noted adjacent to the activity. The greatest advantage of the timescaled logic diagram format is that it combines the advantages of the bar chart and the pure logic diagram. (For a comparison of a pure logic diagram and timescaled logic diagram, see Figures 1-3a, b and 1-4 a, b on pages 22 to 25.) It combines the ease of understanding of the bar chart and the display of interrelationships of the pure logic diagram. As with the bar chart, the length of the activity bar shows the duration of the activity. At a glance, activity durations, floats, and overall project duration are easy to understand in relationship to the timescale (usually appearing at the top of the diagram). The lines connecting the activities show the interrelationships between the activities. Having multiple activities on the same line, rather than a line per activity, makes the display of activity relationships more compact and easier to grasp.

Timescaled Logic Diagrams Dialog Box

The hard copy print of the timescaled logic diagram is accessed through the **Tools** main pull-down menu (Figure 8-1). Click on the **Graphic**

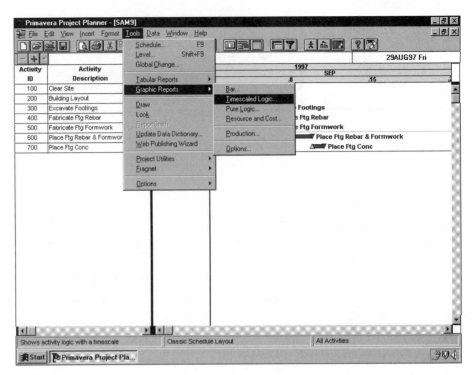

Figure 8-1 Tools Main Pull-Down Menu—Graphic Reports—Timescaled Logic

Reports option and then on **Timescaled Logic....** The **Timescaled Logic Diagrams** title selection box for our project (SAM9) appears (Figure 8-2). The configurations for twelve timescaled logic diagrams (TL-01 to TL-12) are already in the system unless the user has deleted them. These are timescaled logic diagrams that have been configured for other projects. The configurations can be modified and saved for use on your specific project. *Changing the configuration for one project will not affect the configuration used for other projects.*

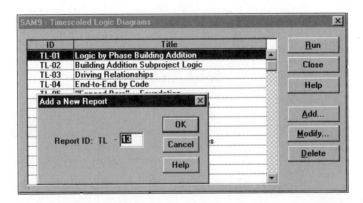

Figure 8-2 Timescaled Logic Diagrams Title Selection Box and Add a New Report Dialog Box

Add a New Report Dialog Box

To add a new timescale report configuration, click on the **Add...** button. The **Add a New Report** dialog box appears with the default of the next sequential number. Click on the **OK** button to accept the new report ID number. The **Timescaled Logic Diagrams** dialog box for timescale configuration TL-13 for our project appears. This dialog box is very similar to the **Bar Charts** dialog box in format and operation. Since most functions work in the same way between the two option boxes, only the differences will be discussed in this chapter.

Content Screen   The first screen to appear with the **Timescaled Logic Diagrams** dialog box is the **Content** screen (Figure 8-3). Compare this screen to Figure 7-30 (p. 163), the **Content** screen for the **Bar Charts** dialog box.

Show early/late schedule   A difference between the two screens is in the **Show early/late schedule** options. The options are **early** for a schedule printout based on early dates and **late** for late finishes.

Draw: Use the **Draw:** options under the **Relationships** section to choose which relationship lines will appear on the hard copy print. The options are **All**, **Driving**, and **None**. Figure 8-4 is an example of a timescaled logic diagram printed with all relationships showing. Figure 8-5 is an example of a printout with no relationships showing.

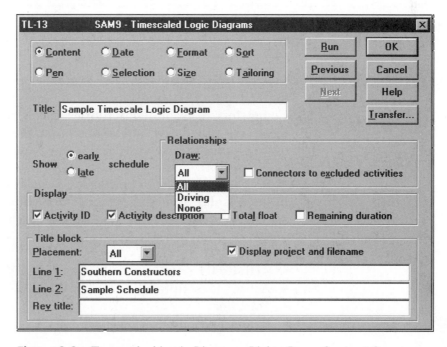

Figure 8-3 Timescaled Logic Diagrams Dialog Box—Content Screen Relationships Options

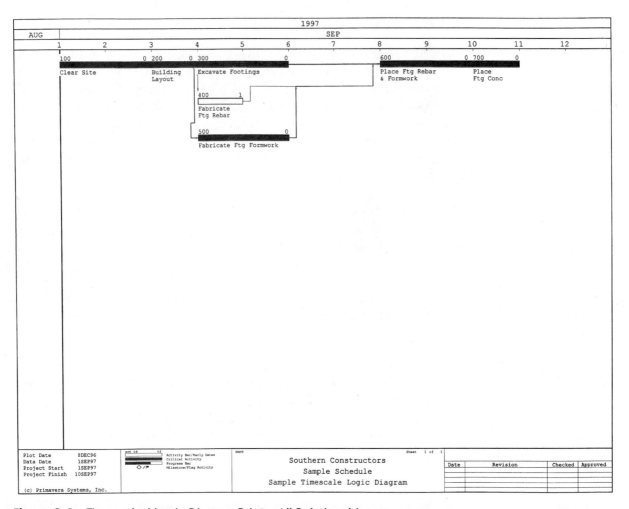

Figure 8-4   Timescaled Logic Diagram Print—All Relationships

Connectors to excluded activities   Another option is to show the driving activity or relationships that determine the critical path. The **Connectors to excluded activities** check box shows a "connector block" for activities with relationships in the graphic but are not part of the graphic themselves.

Display   The next block of options on the **Content** screen is contained in the **Display** section. The four check box options determine the activity information to be printed with each activity. These options are

- **Activity ID**
- **Activity description**
- **Total float**
- **Remaining duration**

For the printout in Figure 8-5, the first three fields are clicked or active. Activity 100 shows activity description "Clear Site" and a

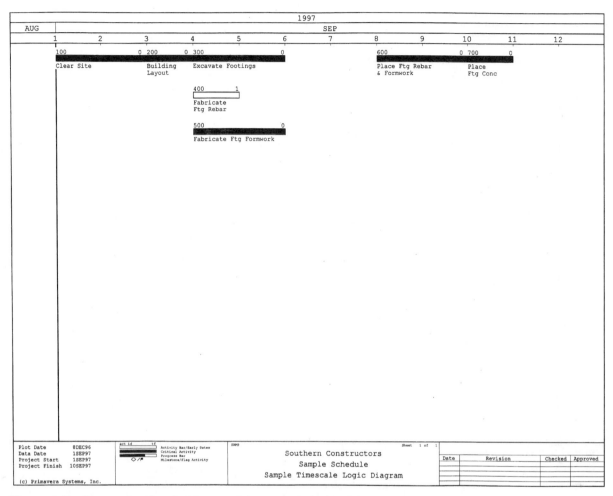

Figure 8-5 Timescaled Logic Diagram Print—No Relationships

total float of 0. In Figure 8-6, the **Total float** field was turned off. Now only activity ID and activity description appear next to the bars.

Pen Screen   The **Pen** screen (Figure 8-7) from the **Timescaled Logic Diagrams** dialog box is almost identical to the **Pen** screen of the **Bar Charts** dialog box (Figure 7-29, p. 163).

Fill Bar field   One option not discussed in Chapter 7 was the **Fill Bar** field. Compare Figure 8-6, which has the **Entire bar** option chosen under **Critical highlighting** (or critical activities), with Figure 8-8, which has the **No fill** option chosen. The actual bar itself is either shaded or unshaded, although when physical progress is input, a portion of the bar is shaded to show percent complete.

Date Screen   The third screen that can be accessed from the **Timescaled Logic Diagrams** dialog box is the **Date** screen (Figure 8-9). Compare this screen to Figure 7-41 (p. 172), the **Date** screen for the **Bar Charts** dialog box. The appearance and options of the two screens are

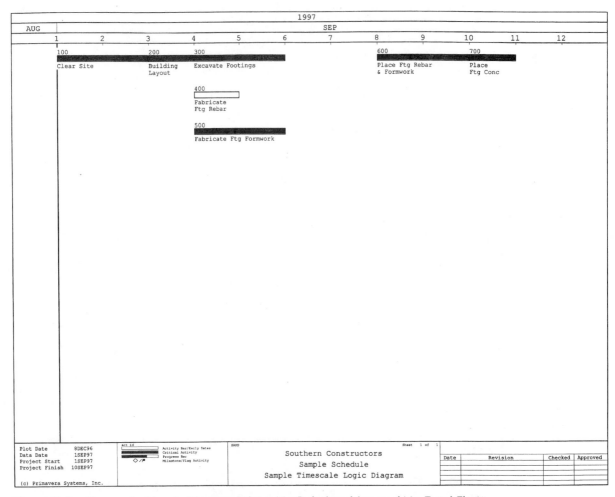

Figure 8-6   Timescaled Logic Diagram Print—No Relationships and No Total Float

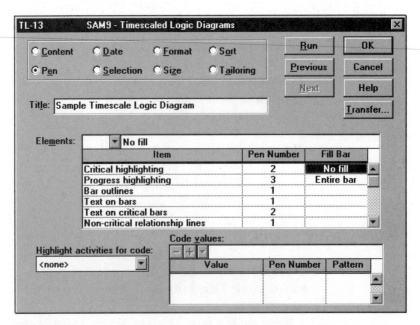

Figure 8-7   Timescaled Logic Diagrams Dialog Box—Pen Screen

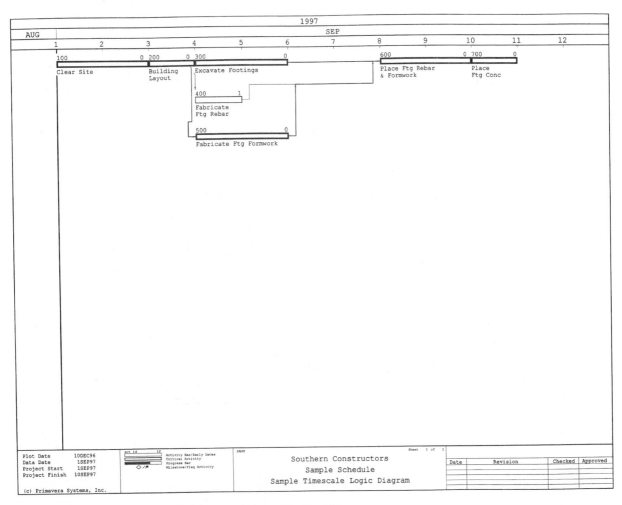

Figure 8-8 Timescaled Logic Diagram Print—No Bar Fill

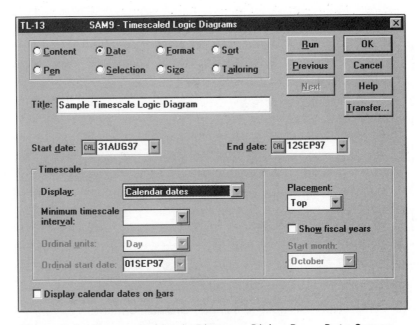

Figure 8-9 Timescaled Logic Diagrams Dialog Box—Date Screen

essentially the same. Figure 8-10 is an example of the ordinal format for a timescaled logic diagram.

Selection Screen   The fourth screen of the **Timescaled Logic Diagrams** dialog box is the **Selection** screen. It is identical in appearance and function to the **Bar Charts** dialog box **Selection** screen (Figure 7-37, p. 168).

Format screen   The fifth screen accessible from the **Timescaled Logic Diagrams** dialog box is the **Format** screen (Figure 8-11). Compare this screen to Figure 7-47 (p. 176), the **Format** screen for bar charts. The **Group by:** and the **Start new page on:** functions and options remain the same between the two screens. The **Orientation:** (landscape and portrait) and the **Horizontal sheets:** (print-to-fit) options also are unchanged.

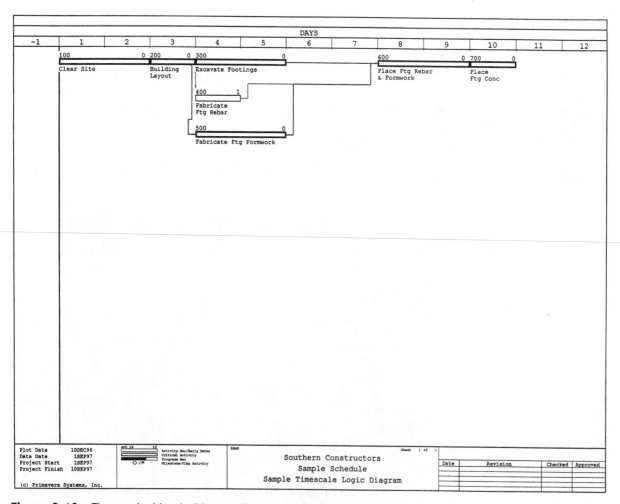

Figure 8-10   Timescaled Logic Diagram Print—Ordinal Format

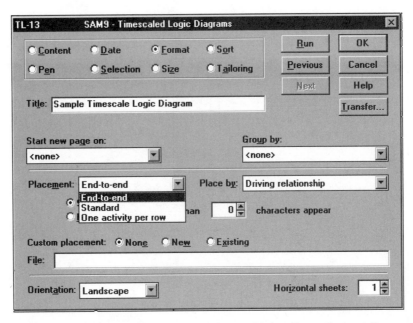

Figure 8-11   Timescaled Logic Diagrams Dialog Box—Format Screen

The primary difference is with the **Placement:** group of options, which are:

- **End-to-end**
- **Standard**
- **One activity per row**

This placement option defines the major difference between bar chart and timescaled logic format. In a bar chart format, each activity represents a horizontal line.

Standard   Figure 8-12 is an example of a schedule printout with the standard option chosen under **Placement:**. Notice the multiple activities per horizontal line and how this organization of the information makes the relationships between the activities much easier to understand than the bar chart format.

End-to-end   Figure 8-13 is an example with the end-to-end option selected under **Placement:**. When this option is chosen, the options **Skip description**, **Break chain**, and **Place by:** become operational or blackened in. In Figure 8-13 the driving relationships, or the critical path, are the deciding factors in activity placement.

One activity per row   Figure 8-14 is an example of the schedule printout of the timescaled logic diagram with the one activity per row option selected under **Placement:**. The row separation under **Size** had to be changed for this print to fit on a single page. This is why the vertical spacing of this print is different in Figures 8-12 and 8-13.

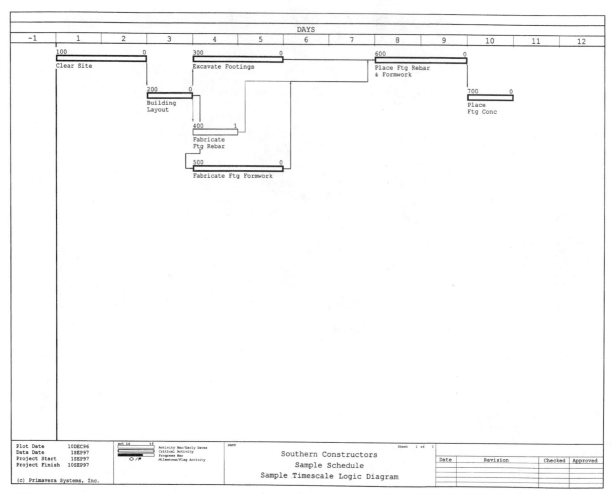

Figure 8-12   Timescaled Logic Diagram Print—Standard Placement

Place by: Use the **Place by:** pull-down menu (Figure 8-15) to change the relationship that determines which activities are placed end-to-end. The choices available are driving relationship or the choices found in the activity codes dictionary. These can be accessed from the **Data** main pull-down menu.

Size Screen   The sixth screen accessible from the **Timescaled Logic Diagrams** dialog box is the **Size** screen (Figure 8-16). Compare this to Figure 7-45 (p. 174), the **Size** screen for the **Bar Charts** dialog box. Figure 8-17 is an example of Figure 8-12 that has been resized to change the way it fits on the paper. The **Point Size** changes are:

Activity bars	8	6
Text on bars	9	7
Row separation	25	10

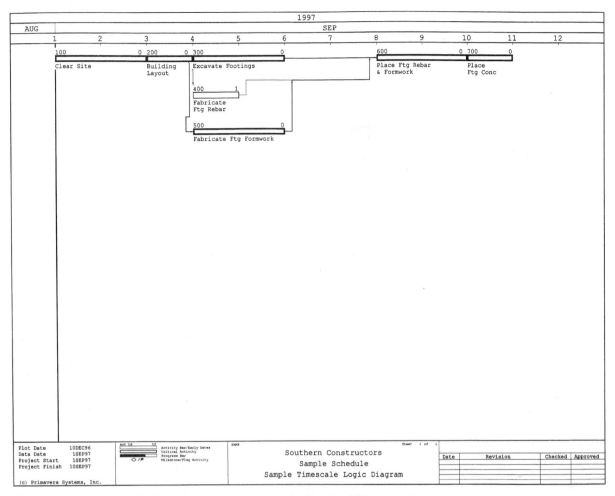

Figure 8-13   Timescaled Logic Diagram Print—End-to-End Placement

Sort Screen   The seventh screen accessible from the **Timescaled Logic Diagrams** dialog box is the **Sort** screen (Figure 8-18). There is no **Sort** screen for the **Bar Charts** dialog box, but, as can be seen by comparing Figure 8-18 and Figure 7-50 (p. 179), **Bar Charts Format** screen, the sorting capabilities are the same. Figure 8-19 is an example of using the activity description as a sort criteria. Figures 8-17 and 8-19 have identical configurations except for the activity description choice as the schedule parameter. Notice where three activities are running concurrently (300, 400, and 500). In Figure 8-17, they are placed by activity ID sequence. In Figure 8-19, they are placed in alphabetical sequence.

Tailoring Screen   The eighth screen accessible from the **Timescaled Logic Diagrams** dialog box is the **Tailoring** screen (Figure 8-20). Compare this figure to Figure 7-51 (p. 179), the **Tailoring** screen for the **Bar Charts** dialog box. The two screens are almost identical in function.

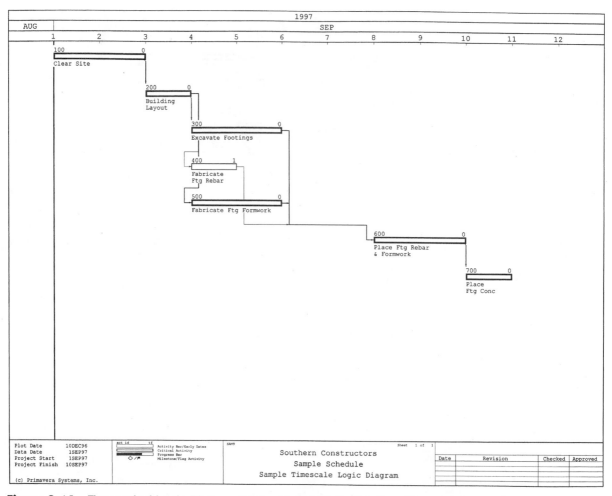

Figure 8-14   Timescaled Logic Diagram Print—One Activity Per Row Placement

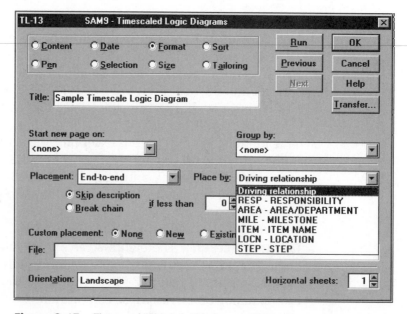

Figure 8-15   Timescaled Logic Diagrams Dialog Box—Format Screen Place By Options

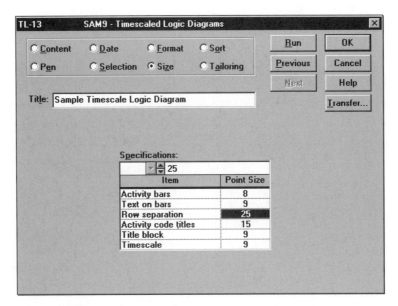

Figure 8-16   Timescaled Logic Diagrams Dialog Box—Size Screen

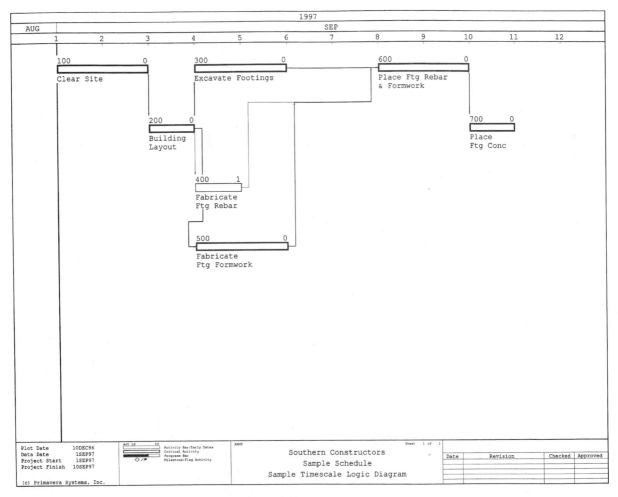

Figure 8-17   Timescaled Logic Diagram Print—Size

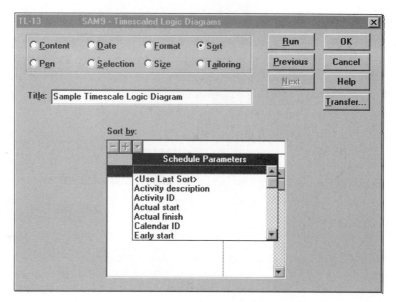

Figure 8-18 Timescaled Logic Diagrams Dialog Box—Sort Screen

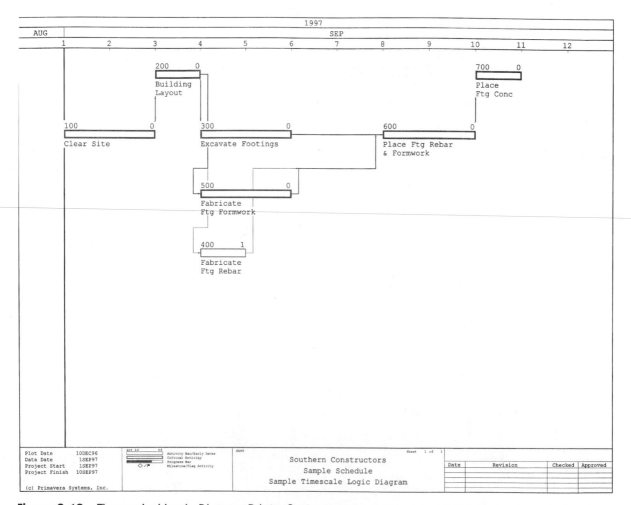

Figure 8-19 Timescaled Logic Diagram Print—Sort

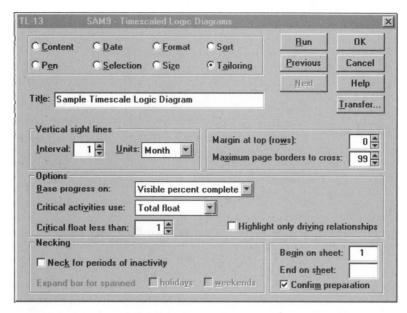

Figure 8-20   Timescaled Logic Diagrams Dialog Box—Tailoring Screen

Timescaled Logic Diagrams Dialog Box Buttons

Now that you have gone through the eight screens of the **Timescaled Logic Diagrams** dialog box, the next step is to use the buttons in the upper right corner of the dialog box. These buttons are constant for all eight screens. Click the **Cancel** button to erase changes to the dialog box. Click the **OK** button to save the configuration for the current project (in this case TL-13, Sample Timescale Logic Diagram) to return to the **Timescaled Logic Diagrams** title selection box (Figure 8-2, p. 188). Then click the **OK** button on the **Add a New Report** dialog box to return to the on-screen bar chart.

Other Buttons   The other buttons in the **Timescaled Logic Diagrams** dialog box (**Run, Previous, Next, OK, Cancel, Help,** and **Transfer...**) and the **Timescaled Logic Diagrams** title selection box (**Close, Help, Add..., Modify..., Delete,** and **Run**) operate the same way as their bar chart counterparts.

PURE LOGIC DIAGRAM

Advantage

As with the timescaled logic diagram, the information about the activity appears with the activity itself, but, in the pure logic diagram, the information appears within the node (activity box). The greatest advantage of the pure logic diagram is that it shows the relationships between the activities in a manner that is clear and easy to understand. The bar chart's format of a line per activity and the timescaled logic diagram's method of making the activity's length correspond to the time necessary to place the activity sometimes make the display of relationships hard to understand. Since all activities have the same size in the pure logic diagram, it makes their positioning on the print more symmetrical and therefore the relationship lines more regular, flowing, and easy to understand. Many schedulers prefer to prepare the rough diagrams in pure logic format and their finished diagrams in bar chart or timescaled logic format.

Pure Logic Diagrams Dialog Box

The print function for the pure logic diagram is accessed through the **Tools** pull-down menu (Figure 8-21). Click on the **Graphic Reports** option and then on **Pure Logic....** The **Pure Logic Diagrams** title selection box for the project SAM9 will appear (Figure 8-22). The configurations for three pure logic diagrams (PL-01 to PL-03) are already in the system. These are pure logic diagrams that have been configured for other projects and can be modified for use on the current project. Changing their configuration does not affect the other projects using the configuration.

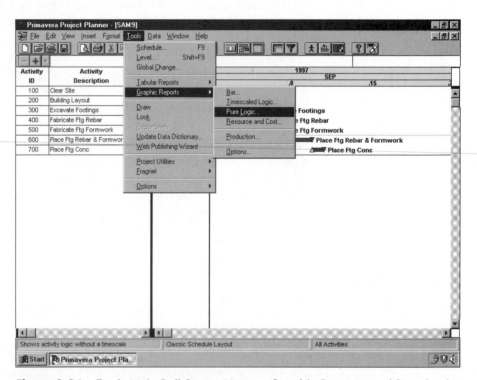

Figure 8-21   Tools Main Pull-Down Menu—Graphic Reports and Pure Logic Options

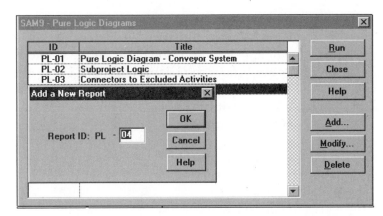

Figure 8-22   Pure Logic Diagrams Title Selection Box and Add a New Report Dialog Box

Add a New Report Dialog Box

To add a new pure logic report configuration, click on the **Add...** button, and the **Add a New Report** dialog box appears with the default of the next sequential number. Click on the **OK** button to accept the new report ID number. The **Pure Logic Diagrams** dialog box for pure logic configuration PL-04 for the project will appear (Figure 8-23).

Content Screen   The first screen, accessible from the **Pure Logic Diagrams** dialog box, is the **Content** screen (Figure 8-23). Compare this screen to Figure 7-30 (p. 163), the **Content** screen for the **Bar Charts** dialog box, and Figure 8-3 (p. 189) for the **Timescaled Logic Diagrams** dialog box.

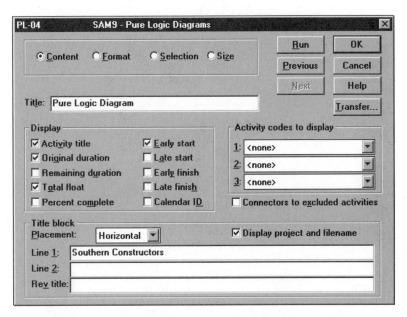

Figure 8-23   Pure Logic Diagrams Dialog Box—Content Screen

Display   The **Content** screen is the main screen in configuring the pure logic diagram. It is used to determine the activity information that will appear within the node on the printout. The check box options appear under the **Display** section. The options are:

- **Acti̲vity title**
- **Original duration**
- **Remaining du̲ration**
- **T̲otal float**
- **Percentage com̲plete**
- **E̲arly start**
- **L̲ate start**
- **Earl̲y finish**
- **Late finis̲h**
- **Calendar I̲D**

Figure 8-24 is a printout of the pure logic diagram run with the information selected in Figure 8-23. Obviously, the more information appearing within the node, the larger the nodes, so fewer nodes can be

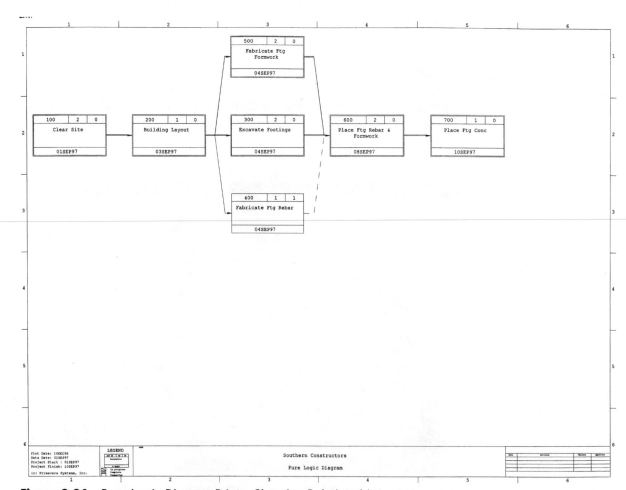

Figure 8-24   Pure Logic Diagram Print—Showing Relationships

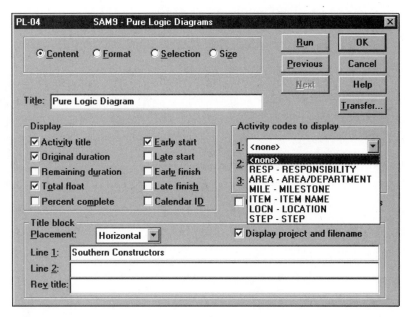

Figure 8-25 Pure Logic Diagrams Dialog Box—Content Screen Activity Codes to Display

placed on each page of the printout. Modify the information appearing in the node to control the number of pages of hard copy print.

Activity codes to display The **Activity codes to display** field (Figure 8-25) can be used to limit the print of activities to those wanted. The sorting or limiting parameters can be found in the activity codes dictionary under the **Data** main pull-down menu. If a printout of only the activities for a particular crew or subcontractor is needed, that printout can be produced.

Title block **Placement:** Under **Title block Placement:**, there are three options: **Horizontal, Vertical**, and **Mask**. Figure 8-24 (p. 204) is an example of horizontal title block placement. Figure 8-26 is an example of vertical title block placement.

Format Screen The second screen accessible with the **Pure Logic Diagrams** dialog box is the **Format** screen (Figure 8-27). Compare this screen to Figure 7-47 (p. 176), the **Format** screen for the **Bar Charts** dialog box, and Figure 8-11 (p. 195), the **Format** screen for the **Timescaled Logic Diagrams** dialog box. The **Group by:** function and option remain the same between the three screens (Figure 8-28). Again, we can group the activities within the printout of the pure logic diagram by activity codes. These groupings are as defined in the activity codes dictionary from the **Data** main pull-down menu.

Page orientation Page orientation (landscape and portrait) is not an option with the pure logic diagram as it was in the other two for-

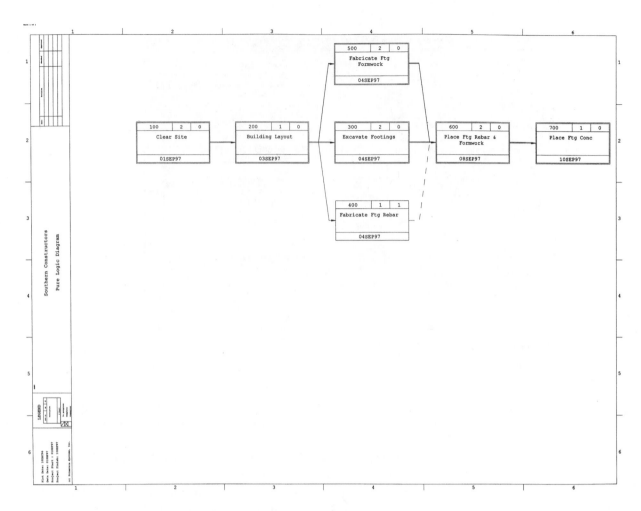

Figure 8-26   Pure Logic Diagram Print—Vertical Title Block Placement

mats. The **Begin on sheet:** and **End on sheet:** options in this screen were part of the **Tailoring** screens of the bar chart and timescaled logic diagram formats and operate in the same manner.

*Critical **a**ctivities use:*   The functions of **Critical a̲ctivities use:** and **Critical fl̲oat:** were in the **Tailoring** screens of the bar charts and timescaled logic diagrams formats and again operate the same way.

Selection Screen   The third screen of the **Pure Logic Diagrams** dialog box is the **Selection** screen (Figure 8-29). This screen is identical in appearance and function to the **Selection** screens for the bar charts (Figure 7-38, p. 170) and the timescaled logic diagrams. This option offers a choice of activities for print that meet certain criteria.

S̲election criteria: The selection criteria for this screen (Figure 8-30) are the same as those for the **Selection** screen for the **Bar Charts** dialog box.

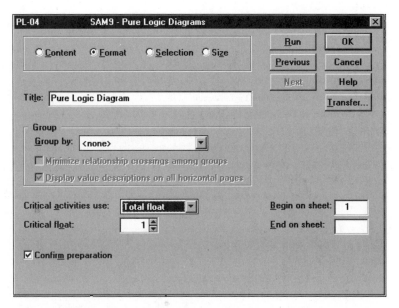

Figure 8-27   Pure Logic Diagrams Dialog Box—Format Screen

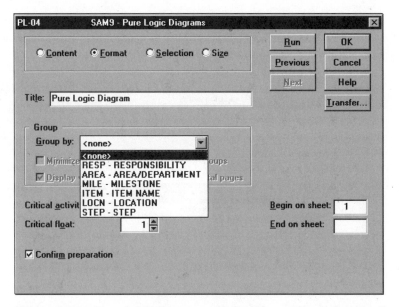

Figure 8-28   Pure Logic Diagrams Dialog Box—Format Screen Group By Options

Size Screen   The fourth screen accessible from the **Pure Logic Diagrams** dialog box is the **Size** screen (Figure 8-31). Compare this to Figure 7-45 (p. 174), the **Size** screen for the bar charts, and Figure 8-16 (p. 199) the **Size** screen for timescaled logic diagrams. The three are similar in function, but all have different items. The **Point Size** fields were changed from the *P3* defaults in Figure 8-31 to the sizes in Figure 8-32 to produce the diagram in Figure 8-33. Compare this format to that of

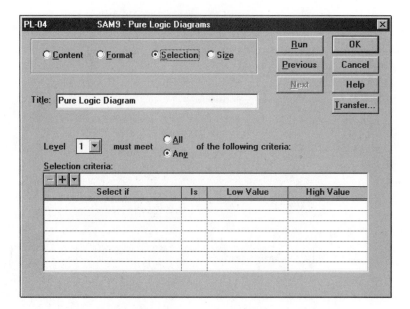

Figure 8-29 Pure Logic Diagrams Dialog Box—Selection Screen

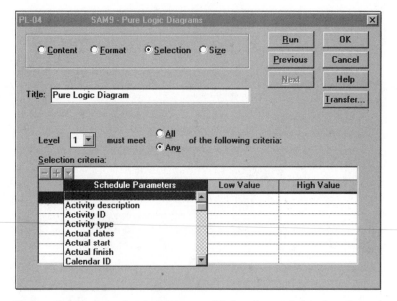

Figure 8-30 Pure Logic Diagrams Dialog Box—Selection Screen
Selection Criteria

Figure 8-26 (p. 206), and the obvious advantages of fitting more information on a single page are readily apparent.

Use the **Horizontal activity separation** and the **Vertical activity separation** rows to determine the distance between activities in points. The actual distance that you select in points is determined by the page size that you choose in the **Print Setup** dialog box. Enter 999 in the **Verti-**

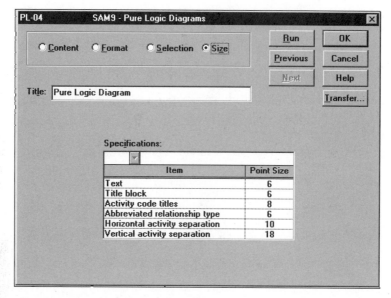

Figure 8-31   Pure Logic Diagrams Dialog Box—Size Screen

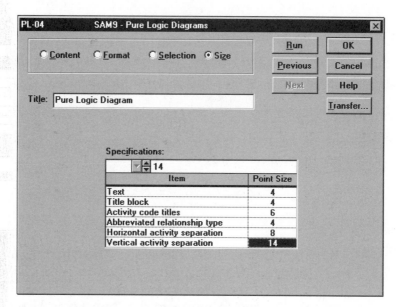

Figure 8-32   Pure Logic Diagrams Dialog Box—Size Screen Specifications

cal activity separation row to fill the page with as many activities as possible.

Pure Logic Diagrams Dialog Box Buttons

Now that you have gone through the four screens of the **Pure Logic Diagrams** dialog box, the next step is to use the buttons in the upper

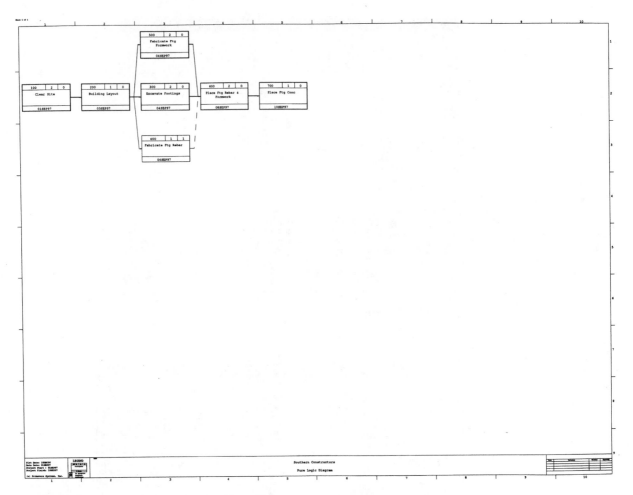

Figure 8-33 Pure Logic Diagram Print—Size

right corner of the dialog box. These buttons are the same for all four screens. Click the **Cancel** button to erase changes to the dialog box. Click on the **OK** button to save the configuration for the particular project and to return to the **Pure Logic Diagrams** title selection box (Figure 8-22, p. 203). Then click the **OK** button to return to the on-screen bar chart.

Other Buttons　The other buttons in the **Pure Logic Diagrams** dialog box (**Run, Previous, Next, OK, Cancel, Help,** and **Transfer...**) and the **Pure Logic Diagrams** title selection box (**Run, Close, Help, Add...,** **Modify...,** and **Delete**) operate the same way as their counterparts using the **Bar...** and **Timescaled Logic...** options under **Graphic Reports** from the **Tools** main pull-down menu.

EXAMPLE PROBLEM: Logic Diagram Prints

Figures 8-34 and 8-35a and b are hard copy prints of logic diagrams for a house put together as an example for student use (see the wood frame house drawings in the Appendix). Figure 8-34 is a hard copy print of the timescaled logic diagram using the **Tools** main pull-down menu option. Figure 8-35a and b is a hard copy print of the pure logic diagram using the **Tools** main pull-down menu option.

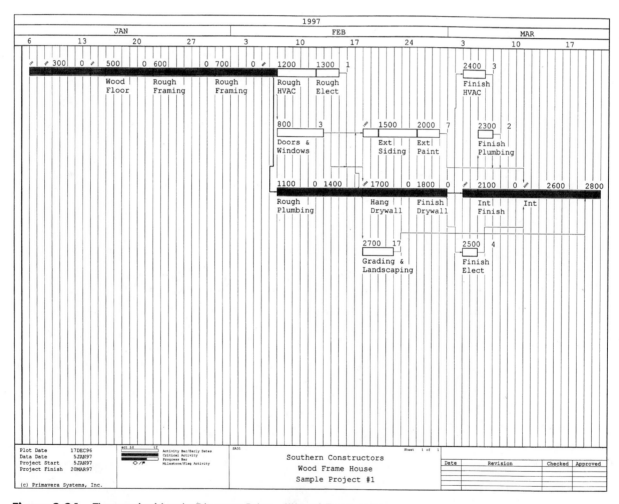

Figure 8-34   Timescaled Logic Diagram Print—Wood Frame House

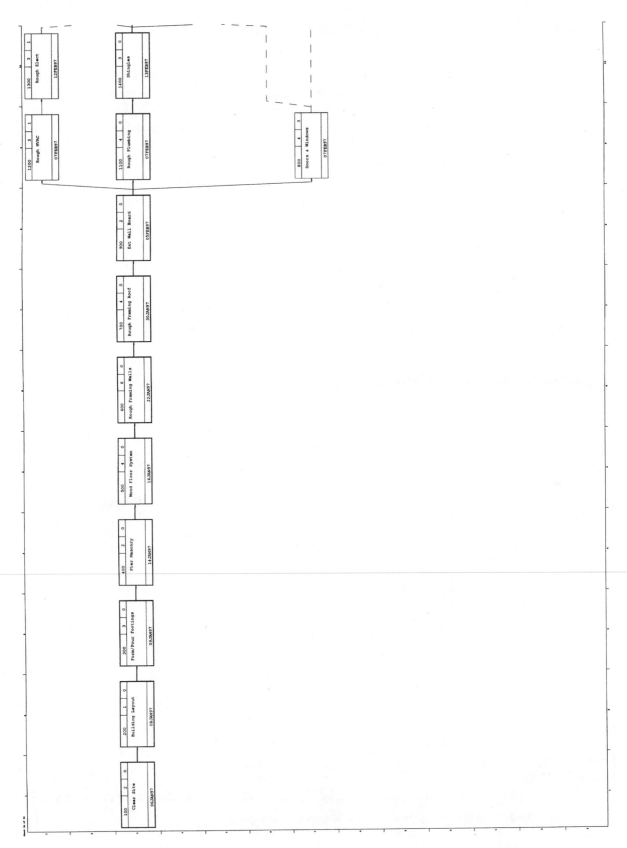

Figure 8-35a   Pure Logic Diagram Print—Wood Frame House, page 1

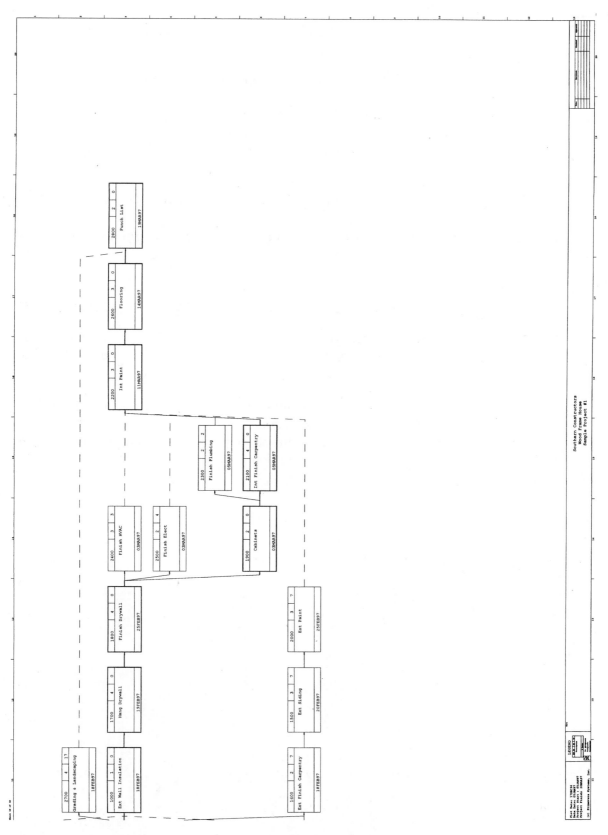

Figure 8-35b   Pure Logic Diagram Print—Wood Frame House, page 2

EXERCISES

1. Small Commercial Concrete Block Building—Logic Diagram Printouts

Prepare logic diagram printouts for the small commercial concrete block building located in the Appendix. This exercise should include the following steps:

1. Prepare a hard copy print of the timescaled logic diagram using the **Tools** main pull-down menu option.
2. Prepare a hard copy print of the pure logic diagram using the **Tools** main pull-down menu option.

2. Large Commercial Building—Logic Diagram Printouts

Prepare logic diagram printouts for the large commercial building located in the Appendix. This exercise should include the following steps:

1. Prepare a hard copy print of the timescaled logic diagram using the **Tools** main pull-down menu option.
2. Prepare a hard copy print of the pure logic diagram using the **Tools** main pull-down menu option.

Tabular Presentation Using *P3*

Objectives

Upon completion of this chapter, the reader should be able to:

• Modify and print a tabular schedule report
• Modify and print a tabular resource report
• Modify and print a tabular cost report
• Use the report writer for custom tabular reports

Sometimes a tabular report is more convenient for transferring information or is a convenient format used to update data. This chapter includes the presentation of the hard copy tabular report in schedule, resource, and cost format.

SCHEDULE REPORTS

Tabular reports can be used in conjunction with diagrams or as stand-alone reports. They are also useful as work sheets when updating the schedule.

Schedule Reports Title Selection Box

You can access the schedule report through the **Tabular Reports** option from the **Tools** main pull-down menu (Figure 9-1). The **Schedule Reports** title selection box will appear (Figure 9-2). As with previous iterations of this dialog box, you can either delete or modify the specifications to existing reports, or add a new report title and specification.

Schedule Reports Dialog Box

Select **SR-01, Classic Schedule Report - Sort by ES, TF**, in the **Schedule Reports** title selection box and click on the **Modify...** button to customize it. The **Schedule Reports** dialog box for SR-01 will appear for report configuration (Figure 9-3).

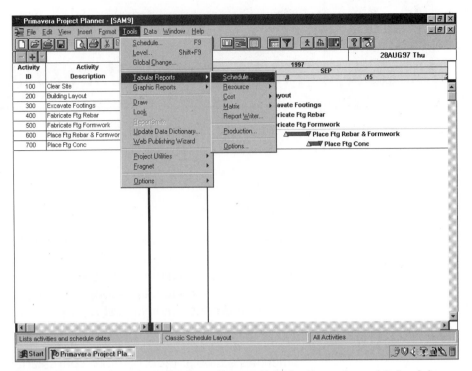

Figure 9-1 Tools Main Pull-Down Menu—Tabular Reports and Schedule Options

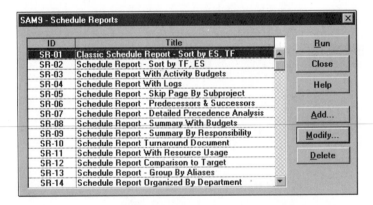

Figure 9-2 Schedule Reports Title Selection Box

Content Screen   The first screen to appear in the **Schedule Reports** dialog box is the **Content** screen.

**Include the following data:**   **Content Code** is used to define the fields that will appear on the tabular report. In Figure 9-3, under the **Include the following data:** section, only the **Activity code line** is selected for print. Figure 9-4 is a printout of the schedule tabular report with only this option selected.

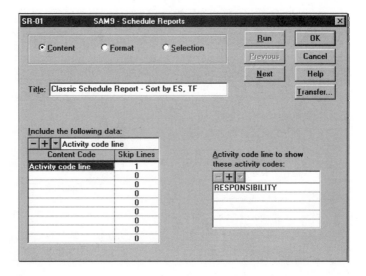

Figure 9-3 Schedule Reports Dialog Box—Content Screen

```
-------------------------------------------------------------------------------------------------
Southern Constructors                    PRIMAVERA PROJECT PLANNER              Sample Schedule

REPORT DATE 18DEC96  RUN NO.   53                                    START DATE  1SEP97  FIN DATE 10SEP97
              15:13
Classic Schedule Report - Sort by ES, TF                            DATA DATE   1SEP97  PAGE NO.   1


----- -----  ---- ---- - ---  ---------- ------------------------------------ -------- -------- -------- -------- -----
ACTIVITY     ORIG REM                              ACTIVITY DESCRIPTION         EARLY    EARLY    LATE     LATE   TOTAL
   ID        DUR  DUR  %  CODE                                                 START    FINISH   START    FINISH FLOAT
----- -----  ---- ---- - ---  ---------- ------------------------------------ -------- -------- -------- -------- -----
       100     2   2   0       Clear Site                                      1SEP97   2SEP97   1SEP97   2SEP97    0

       200     1   1   0       Building Layout                                 3SEP97   3SEP97   3SEP97   3SEP97    0

       300     2   2   0       Excavate Footings                              4SEP97   5SEP97   4SEP97   5SEP97    0

       500     2   2   0       Fabricate Ftg Formwork                         4SEP97   5SEP97   4SEP97   5SEP97    0

       400     1   1   0       Fabricate Ftg Rebar                            4SEP97   4SEP97   5SEP97   5SEP97    1

       600     2   2   0       Place Ftg Rebar & Formwork                     8SEP97   9SEP97   8SEP97   9SEP97    0

       700     1   1   0       Place Ftg Conc                                 10SEP97  10SEP97  10SEP97  10SEP97   0
```

Figure 9-4 Classic Schedule Report Print—Sort by ES,TF

> ***Activity code line*** The columns of the **Activity code line** can be turned on and off using the **Format** screen. Other activity information besides the **Activity code line** information can be included in this report. The other items of information that can be added are the **Schedule Report Elements** shown in Table 9-1 and Figure 9-5.

Resource line	Detailed predecessor
Workperiod	Detailed successor
All activity codes	Predecessor analysis
Log record	Successor analysis
Constrained date	Update line
Predecessor activity	Budgeted cost line
Successor activity	Free float line

Table 9-1 Schedule Report Elements

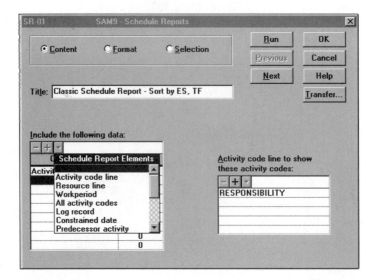

Figure 9-5 Schedule Reports Dialog Box—Content Screen Schedule Report Elements

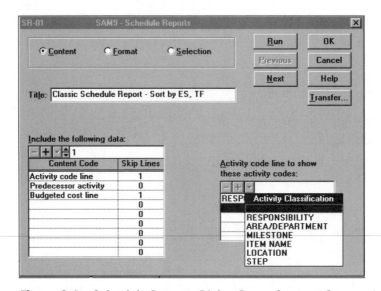

Figure 9-6 Schedule Reports Dialog Box—Content Screen Activity Codes

Skip lines   In Figure 9-6, **Predecessor activity** and **Budgeted cost line** have been added to the **Content Code** field. A 1 was added in the **Skip Lines** field after the **Budgeted cost line** to create a separator between activities. Figure 9-7 is the printout of the schedule tabular report run with this configuration. For Activity 200, the predecessor Activity 100 and the budgeted cost of $300.00 were added.

Activity code line to show these activity codes: In the **Content** screen, click on this field to select codes that will appear in the Code column of the schedule report. To find and modify the choices for this field,

```
--------------------------------------------------------------------------------------------------
Southern Constructors                  PRIMAVERA PROJECT PLANNER          Sample Schedule

REPORT DATE 18DEC96  RUN NO.   55                              START DATE  1SEP97  FIN DATE 10SEP97
              15:20
Classic Schedule Report - Sort by ES, TF                      DATA DATE   1SEP97  PAGE NO.     1


----- -----  ---- ----  - ---  ----------  -------------------------------------  -------- -------- -------- -------- -----
ACTIVITY     ORIG REM                                ACTIVITY DESCRIPTION          EARLY    EARLY    LATE     LATE     TOTAL
      ID     DUR  DUR    %  CODE                         BUDGET      EARNED         START    FINISH   START    FINISH   FLOAT
----- -----  ---- ----  - ---  ----------  -------------------------------------  -------- -------- -------- -------- -----
       100    2    2     0        Clear Site                                       1SEP97   2SEP97   1SEP97   2SEP97     0
                                                       1000.00       .00

       200    1    1     0        Building Layout                                  3SEP97   3SEP97   3SEP97   3SEP97     0
                               PRED ACT.IDS ,*     100,
                                                        300.00       .00

       300    2    2     0        Excavate Footings                                4SEP97   5SEP97   4SEP97   5SEP97     0
                               PRED ACT.IDS ,*     200,
                                                        400.00       .00

       500    2    2     0        Fabricate Ftg Formwork                           4SEP97   5SEP97   4SEP97   5SEP97     0
                               PRED ACT.IDS ,*     200,
                                                        600.00       .00

       400    1    1     0        Fabricate Ftg Rebar                              4SEP97   4SEP97   5SEP97   5SEP97     1
                               PRED ACT.IDS ,*     200,
                                                        500.00       .00

       600    2    2     0        Place Ftg Rebar & Formwork                       8SEP97   9SEP97   8SEP97   9SEP97     0
                               PRED ACT.IDS ,*     300,     400,*     500,
                                                        500.00       .00

       700    1    1     0        Place Ftg Conc                                   10SEP97  10SEP97  10SEP97  10SEP97    0
                               PRED ACT.IDS ,*     600,
                                                        300.00       .00

                                                     ============ ============
       REPORT TOTAL                                     3600.00       .00
```

Figure 9-7   Classic Schedule Report Print—Sort by ES, TF—Data

access the activity codes dictionary from the **Data** main pull-down menu. The standard *P3* default choices for activity codes are:

- Responsibility
- Area/Department
- Milestone
- Item Name
- Location
- Step

Format Screen   The second screen to appear with the **Schedule Reports** dialog box is the **Format** screen (Figure 9-8).

Skip line on:, Skip page on:, and Summarize on: In these fields, you can select criteria for sorting activities to print or skip lines or pages between the different sort criteria. Each party will get only relevant information in relation to his or her own part of the schedule. The choices for these fields (Figure 9-9) can be found and modified in the activity codes dictionary, which is accessed from the **Data** main pull-down menu.

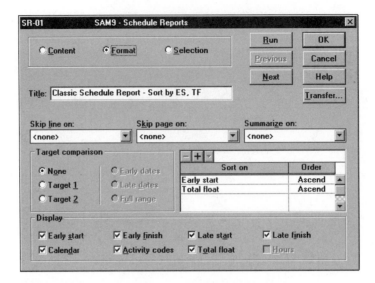

Figure 9-8   Schedule Reports Dialog Box—Format Screen

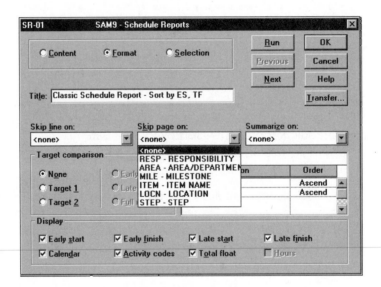

Figure 9-9   Schedule Reports Dialog Box—Format Screen Skip Page On Options

Target comparison   The next section of the **Format** screen is **Target comparison**. Use these options to compare the present schedule (usually updated) to a target (baseline) schedule. Of the three options, **None**, **Target 1**, and **Target 2**, the **None** option is chosen in Figure 9-8.

Early dates, Late dates, and Full range   Selecting **None** causes the field choices of **Early dates**, **Late dates**, and **Full range** to be grayed out, or nonoperational. If a project does have associated target schedules, selecting **Target 1** or **Target 2** causes these options fields to become black, or operational (Figure 9-10). Figure 9-11 is a printout of the tabular schedule with **Target 1** and **Early dates** chosen.

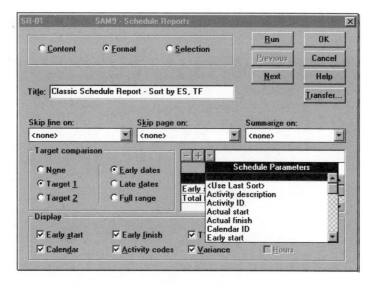

Figure 9-10   Schedule Reports Dialog Box—Format Screen Sort On Options

Variance   Compare Figure 9-11, tabular report with target comparison, with Figure 9-4 (p. 217), tabular report with no target comparison. The Total Float column in Figure 9-4 is replaced with the Var., or Variance, column, which is used to compare actual progress (from the updated schedule) to the target (baseline) schedule.

Sort on   Use the **Sort on** field of the **Format** screen to organize the hard copy print. The sort options for this field, shown in Figure 9-12, are the same as:

- **Sort on** field of the **Format** screen for the **Bar Charts** dialog box (Figure 7-50, p. 179)
- **Sort by:** field of the **Sort** screen for the **Timescaled Logic Diagrams** dialog box (Figure 8-18, p. 200)

```
------------------------------------------------------------------------------------------------------------
Southern Constructors                  PRIMAVERA PROJECT PLANNER              Sample Schedule

REPORT DATE 18DEC96  RUN NO.   58                                       START DATE  1SEP97  FIN DATE 10SEP97
            15:31
Classic Schedule Report - Sort by ES, TF                               DATA DATE   1SEP97  PAGE NO.    1


----- -----  ---- ---- - --- ----------  -------------------------------------------- -------- -------- -------- -------- -----
ACTIVITY    TAR  CUR                                                     CURRENT  EARLY    TARGET   EARLY
  ID        DUR  DUR   %  CODE              ACTIVITY DESCRIPTION          START    FINISH   START    FINISH   VAR.
----- -----  ---- ---- - --- ----------  -------------------------------------------- -------- -------- -------- -------- -----
    100       2    2   0          Clear Site                            1SEP97   2SEP97   1SEP97   2SEP97    0

    200       1    1   0          Building Layout                       3SEP97   3SEP97   3SEP97   3SEP97    0

    300       2    2   0          Excavate Footings                     4SEP97   5SEP97   4SEP97   5SEP97    0

    500       2    2   0          Fabricate Ftg Formwork                4SEP97   5SEP97   4SEP97   5SEP97    0

    400       1    1   0          Fabricate Ftg Rebar                   4SEP97   4SEP97   8SEP97   8SEP97    2

    600       2    2   0          Place Ftg Rebar & Formwork            8SEP97   9SEP97   9SEP97  10SEP97    1

    700       1    1   0          Place Ftg Conc                       10SEP97  10SEP97  11SEP97  11SEP97    1
```

Figure 9-11   Classic Schedule Report Print—Sort by ES, TF—Sort On

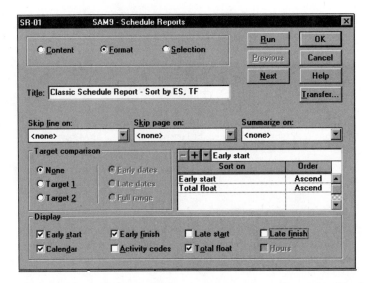

Figure 9-12   Schedule Reports Dialog Box—Format—Display

Display   In the **Display** section of the **Format** screen you can turn on and off the columns that appear in Figure 9-4 (p. 217). The columns contain the information that appears when the **Activity code line** in the **Content** screen is selected (Figure 9-3, p. 217). The **Display** options are:

- **Early start**
- **Early finish**
- **Late start**
- **Late finish**
- **Calendar**
- **Activity codes**
- **Total float**
- **Hours**

Figure 9-13 is a tabular schedule report printed as configured with the options shown in Figure 9-12. The **Late start, Late finish,** and **Activity codes** check boxes have been turned off. Compare Figure 9-13 to Figure 9-4 (p. 217).

Selection Screen   The last screen to appear with the **Schedule Reports** dialog box is the **Selection** screen (Figure 9-14). Use this screen to organize the hard copy print.

Selection criteria: The sort options for this field are the same as the **Sort on** field of the **Format** screen for the **Bar Charts** dialog box (Figure 7-50, p. 179) and the **Sort by:** field of the **Sort** screen for the **Timescaled Logic Diagrams** dialog box (Figure 8-18, p. 200). The **Selection criteria:** options are shown in Table 7-2 (p. 169).

Although the choices are the same for **Selection criteria:** under the **Selection** screen as the **Sort on** field of the **Format** screen, they are

```
-------------------------------------------------------------------------------------------------
Southern Constructors                    PRIMAVERA PROJECT PLANNER              Sample Schedule

REPORT DATE 18DEC96  RUN NO.   62                                    START DATE  1SEP97  FIN DATE 10SEP97
            15:44
Classic Schedule Report - Sort by ES, TF                            DATA DATE   1SEP97  PAGE NO.    1


----- -----  ---- ---- - ---  ----------  ------------------------------------  -------- --------  -----
ACTIVITY     ORIG REM                                                           EARLY    EARLY      TOTAL
   ID        DUR  DUR   %                  ACTIVITY DESCRIPTION                  START    FINISH     FLOAT
----- -----  ---- ---- - ---  ----------  ------------------------------------  -------- --------  -----
      100      2    2   0      Clear Site                                       1SEP97   2SEP97         0

      200      1    1   0      Building Layout                                  3SEP97   3SEP97         0

      300      2    2   0      Excavate Footings                               4SEP97   5SEP97         0

      500      2    2   0      Fabricate Ftg Formwork                          4SEP97   5SEP97         0

      400      1    1   0      Fabricate Ftg Rebar                             4SEP97   4SEP97         1

      600      2    2   0      Place Ftg Rebar & Formwork                      8SEP97   9SEP97         0

      700      1    1   0      Place Ftg Conc                                 10SEP97  10SEP97         0
```

Figure 9-13   Classic Schedule Report Print—Sort by ES, TF—Display

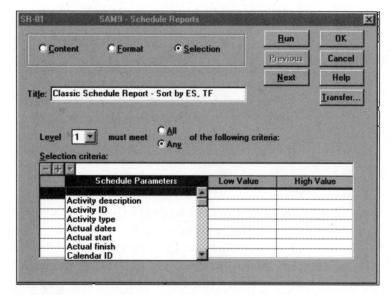

Figure 9-14   Schedule Reports Dialog Box—Selection Screen Selection Criteria

used for different purposes. Under the **Format** screen, only the sequence of print is selected. With the **Selection** screen, you can choose various ranges of activities to print; for example, activities for the next 30 days or activities that are behind schedule.

RESOURCE REPORTS

The tabular resource report is accessed by selecting **Resource** from the **Tabular Reports** option from the **Tools** main pull-down menu (Figure 9-15). There are five different options under **Resource: Control...**, **Productivity...**, **Earned Value (Units)...**, **Tabular...**, and **Loading...**.

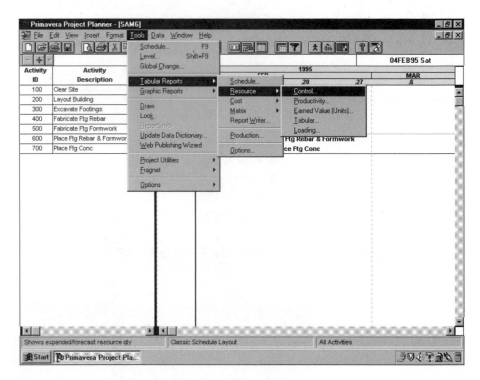

Figure 9-15   Tools Main Pull-Down Menu—Tabular Reports and Resource Control Options

Resource Control Report

Resource Control Reports Dialog Box   Figure 9-16 is an example of a resource schedule in tabular format. This report is particularly useful in controlling and monitoring resources. Select the **Control...** option from the **Resource** option of **Tabular Reports** from the **Tools** main pull-down menu. The **Resource Control Reports** title selection box will appear (Figure 9-17). As with other previous iterations of this selection, you can either delete or modify the specifications to existing reports to meet present project needs or add a new report title and specification. Select **RC-01, Resource Control - Detail By Activity**, and click on the **Modify...** button to customize it as needed. The **Resource Control Reports** dialog box for RC-01 will appear for configuring the current project (Figure 9-18).

Resource Selection Screen   The first screen to appear with the **Resource Control Reports** dialog box is the **Resource Selection** screen, which has the **Res Selection** option selected. This screen enables the user to select which resources will appear in the resource control report.

To select a resource for this report, click on the **Profile if** field. Choose from the options of GT (greater than), LT (less than), WR (within range), or NR (not within range). Then define the resource in the **Low Value Resource** field. This is done by clicking on the field and then on the

```
-----------------------------------------------------------------------------------------------------------------
Southern Constructors                     PRIMAVERA PROJECT PLANNER            Resource Refine

REPORT DATE  29DEC96  RUN NO.  39          RESOURCE CONTROL ACTIVITY REPORT     START DATE 1SEP97  FIN DATE 10SEP97
             7:59
Resource Control - Detail By Activity                                          DATA DATE  1SEP97   PAGE NO.    1
-----------------------------------------------------------------------------------------------------------------

                       COST      ACCOUNT   UNIT            PCT    ACTUAL     ACTUAL    ESTIMATE TO
ACTIVITY ID RESOURCE  ACCOUNT   CATEGORY  MEAS   BUDGET    CMP   TO DATE  THIS PERIOD   COMPLETE   FORECAST   VARIANCE
---------- --------- --------- ---------- ----  --------  ----- -------- -----------  ----------- --------- ----------
       100 Clear Site
           RD    2 ES 1SEP97 EF 2SEP97 LS 1SEP97 LF 2SEP97 TF   0

           CARP 1                         Days    2.00    .0       .00        .00        2.00      2.00       .00
                                                -------- ----- -------- ----------- ----------- --------- ----------
           TOTAL :                               2.00    .0       .00        .00        2.00      2.00       .00

       200 Layout Building
           RD    1 ES 3SEP97 EF 3SEP97 LS 3SEP97 LF 3SEP97 TF   0

           CARP 2                         Days    1.00    .0       .00        .00        1.00      1.00       .00
           LAB 1                          Days    1.00    .0       .00        .00        1.00      1.00       .00
                                                -------- ----- -------- ----------- ----------- --------- ----------
           TOTAL :                               2.00    .0       .00        .00        2.00      2.00       .00

       300 Excavate Footings
           RD    2 ES 4SEP97 EF 5SEP97 LS 4SEP97 LF 5SEP97 TF   0

           LAB 1                          Days    2.00    .0       .00        .00        2.00      2.00       .00
           EQUIP OP                       Days    2.00    .0       .00        .00        2.00      2.00       .00
                                                -------- ----- -------- ----------- ----------- --------- ----------
           TOTAL :                               4.00    .0       .00        .00        4.00      4.00       .00

       400 Fabricate Ftg Rebar
           RD    1 ES 4SEP97 EF 4SEP97 LS 5SEP97 LF 5SEP97 TF   1

           LAB 1                          Days    1.00    .0       .00        .00        1.00      1.00       .00
           LAB 2                          Days    1.00    .0       .00        .00        1.00      1.00       .00
                                                -------- ----- -------- ----------- ----------- --------- ----------
           TOTAL :                               2.00    .0       .00        .00        2.00      2.00       .00

       500 Fabricate Ftg Formwork
           RD    2 ES 4SEP97 EF 5SEP97 LS 4SEP97 LF 5SEP97 TF   0

           CARP 2                         Days    2.00    .0       .00        .00        2.00      2.00       .00
           LAB 2                          Days    2.00    .0       .00        .00        2.00      2.00       .00
                                                -------- ----- -------- ----------- ----------- --------- ----------
           TOTAL :                               4.00    .0       .00        .00        4.00      4.00       .00

       600 Place Ftg Rebar & Formwork
           RD    2 ES 8SEP97 EF 9SEP97 LS 8SEP97 LF 9SEP97 TF   0

           EQUIP OP                       Days    2.00    .0       .00        .00        2.00      2.00       .00
           LAB 2                          Days    2.00    .0       .00        .00        2.00      2.00       .00
                                                -------- ----- -------- ----------- ----------- --------- ----------
           TOTAL :                               4.00    .0       .00        .00        4.00      4.00       .00

       700 Place Ftg Conc
           RD    1 ES 10SEP97 EF 10SEP97 LS 10SEP97 LF 10SEP97 TF 0

           EQUIP OP                       Days    1.00    .0       .00        .00        1.00      1.00       .00
           LAB 1                          Days    1.00    .0       .00        .00        1.00      1.00       .00
           LAB 2                          Days     .00    .0       .00        .00         .00       .00       .00
                                                -------- ----- -------- ----------- ----------- --------- ----------
           TOTAL :                               2.00    .0       .00        .00        2.00      2.00       .00

                                                -------- ----- -------- ----------- ----------- --------- ----------
                          REPORT TOTALS          20.00   .0       .00        .00       20.00     20.00       .00
                                                ======== ===== ======== =========== =========== ========= ==========
```

Figure 9-16 Resource Control Activity Report Print—Detail by Activity

down arrow under **Res_ource Selection:**. A selection menu will appear for all of the resources as input in the resources dictionary. Another option is to fill the **Low Value Resource** field with question marks (???????). By using this wildcard, all resources within the resources dictionary will be selected.

Figure 9-16 is the RC-01 report run selecting the resources as configured in Figure 9-18. Notice that Activity 200, Layout Building, has a budget of 1 day for CARP 2 and 1 day for LAB 1. Since no work has been completed on this activity, the values for estimate to complete and forecast are the same as the budget.

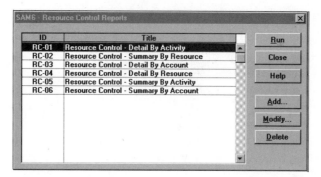

Figure 9-17   Resource Control Reports Title Selection Box

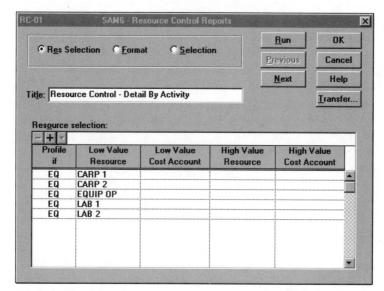

Figure 9-18   Resource Control Reports Dialog Box—Resource Selection
Screen

Format Screen   The second screen to appear with the **Resource Control Reports** dialog box is the **Format** screen (Figure 9-19). This **Format** screen is very similar in function and appearance to the other **Format** screens we have used. This screen enables us to organize the hard copy print by criteria. Examples are sorts by contract (subcontractor) or responsibility (crew). The **Report organized by:** field has three options: **Activity ID, Resource,** and **Cost Account**.

Sort on   The **Sort on** field under the **Sort by:** option of the **Format** screen organizes the hard copy print. The sort options for this field are the same as the **Sort on** fields of the **Format** screens for the **Schedule Reports** dialog box and the **Bar Charts** dialog box. It is also the same as the **Sort by:** field of the **Sort** screen for the **Time-scaled Logic Diagrams** dialog box.

Selection Screen   The third screen to appear with the **Resource Control Reports** dialog box is the **Selection** screen (Figure 9-20). This **Selec-**

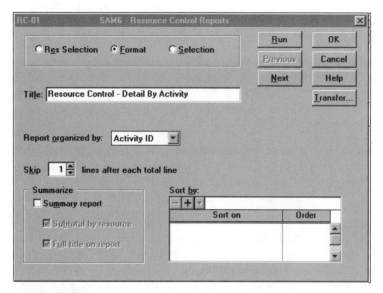

Figure 9-19   Resource Control Reports Dialog Box—Format Screen

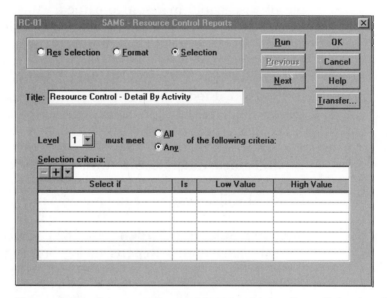

Figure 9-20   Resource Control Reports Dialog Box—Selection Screen

tion screen is very similar in function and appearance to the other
Selection screens. With this screen you can choose to print only the
information meeting some criteria to customize reports for specialized
distribution. Examples are sorts by contract (subcontractor) or responsibility (crew).

Selection criteria: The selection criteria in the **Selection** screen are the
same as the selection criteria for the **Sort on** field in the **Format**
screen. Although the choices are the same, they are used for different
purposes. Under the **Format** screen, you can only select the sequence

of print. Under the **Selection** screen, you can print ranges of selected activities; for example, only activities that are behind schedule.

Earned Value Report

The earned value report shows a detailed picture of performance in terms of the budgeted cost for work performed (BCWP), or earned value, the actual cost for work performed (ACWP), and the budgeted cost for work scheduled (BCWS). *P3* calculates earned value as the product of the percent complete and the budget. *P3* uses the resource percent complete if it is specified for a resource; otherwise *P3* uses the schedule percent complete for the activity. The budgeted amount on budgets can be based on either the current or the target schedule. Specify your choice in the **Earned Value Calculations** dialog box. To access the **Earned Value Calculations** dialog box, select the **Tools** main pull-down menu, then select **Options** and **Earned Value....**

Earned Value (Units) Reports Dialog Box   Figure 9-21 is an example of a schedule in resource earned value tabular format. Earned value represents the value of the work performed rather than the actual work performed. This report is particularly useful in calculating physical progress. Select the **Earned Value (Units)...** option from the **Resource** option of **Tabular Reports** from the **Tools** main pull-down menu. The **Earned Value (units) Reports** title selection box will appear (Figure 9-22). As with previous iterations of this dialog box, you can either delete or modify the specifications to existing reports to meet present project needs or add a new report title and specification. Select **RE-01**, **Earned Value Report - Units**, and click on the **Modify...** button to customize it as needed. The **Earned Value (units) Reports** dialog box for RE-01 will appear for configuring the current project (Figure 9-23).

Resource Selection Screen   The first screen to appear with the **Earned Value (units) Reports** dialog box is the **Resource Selection**

```
-------------------------------------------------------------------------------------------------------------
Southern Constructors                    PRIMAVERA PROJECT PLANNER              Resource Refine

REPORT DATE  29DEC96  RUN NO.   42          EARNED VALUE REPORT - QUANTITY        START DATE  1SEP97  FIN DATE 10SEP97
             8:01
Earned Value Report - Units                                                      DATA DATE   1SEP97  PAGE NO.    1

-------------------------------------------------------------------------------------------------------------
   COST                        PCT  .........CUMULATIVE TO DATE.........  ........VARIANCE.........  ......AT COMPLETION......
  ACCOUNT    RESOURCE ACTIVITY ID CMP    ACWP         BCWP         BCWS        COST      SCHEDULE      BUDGET      ESTIMATE
----------  -------- ----------- ---  ------------ ------------ ------------  ----------- ----------  ----------- -----------

            EQUIP OP - Equipment Operator

            EQUIP OP      300    .0      .00          .00          .00          .00         .00        2.00        2.00
            EQUIP OP      600    .0      .00          .00          .00          .00         .00        2.00        2.00
            EQUIP OP      700    .0      .00          .00          .00          .00         .00        1.00        1.00
                                ----  ------------ ------------ ------------  ----------- ----------  ----------- -----------
            EQUIP OP    TOTAL    .0      .00          .00          .00          .00         .00        5.00        5.00
=============================================================================================================


            REPORT     TOTALS   .0      .00          .00          .00          .00         .00        5.00        5.00
=============================================================================================================
```

Figure 9-21   Earned Value Quantity Report Print—Units

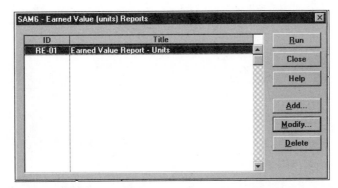

Figure 9-22   Earned Value (Units) Reports Title Selection Box

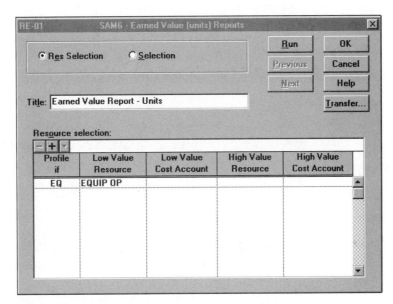

Figure 9-23   Earned Value (Units) Reports Dialog Box—Resource Selection Screen

screen. Use this screen to select the resources that will appear in the earned value report. The options for this field are defined in the resources dictionary under the **Data** main pull-down menu. Figure 9-21 (p. 228) is the RE-01 report run selecting the resources as configured in Figure 9-23. Notice that the resource selected, EQUIP OP, appears in Activity 300, 600, and 700. In Figure 9-21 (p. 228), ACWP is actual *quantity* for work performed, BCWP is budgeted *quantity* for work performed, and BCWS is budgeted *quantity* for work scheduled. These abbreviation are deceptive. *P3* uses the abbreviation C for cost, but since an earned value resource report is being produced, *P3* returns quantities. The scheduled budget reflects the amount of work that should have been finished according to the target plan.

Selection Screen   The second screen to appear with the **Earned Value (units) Reports** dialog box is the **Selection** screen (Figure 9-24). This **Selection** screen is similar in function and appearance to the **Selection**

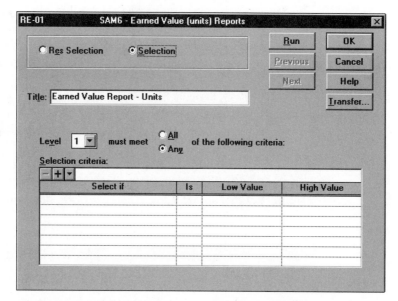

Figure 9-24   Earned Value (Units) Reports Dialog Box—Selection Screen

screens described previously. The options for this screen were covered earlier in this chapter under the **Selection** screen for the tabular schedule report.

Tabular Resource Report

Tabular Resource Reports Dialog Box   Figure 9-25 is an example of a schedule in tabular resource format. This report is particularly useful in planning for and tracking a particular resource. Select the **Tabular...** option of the **Resource** options of **Tabular Reports** from the **Tools** main pull-down menu. The **Tabular Resource Reports** title selection box will appear (Figure 9-26). As with previous iterations of this dialog box, you

```
-----------------------------------------------------------------------------------------------------
Southern Constructors                  PRIMAVERA PROJECT PLANNER            Resource Refine

REPORT DATE  29DEC96  RUN NO.   55           TABULAR RESOURCE REPORT-DAILY       START DATE  1SEP97  FIN DATE 10SEP97
             9:04
Tabular Resource Use - Weekly                                                    DATA DATE   1SEP97   PAGE NO.    1
-----------------------------------------------------------------------------------------------------
   PERIOD     ----AVAILABLE----      -----EARLY SCHEDULE----     -----LATE SCHEDULE-----      ---TARGET 1 SCHEDULE---
  BEGINNING   NORMAL   MAXIMUM        USAGE     CUMULATIVE        USAGE     CUMULATIVE         USAGE     CUMULATIVE
-----------  -------  -------        -------   ------------      -------   ------------       -------   ------------
             EQUIP OP - Equipment Operator                    UNIT OF MEASURE = Days

  31AUG97        5        7            .00         .00           .00         .00              .00         .00
***DATA DATE***
   1SEP97        5        7            .00         .00           .00         .00              .00         .00
   2SEP97        5        7            .00         .00           .00         .00              .00         .00
   3SEP97        5        7            .00         .00           .00         .00              .00         .00
   4SEP97        5        7           2.00        2.00           .00         .00              .00         .00
   5SEP97        5        7            .00        2.00           .00         .00              .00         .00
   6SEP97        5        7            .00        2.00           .00         .00              .00         .00
   7SEP97        5        7            .00        2.00           .00         .00              .00         .00
   8SEP97        5        7           2.00        4.00           .00         .00              .00         .00
   9SEP97        5        7            .00        4.00           .00         .00              .00         .00
  10SEP97        5        7           1.00        5.00           .00         .00              .00         .00
  11SEP97     9999999  9999999         .00        5.00           .00         .00              .00         .00
```

Figure 9-25   Tabular Resource Report Daily—Print

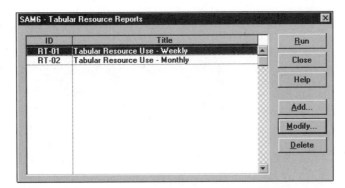

Figure 9-26   Tabular Resource Reports Title Selection Box

can either delete or modify the specifications to existing reports to meet present project needs or add a new report title and specification. Select **RT-01, Tabular Resource Use - Weekly**, and click on the **Modify...** button to customize it. The **Tabular Resource Reports** dialog box for RT-01 will appear (Figure 9-27).

Resource Selection Screen   The first screen to appear with the **Tabular Resource Reports** dialog box is the **Resource Selection** screen for selecting the resources that will appear in the tabular resource report. Figure 9-25 is the RT-01 report run selecting the resources as configured in Figure 9-27. Notice that for EQUIP OP the report provides both daily and cumulative usage.

Selection Screen   The second screen to appear with the **Tabular Resource Reports** dialog box is the **Selection** screen (Figure 9-28), which is similar in function and appearance to the other **Selection**

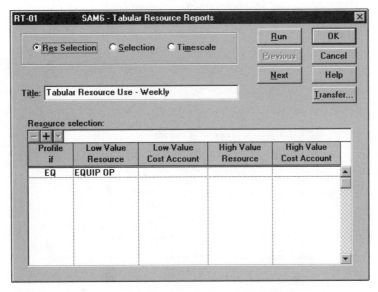

Figure 9-27   Tabular Resource Reports Dialog Box—Resource Selection Screen

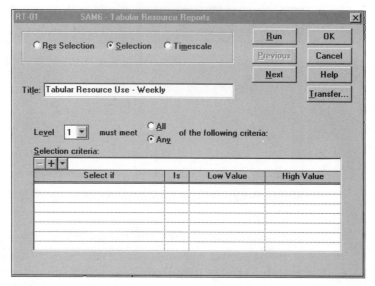

Figure 9-28 Tabular Resource Reports Dialog Box—Selection Screen

screens. Use it to customize reports for narrow distribution. For the options for this field, see the **Selection** screen section for the tabular schedule report (pp. 222–223).

Timescale Screen The last screen to appear with the **Tabular Resource Reports** dialog box is the **Timescale** screen (Figure 9-29). As with the bar chart and timescaled logic diagrams, you can designate the starting and ending times of this tabular report. Figure 9-25 (p. 230) is the RT-01 report run selecting the resources as configured in Figure 9-29. Notice that the **Timescale units:** is Day and the **Days shown per week:** is 7. Excluding weekends, since they are nonwork periods, would produce a cleaner report.

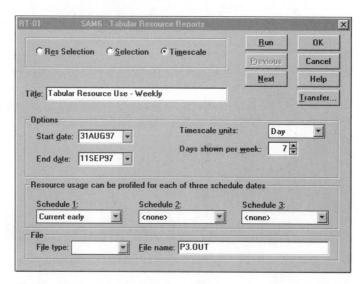

Figure 9-29 Tabular Resource Reports Dialog Box—Timescale Screen

Cost Reports

Tracking Costs by Activity

The third major category of tabular reports is the cost reports. See the **Cost** option of **Tabular Reports** from the **Tools** main pull-down menu (Figure 9-30). The schedule and resource reports were the first two categories. Figure 9-31 is an example of a schedule in tabular cost format. This form is particularly useful in tracking costs by activity. For Activity 600, Place Ftg Rebar & Formwork, Cost account 3158 has a budget of $200, an estimate to complete of $200, and a forecast of $200. When actual costs are input, the project can be monitored and controlled.

Cost Series of Reports

The cost series of reports is accessed from the **Tabular Reports** option from the **Tools** main pull-down menu. There are five types of reports included in the **Cost** option. They are control; cost, price, and rates; earned value; tabular; and loading.

Control Control Report

Cost Control Reports Dialog Box   Figure 9-31 is an example of a cost control tabular report. Select the **Control...** option of **Cost** of **Tabular**

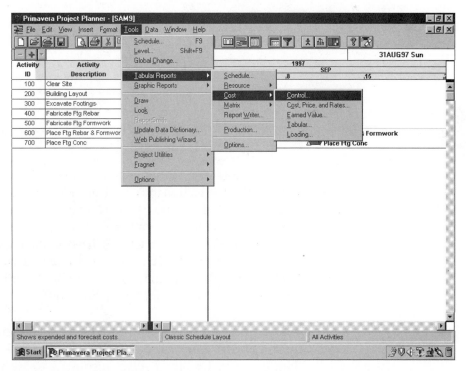

Figure 9-30   Tools Main Pull-Down Menu—Tabular Reports Cost and Control Options

```
--------------------------------------------------------------------------------------------
Southern Constructors            PRIMAVERA PROJECT PLANNER        Sample Schedule

REPORT DATE  29DEC96 RUN NO.  66     COST CONTROL ACTIVITY REPORT   START DATE 1SEP97  FIN DATE 10SEP97
             9:16
Cost Control - Detailed by Activity                               DATA DATE 1SEP97   PAGE NO.   1

--------------------------------------------------------------------------------------------
                    COST    ACCOUNT UNIT          PCT   ACTUAL    ACTUAL   ESTIMATE TO
ACTIVITY ID RESOURCE ACCOUNT CATEGORY MEAS  BUDGET  CMP  TO DATE  THIS PERIOD COMPLETE  FORECAST   VARIANCE

    100 Clear Site
        RD   2 ES 1SEP97  EF 2SEP97 LS 1SEP97 LF 2SEP97 TF  0
             2104    T Tot Cost      1000.00  .0     .00      .00  1000.00  1000.00      .00
        TOTAL :                      1000.00  .0     .00      .00  1000.00  1000.00      .00

    200 Building Layout
        RD   1 ES 3SEP97  EF 3SEP97 LS 3SEP97 LF 3SEP97 TF  0
             1306    T Tot Cost       300.00  .0     .00      .00   300.00   300.00      .00
        TOTAL :                       300.00  .0     .00      .00   300.00   300.00      .00

    300 Excavate Footings
        RD   2 ES 4SEP97  EF 5SEP97 LS 4SEP97 LF 5SEP97 TF  0
             2204    T Tot Cost       400.00  .0     .00      .00   400.00   400.00      .00
        TOTAL :                       400.00  .0     .00      .00   400.00   400.00      .00

    400 Fabricate Ftg Rebar
        RD   1 ES 4SEP97  EF 4SEP97 LS 5SEP97 LF 5SEP97 TF  1
             3217    T Tot Cost       500.00  .0     .00      .00   500.00   500.00      .00
        TOTAL :                       500.00  .0     .00      .00   500.00   500.00      .00

    500 Fabricate Ftg Formwork
        RD   2 ES 4SEP97  EF 5SEP97 LS 4SEP97 LF 5SEP97 TF  0
             3158    T Tot Cost       600.00  .0     .00      .00   600.00   600.00      .00
        TOTAL :                       600.00  .0     .00      .00   600.00   600.00      .00

    600 Place Ftg Rebar & Formwork
        RD   2 ES 8SEP97  EF 9SEP97 LS 8SEP97 LF 9SEP97 TF  0
             3158             200.00  .0     .00      .00   200.00   200.00      .00
             3217             300.00  .0     .00      .00   300.00   300.00      .00
        TOTAL :              500.00  .0     .00      .00   500.00   500.00      .00

    700 Place Ftg Conc
        RD   1 ES 10SEP97 EF 10SEP97 LS 10SEP97 LF 10SEP97 TF  0
             3130             300.00  .0     .00      .00   300.00   300.00      .00
        TOTAL :              300.00  .0     .00      .00   300.00   300.00      .00

            REPORT TOTALS   3600.00  .0     .00      .00  3600.00  3600.00      .00
==============================================================================================
```

Figure 9-31 Cost Control Activity Print—Detailed by Activity

Reports from the **Tools** main pull-down menu. The **Cost Control Reports** title selection box will appear (Figure 9-32). As with similar uses of this dialog box, you can delete or modify the specifications to existing reports to meet present project needs or add a new report title and specification. Select **CC-07, Cost Control - Detailed by Activity**, and click on the **Modify...** button to customize it to your needs. The **Cost Control Reports** dialog box for CC-07 will appear for configuring the project we are presently working on (Figure 9-33).

Resource Selection Screen The first screen to appear with the **Cost Control Reports** dialog box is the **Resource Selection** screen (Figure 9-33). Use it to select resources to include in the cost control report. The options for this field are defined in the cost accounts dictio-

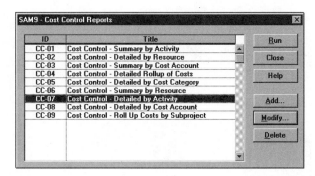

Figure 9-32   Cost Control Reports Title Selection Box

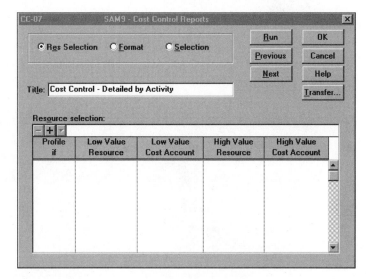

Figure 9-33   Cost Control Reports Dialog Box—Resource Selection Screen

nary under the **Data** main pull-down menu. Figure 9-31 (p. 230) is the CC-07 report run selecting the resources as configured in Figure 9-33. Notice that all accounts were selected in the figure since no criteria limitations were placed on their selection.

Format Screen   The second screen to appear with the **Cost Control Reports** dialog box is the **Format** screen (Figure 9-34). This **Format** screen is similar in function and appearance to the earlier **Format** screens. It allows sorting the hard copy print by criteria for specialized distribution of reports. Examples are sorts by contract (subcontractor) or responsibility (crew).

Selection Screen   The third screen to appear with the **Cost Control Reports** dialog box is the **Selection** screen (Figure 9-35), which is similar in function and appearance to the other **Selection** screens. For options for this field, see Table 7-2 (p. 169). The selection criteria for the **Sort on** field in the **Format** screen and the **Selection Criteria:** in the **Selection** screen are the same.

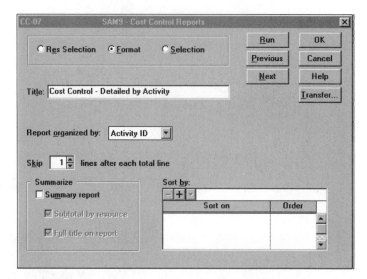

Figure 9-34 Cost Control Reports Dialog Box—Format Screen

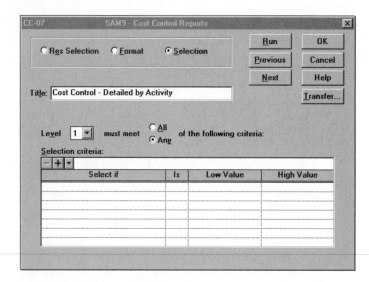

Figure 9-35 Cost Control Reports Dialog Box—Selection Screen

Earned Value Report

Earned Value Reports Dialog Box Figure 9-36 is an example of an earned value tabular cost report. Select the **Earned Value...** of the **Cost** options of **Tabular Reports** from the **Tools** main pull-down menu. The **Earned Value Reports** title selection box will appear (Figure 9-37). As with previous iterations of this dialog box, use it either to delete or to modify the specifications to existing reports to meet present project needs or to add a new report title and specification. Select **CE-01, Earned Value Report - Cost**, and click on the **Modify...** button to customize it as needed. The **Earned Value Reports** dialog box for CE-01 will appear for configuring the present project (Figure 9-38).

```
--------------------------------------------------------------------------------------------
Southern Constructors                    PRIMAVERA PROJECT PLANNER          Sample Schedule

REPORT DATE  29DEC96  RUN NO.   69          EARNED VALUE REPORT - COST       START DATE  1SEP97   FIN DATE 10SEP97
             10:05
Earned Value Report - Cost                                                   DATA DATE   1SEP97   PAGE NO.    1

--------------------------------------------------------------------------------------------
   COST                        PCT  .........CUMULATIVE TO DATE.........  ........VARIANCE.........  ......AT COMPLETION......
  ACCOUNT    RESOURCE ACTIVITY ID  CMP   ACWP         BCWP         BCWS       COST      SCHEDULE      BUDGET      ESTIMATE
-----------  -------- ----------- ----- ------------ ------------ ------------ ------------ ------------ ------------ ------------

                      ????????????

1306         T              200    .0     .00          .00          .00         .00          .00       300.00       300.00
2104         T              100    .0     .00          .00          .00         .00          .00      1000.00      1000.00
2204         T              300    .0     .00          .00          .00         .00          .00       400.00       400.00
3130                        700    .0     .00          .00          .00         .00          .00       300.00       300.00
3158                        600    .0     .00          .00          .00         .00          .00       200.00       200.00
3158         T              500    .0     .00          .00          .00         .00          .00       600.00       600.00
3217                        600    .0     .00          .00          .00         .00          .00       300.00       300.00
3217         T              400    .0     .00          .00          .00         .00          .00       500.00       500.00
                                  ----- ------------ ------------ ------------ ------------ ------------ ------------ ------------
????????????          TOTAL        .0     .00          .00          .00         .00          .00      3600.00      3600.00
============================================================================================

             REPORT    TOTALS      .0     .00          .00          .00         .00          .00      3600.00      3600.00
============================================================================================
```

Figure 9-36   Earned Value Cost Report Print

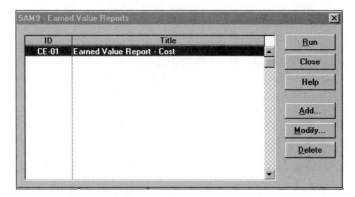

Figure 9-37   Earned Value Reports Title Selection Box

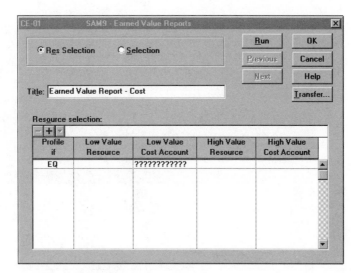

Figure 9-38   Earned Value Reports Dialog Box—Resource Selection Screen

Resource Selection Screen   The first screen to appear with the **Earned Value Reports** dialog box is the **Resource Selection** screen (Figure 9-38). This screen enables the user to select resources to appear in the earned value report. Figure 9-36 (p. 237) is the CE-01 report run using the configuration in Figure 9-38. Notice that all accounts were selected since the **Low Value Cost Account** field was filled with ?s. There were no criteria limitations placed on their selection.

Selection Screen   The second screen under the **Earned Value Reports** dialog box is the **Selection** screen (Figure 9-39). It is very similar in function and appearance to the **Selection** screens described previously.

Tabular Cost Report

Tabular Cost Reports Dialog Box   Figure 9-40 is an example of a tabular cost report. Select the **Tabular...** from the **Cost** options of **Tabular Reports** from the **Tools** main pull-down menu. The **Tabular Cost Reports** title selection box will appear (Figure 9-41). Use this box to delete or modify the specifications to existing reports to meet present project needs or add a new report title and specification. Select **CT-01, Tabular Cost - Monthly Project Cash Flow**, and click on the **Modify...** button to customize it to meet your needs. The **Tabular Cost Reports** dialog box for CT-01 will appear for configuring the present project (Figure 9-42).

Resource Selection Screen   The first screen to appear with the **Tabular Cost Reports** dialog box is the **Resource Selection** screen (Figure 9-42) for choosing the accounts to print in the tabular cost report. Figure 9-40 is the CT-01 report run using the configuration in Figure 9-42.

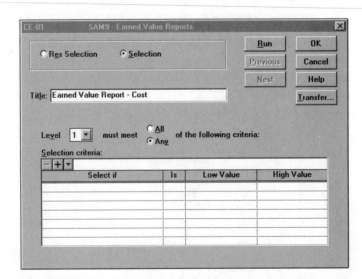

Figure 9-39   Earned Value Reports Dialog Box—Selection Screen

```
Southern Constructors                      PRIMAVERA PROJECT PLANNER          Sample Schedule

REPORT DATE  29DEC96  RUN NO.   77              TABULAR COST REPORT - DAILY        START DATE  1SEP97  FIN DATE 10SEP97
             10:22
Tabular Cost - Monthly Project Cash Flow                                           DATA DATE  1SEP97   PAGE NO.    1

------------------------------------------------------------------------------------------------------------------------
    PERIOD          -----EARLY SCHEDULE----            -----LATE SCHEDULE-----          ---TARGET 1 SCHEDULE---
  BEGINNING          USAGE     CUMULATIVE               USAGE     CUMULATIVE             USAGE      CUMULATIVE
 ------------       --------  ------------             --------  ------------           --------   ------------
            COST ACCOUNT = ???????????? -

  31AUG97               .00         .00                   .00        .00                   .00         .00
***DATA DATE***
   1SEP97            1000.00     1000.00                  .00        .00                   .00         .00
   2SEP97               .00     1000.00                   .00        .00                   .00         .00
   3SEP97             300.00     1300.00                  .00        .00                   .00         .00
   4SEP97            1500.00     2800.00                  .00        .00                   .00         .00
   5SEP97               .00     2800.00                   .00        .00                   .00         .00
   6SEP97               .00     2800.00                   .00        .00                   .00         .00
   7SEP97               .00     2800.00                   .00        .00                   .00         .00
   8SEP97             500.00     3300.00                  .00        .00                   .00         .00
   9SEP97               .00     3300.00                   .00        .00                   .00         .00
  10SEP97             300.00     3600.00                  .00        .00                   .00         .00
  11SEP97               .00     3600.00                   .00        .00                   .00         .00
```

Figure 9-40 Monthly Project Cash Flow Tabular Cost Report Print—Daily

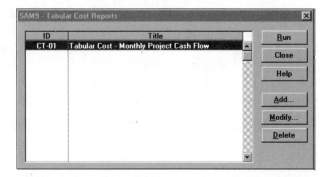

Figure 9-41 Tabular Cost Reports Title Selection Box

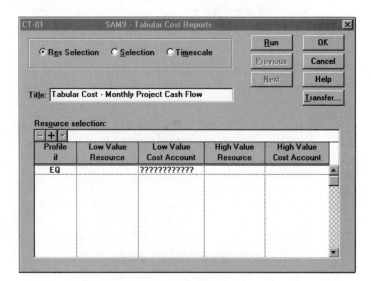

Figure 9-42 Tabular Cost Reports Dialog Box—Resource Selection Screen

Notice that all accounts were selected, since the **Low Value Cost Account** field was filled with ?s; no criteria limitations were placed on their selection. The plan is for $1,000 to be spent on 1SEP97 (September

1, 1997). The cumulative cost of $3,600 will be spent by 11SEP97, the end of this schedule.

Selection Screen   The second screen to appear with the **Tabular Cost Reports** dialog box is the **Selection** screen (Figure 9-43).

Timescale Screen   The **Timescale** screen (Figure 9-44) has **Start date:** for choosing the beginning date of the tabular report, in this case 31AUG97. The **End date:** is 11SEP97. The **Timescale units:** used for Figure 9-40 (p. 239) was the day; either the week or the month might be a more appropriate unit for cash flow purposes for longer projects.

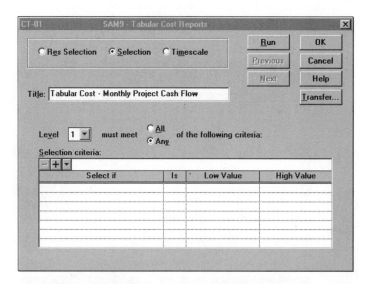

Figure 9-43   Tabular Cost Reports Dialog Box—Selection Screen

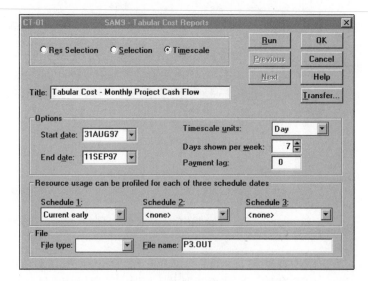

Figure 9-44   Tabular Cost Reports Dialog Box—Timescale Screen

REPORT WRITER

Types of Reports

Report Writer allows you to customize many of the default report types, include calculated data, or start a new report from scratch. To access Report Writer, select **Report Writer...** from the **Tabular Reports** option from the **Tools** main pull-down menu. With this option selected, the **Report Writer Reports** title selection box will appear (Figure 9-45). The default reports within report writer are shown in Figure 9-45.

Report Writer Reports Dialog Box

Select **RW-01, Classic Schedule Report,** and click on the **Modify...** button to customize it. The **Report Writer Reports** dialog box for RW-01 will appear (Figure 9-46). Figure 9-47 is a hard copy of the RW-01 report before modification.

Arithmetic Screen   The first screen to appear with the **Report Writer Reports** dialog box is the **Arithmetic** screen (Figure 9-46). You can use it to specify calculated data for report columns as identified by temporary variables from the **Content** screen. Click on the cell, then click the right mouse button to select from a list of data items. Calculate data by using if/then logic statements. If statements define the condition under which certain events should occur. Then statements initiate the action or calculation that you want to take place. The Else statements establish a condition that occurs when the If statement is false.

Selection Screen   The second screen to appear with the **Report Writer Reports** dialog box is the **Selection** screen (Figure 9-48). Use it to organize the printout as needed to represent the information effectively. The sort options for the **Selection criteria** field are the same as the, **Sort on**

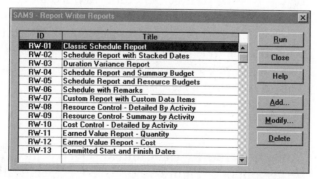

Figure 9-45   Report Writer Reports Title Selection Box

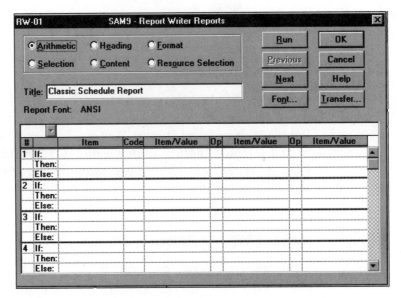

Figure 9-46 Report Writer Reports Dialog Box—Arithmetic Screen

```
Southern Constructors                    PRIMAVERA PROJECT PLANNER        Sample Schedule

REPORT DATE 29DEC96  RUN NO.   80          -----Project Schedule-----     START DATE 01SEP97  FIN DATE 10SEP97
          10:29
Classic Schedule Report                                                  DATA DATE  01SEP97  PAGE NO.    1

ACTIVITY    ORIG REM  CAL          ACTIVITY                              EARLY    EARLY    LATE     LATE     TOTAL
ID          DUR  DUR  ID   %  CODE DESCRIPTION                           START    FINISH   START    FINISH   FLOAT
--------    ---- ---- ---  -- ---- -----------                          -------- -------- -------- -------- -----
     100     2    2    1   0       Clear Site                            01SEP97  02SEP97  01SEP97  02SEP97    0
     200     1    1    1   0       Building Layout                       03SEP97  03SEP97  03SEP97  03SEP97    0
     400     1    1    1   0       Fabricate Ftg Rebar                   04SEP97  04SEP97  05SEP97  05SEP97    1
     300     2    2    1   0       Excavate Footings                     04SEP97  05SEP97  04SEP97  05SEP97    0
     500     2    2    1   0       Fabricate Ftg Formwork                04SEP97  05SEP97  04SEP97  05SEP97    0
     600     2    2    1   0       Place Ftg Rebar & Formwork            08SEP97  09SEP97  08SEP97  09SEP97    0
     700     1    1    1   0       Place Ftg Conc                        10SEP97  10SEP97  10SEP97  10SEP97    0
```

Figure 9-47 Project Schedule Print—Classic Schedule Report

field of the **Format** screen for the **Bar Charts** dialog box and the **Sort by:** field of the **Sort** screen for the **Timescaled Logic Diagrams** dialog box.

Heading Screen The next screen for the **Report Writer Reports** dialog box is the **Heading** screen (Figure 9-49). Figure 9-47 is a hard copy of the RW-01, Classic Schedule Report, as configured in Figure 9-49. Notice in Figure 9-47 that the title Southern Constructors is printed on the first line and is aligned to the left. The second header entry is Sample Schedule. It is printed on the first line and is 96 spaces from the left. The project name is not printed, since there is a 0 in the **Row** column. The project start date, project data date, and the project finish date are listed. The report title Classic Schedule Report is printed on the fifth line and is aligned to the left.

Content Screen The **Content** screen is the power behind the Report Writer options (Figure 9-50). Use its **Specifications:** fields to indicate which columns are to be printed, the column title, column width, data

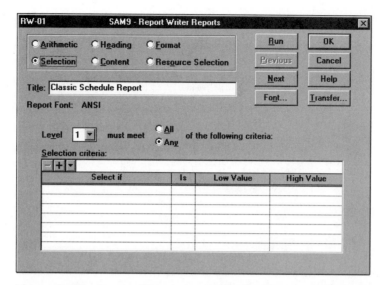

Figure 9-48   Report Writer Reports Dialog Box—Selection Screen

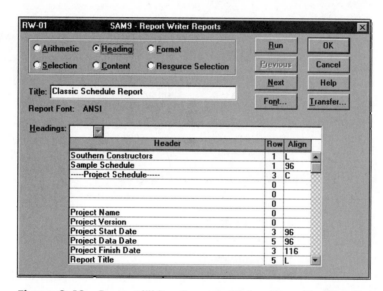

Figure 9-49   Report Writer Reports Dialog Box—Heading Screen

to be placed in the field, alignment, and wrapping characteristics. Figure 9-47 (p. 242) is a hard copy of the RW-01, Classic Schedule Report, as configured in Figure 9-50. Notice the title for **Column 1**, ACTIVITY ID, can appear in three fields: **Line 1 title**, **Line 2 title**, and **Line 3 title**.

Data item   To define the information to be placed in the column, click on the **Data item** field and click the down arrow to bring up the **Data Item** dialog box (Figure 9-51). The major options under **Data item** are:

- **Activity data**
- **Constraints**
- **Resource/cost data**

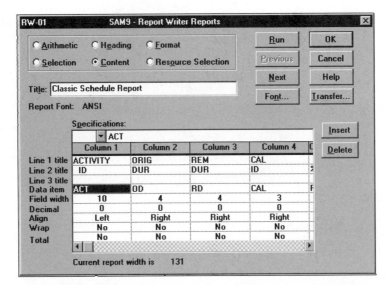

Figure 9-50   Report Writer Reports Dialog Box—Content Screen

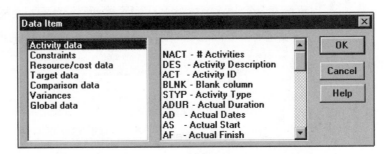

Figure 9-51   Data Item Dialog Box

- Target data
- Comparison data
- Variances
- Global data

Activity data   For each of the classifications on the left half of the **Data Item** dialog box, there are options on the right half of the dialog box. The **Activity data** options are given in Table 9-2.

Constraints   The **Constraints** options of the **Data Item** dialog box are given in Table 9-3.

Resource/cost data   The **Resource/cost data** options of the **Data Item** dialog box are given in Table 9-4.

Global data   The **Global data** options are:

SD	-	Project Start
DD	-	Data Date
FD	-	Project Finish

NACT	-	# Activities
DES	-	Activity Description
ACT	-	Activity ID
BLNK	-	Blank column
ATYP	-	Activity Type
ADUR	-	Actual Duration
AD	-	Actual Dates
AS	-	Actual Start
AF	-	Actual Finish
CAL	-	Calendar ID
ED	-	Early Dates
ES	-	Early Start
EF	-	Early Finish
FFL	-	Free Float
LD	-	Late Dates
LS	-	Late Start
LF	-	Late Finish
LTYP	-	Leveling Type
MEM	-	Log Records
OD	-	Original Duration
PCT	-	Percent Complete
RD	-	Remaining Duration
RSM	-	Resume Date
SUS	-	Suspended Date
TF	-	Total Float
WBS	-	Work Breakdown Structure
WBST	-	WBS Title
RESP	-	RESPONSIBILITY
AREA	-	AREA/DEPARTMENT
MILE	-	MILESTONE
ITEM	-	ITEM NAME
LOCN	-	LOCATION
STEP	-	STEP
PLST	-	Planned Start
PLFN	-	Planned Finnish
SPEC	-	Specification
SUBM	-	Submittal

Table 9-2   Data Item—Activity Data Options

SNE	-	ES Constraint
SNL	-	LS Constraint
FNE	-	EF Constraint
FNL	-	LF Constraint
ON	-	Start On
MS	-	Mandatory Start
MF	-	Mandatory Finish
ZTF	-	Zero total float
ZFF	-	Zero free float
XF	-	Expected Finish
CON	-	Constraint exists

Table 9-3   Data Item—Constraints Options

Field width   The **Field width** field determines the width of each column in number of characters. For example, Column 1 of Figure 9-47 (p. 242) has a field width of 10 characters (see Figure 9-50, p. 244).

Align   The **Align** field determines whether the characters in the field are aligned to the left or right of or centered on the field. Column 1 of Figure 9-47 (p. 242) has the **Align** field set to left.

Wrap   The **Wrap** field options are Yes and No. This function wraps the print to another line in the column if the number of characters in the

CTD	-	Cost to date
CTP	-	Cost this period
QTD	-	Qty. to date
QTP	-	Qty. this period
BC	-	Budgeted cost
BQ	-	Budgeted quantity
CVAR	-	Completion variance (cost)
QVAR	-	Completion variance (quantity)
ACC	-	Cost account (11)
ACX	-	Cost account (12)
CAT	-	Cost category
CAC	-	Cost at completion
CTC	-	Cost to complete
CV	-	Cost variance (BCWP - ACWP)
CVQ	-	Cost variance units (BQWP - AQWP)
RDRV	-	Res. driving flag
BCWP	-	Earned value (cost)
BQWP	-	Earned value (units)
QTC	-	Qty. to complete
QAC	-	Qty. at completion
RPCT	-	Res. percent comp.
RES	-	Resource
RID	-	Resource curve
RDUR	-	Resource duration
RUM	-	Res. unit of measure
RLAG	-	Resource lag
BCWS	-	Sched. budget cost
BQWS	-	Sched. budget qty.
SV	-	Schedule variance (BCWP - BCWS)
SVQ	-	Schedule variance (BQWP - BQWS)
UPT	-	Units/time period
NRES	-	Number of resources
ORBC	-	Orig Budget Cost
ORBQ	-	Orig Budget Qty

Table 9-4 Data Item—Resource/Cost Data Options

Field width is not sufficient for the entry. This function is turned off in Figure 9-50 (p. 244).

Total The **Total** field options are Yes and No. This function indicates whether subtotals are displayed within a column. This function is turned off in Figure 9-50 (p. 244).

Format Screen The fifth screen to appear with the **Report Writer Reports** dialog box is the **Format** screen (Figure 9-52). The functions and operations of this screen are very similar to previously discussed

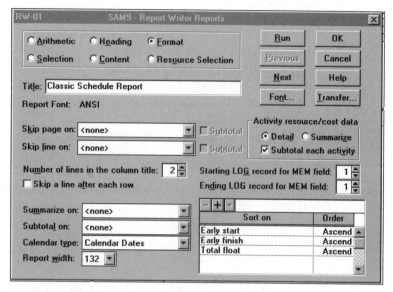

Figure 9-52   Report Writer Reports Dialog Box—Format Screen

Format screens. Figure 9-47 (p. 242) is a hard copy of the RW-01, Classic Schedule Report, as configured in Figure 9-52.

Resource Selection Screen   The last screen to appear with the **Report Writer Reports** dialog box is the **Resource Selection** screen (Figure 9-53). This screen selects the resources that will appear in the resource control report. The functions and operations of this screen are very similar to previously discussed **Resource Selection** screens. Here no particular resource selection criteria was input to modify the report.

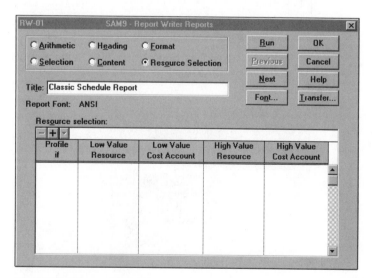

Figure 9-53   Report Writer Reports Dialog Box—Resource Selection Screen

PRODUCTION

Run Multiple Reports

P3's production feature offers the convenience of running multiple reports with one execution command. If at the end of each month or update period you wish to run a series of the same reports, you can save time by selecting a group of reports. To do this, select **Production...** from the **Tabular Reports** option from the **Tools** main pull-down menu. Selecting this option causes the **Production** reports title selection box to appear (Figure 9-54). Whenever a report is included or added to any of the reports discussed in this chapter, it becomes a part of the production reports.

Each report series can be identified by a code. In Figure 9-54, three reports have been identified as part of Series A. To run a series of reports, click on the **Run Series...** button. The **Run a Series** dialog box will appear (Figure 9-55). In Figure 9-55 Series A has been selected. Click the OK button, and all three reports, or the entire series, will be executed.

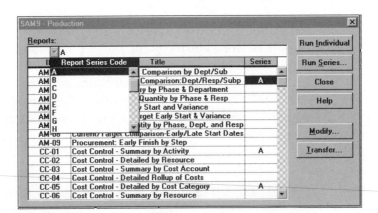

Figure 9-54   Production Reports Title Selection Box Report Series Code

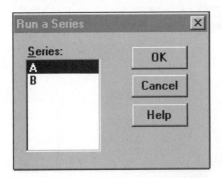

Figure 9-55   Run a Series Dialog Box

EXAMPLE PROBLEM: Tabular Reports

Figures 9-56 to 9-58 are hard copy printouts of tabular reports for a house put together as an example for student use (see the wood frame house drawings in the Appendix).

Figure 9-56 is a hard copy print of the tabular classic schedule report sorted by ES and TF. Figure 9-57 a to d is the tabular resource control report detailed by activity. Figure 9-58 a to d is the tabular cost control report detailed by activity.

```
-------------------------------------------------------------------------------------------------------------
Student Constructors                    PRIMAVERA PROJECT PLANNER                Sample Project #1

REPORT DATE 30DEC96  RUN NO.   28                              START DATE  5JAN97  FIN DATE 20MAR97
              6:23
Classic Schedule Report - Sort by ES, TF                      DATA DATE   5JAN97  PAGE NO.    1
```

ACTIVITY ID	ORIG DUR	REM DUR	%	ACTIVITY DESCRIPTION	EARLY START	EARLY FINISH	LATE START	LATE FINISH	TOTAL FLOAT
100	2	2	0	Clear Site	6JAN97	7JAN97	6JAN97	7JAN97	0
200	1	1	0	Building Layout	8JAN97	8JAN97	8JAN97	8JAN97	0
300	3	3	0	Form/Pour Footings	9JAN97	13JAN97	9JAN97	13JAN97	0
400	2	2	0	Pier Masonry	14JAN97	15JAN97	14JAN97	15JAN97	0
500	4	4	0	Wood Floor System	16JAN97	21JAN97	16JAN97	21JAN97	0
600	6	6	0	Rough Framing Walls	22JAN97	29JAN97	22JAN97	29JAN97	0
700	4	4	0	Rough Framing Roof	30JAN97	4FEB97	30JAN97	4FEB97	0
900	2	2	0	Ext Wall Board	5FEB97	6FEB97	5FEB97	6FEB97	0
1100	4	4	0	Rough Plumbing	7FEB97	12FEB97	7FEB97	12FEB97	0
1200	3	3	0	Rough HVAC	7FEB97	11FEB97	10FEB97	12FEB97	1
800	4	4	0	Doors & Windows	7FEB97	12FEB97	12FEB97	17FEB97	3
1300	3	3	0	Rough Elect	12FEB97	14FEB97	13FEB97	17FEB97	1
1400	3	3	0	Shingles	13FEB97	17FEB97	13FEB97	17FEB97	0
1000	1	1	0	Ext Wall Insulation	18FEB97	18FEB97	18FEB97	18FEB97	0
1600	2	2	0	Ext Finish Carpentry	18FEB97	19FEB97	27FEB97	28FEB97	7
2700	4	4	0	Grading & Landscaping	18FEB97	21FEB97	13MAR97	18MAR97	17
1700	4	4	0	Hang Drywall	19FEB97	24FEB97	19FEB97	24FEB97	0
1500	3	3	0	Ext Siding	20FEB97	24FEB97	3MAR97	5MAR97	7
1800	4	4	0	Finish Drywall	25FEB97	28FEB97	25FEB97	28FEB97	0
2000	3	3	0	Ext Paint	25FEB97	27FEB97	6MAR97	10MAR97	7
1900	2	2	0	Cabinets	3MAR97	4MAR97	3MAR97	4MAR97	0
2400	3	3	0	Finish HVAC	3MAR97	5MAR97	6MAR97	10MAR97	3
2500	2	2	0	Finish Elect	3MAR97	4MAR97	7MAR97	10MAR97	4
2100	4	4	0	Int Finish Carpentry	5MAR97	10MAR97	5MAR97	10MAR97	0
2300	2	2	0	Finish Plumbing	5MAR97	6MAR97	7MAR97	10MAR97	2
2200	3	3	0	Int Paint	11MAR97	13MAR97	11MAR97	13MAR97	0
2600	3	3	0	Flooring	14MAR97	18MAR97	14MAR97	18MAR97	0
2800	2	2	0	Punch List	19MAR97	20MAR97	19MAR97	20MAR97	0

Figure 9-56 Classic Schedule Report Print—Sort by ES, TF—Wood Frame House

250   Chapter 9

```
------------------------------------------------------------------------------------------------------------------------
Student Constructors                      PRIMAVERA PROJECT PLANNER            Sample Project #1

REPORT DATE  30DEC96  RUN NO.   32        RESOURCE CONTROL ACTIVITY REPORT     START DATE  5JAN97  FIN DATE 20MAR97
             6:28
Resource Control - Detail By Activity                                         DATA DATE  5JAN97   PAGE NO.   1
------------------------------------------------------------------------------------------------------------------------

                         COST     ACCOUNT  UNIT            PCT    ACTUAL     ACTUAL     ESTIMATE TO
ACTIVITY ID RESOURCE    ACCOUNT   CATEGORY MEAS   BUDGET   CMP   TO DATE  THIS PERIOD   COMPLETE    FORECAST   VARIANCE
----------- --------    -------   -------- ----  --------- ----- -------- -----------  ----------- ---------- ----------

      100  Clear Site
           RD   2 ES  6JAN97  EF  7JAN97  LS  6JAN97  LF  7JAN97  TF   0

           SUB               S SUBCNT'R        .00    .0       .00        .00         .00        .00        .00
                                          -----------  -----  -----------  -----------  -----------  ----------  ----------
           TOTAL :                               .00    .0       .00        .00         .00        .00        .00

      200  Building Layout
           RD   1 ES  8JAN97  EF  8JAN97  LS  8JAN97  LF  8JAN97  TF   0

           CARPENTR          L LABOR    day      .10    .0       .00        .00         .10        .10        .00
           CRPN FOR          L LABOR    day      .10    .0       .00        .00         .10        .10        .00
           LAB CL 1          L LABOR    day      .20    .0       .00        .00         .20        .20        .00
           LAB CL 2          L LABOR    day      .20    .0       .00        .00         .20        .20        .00
           MATERIAL          M MATERIAL          .00    .0       .00        .00         .00        .00        .00
           SUB               S SUBCNT'R          .00    .0       .00        .00         .00        .00        .00
                                          -----------  -----  -----------  -----------  -----------  ----------  ----------
           TOTAL :                               .60    .0       .00        .00         .60        .60        .00

      300  Form/Pour Footings
           RD   3 ES  9JAN97  EF 13JAN97  LS  9JAN97  LF 13JAN97  TF   0

           CARPENTR          L LABOR    day      .90    .0       .00        .00         .90        .90        .00
           CRPN FOR          L LABOR    day      .90    .0       .00        .00         .90        .90        .00
           LAB CL 1          L LABOR    day     1.79    .0       .00        .00        1.79       1.79        .00
           LAB CL 2          L LABOR    day     1.79    .0       .00        .00        1.79       1.79        .00
           MATERIAL          M MATERIAL          .00    .0       .00        .00         .00        .00        .00
           SUB               S SUBCNT'R          .00    .0       .00        .00         .00        .00        .00
                                          -----------  -----  -----------  -----------  -----------  ----------  ----------
           TOTAL :                              5.38    .0       .00        .00        5.38       5.38        .00

      400  Pier Masonry
           RD   2 ES 14JAN97  EF 15JAN97  LS 14JAN97  LF 15JAN97  TF   0

           CARPENTR          L LABOR    day      .51    .0       .00        .00         .51        .51        .00
           CRPN FOR          L LABOR    day      .26    .0       .00        .00         .26        .26        .00
           CRPN HLP          L LABOR    day      .51    .0       .00        .00         .51        .51        .00
           LAB CL 1          L LABOR    day     1.03    .0       .00        .00        1.03       1.03        .00
           LAB CL 2          L LABOR    day     1.03    .0       .00        .00        1.03       1.03        .00
           MASON             L LABOR    day     2.06    .0       .00        .00        2.06       2.06        .00
           MATERIAL          M MATERIAL          .00    .0       .00        .00         .00        .00        .00
                                          -----------  -----  -----------  -----------  -----------  ----------  ----------
           TOTAL :                              5.40    .0       .00        .00        5.40       5.40        .00

      500  Wood Floor System
           RD   4 ES 16JAN97  EF 21JAN97  LS 16JAN97  LF 21JAN97  TF   0

           CARPENTR          L LABOR    day     6.68    .0       .00        .00        6.68       6.68        .00
           CRPN FOR          L LABOR    day     3.34    .0       .00        .00        3.34       3.34        .00
           CRPN HLP          L LABOR    day     6.68    .0       .00        .00        6.68       6.68        .00
           MATERIAL          M MATERIAL          .00    .0       .00        .00         .00        .00        .00
                                          -----------  -----  -----------  -----------  -----------  ----------  ----------
           TOTAL :                             16.70    .0       .00        .00       16.70      16.70        .00

      600  Rough Framing Walls
           RD   6 ES 22JAN97  EF 29JAN97  LS 22JAN97  LF 29JAN97  TF   0

           CARPENTR          L LABOR    day     6.55    .0       .00        .00        6.55       6.55        .00
           CRPN FOR          L LABOR    day     3.28    .0       .00        .00        3.28       3.28        .00
           CRPN HLP          L LABOR    day     6.55    .0       .00        .00        6.55       6.55        .00
           MATERIAL          M MATERIAL          .00    .0       .00        .00         .00        .00        .00
                                          -----------  -----  -----------  -----------  -----------  ----------  ----------
           TOTAL :                             16.38    .0       .00        .00       16.38      16.38        .00

      700  Rough Framing Roof
           RD   4 ES 30JAN97  EF  4FEB97  LS 30JAN97  LF  4FEB97  TF   0

           CARPENTR          L LABOR    day     7.51    .0       .00        .00        7.51       7.51        .00
           CRPN FOR          L LABOR    day     3.72    .0       .00        .00        3.72       3.72        .00
```

Figure 9-57a Resource Control Activity Report Print—Wood Frame House, page 1

```
-----------------------------------------------------------------------------------------------------
Student Constructors              PRIMAVERA PROJECT PLANNER        Sample Project #1

REPORT DATE  30DEC96  RUN NO.  32    RESOURCE CONTROL ACTIVITY REPORT   START DATE  5JAN97  FIN DATE 20MAR97
             6:28
Resource Control - Detail By Activity                            DATA DATE  5JAN97   PAGE NO.   2
-----------------------------------------------------------------------------------------------------

                       COST    ACCOUNT UNIT          PCT   ACTUAL     ACTUAL   ESTIMATE TO
ACTIVITY ID RESOURCE  ACCOUNT CATEGORY MEAS  BUDGET  CMP  TO DATE  THIS PERIOD  COMPLETE  FORECAST  VARIANCE
----------- --------  ------- -------- ----  ------  ---  -------  -----------  --------  --------  --------
            CRPN HLP   L LABOR   day    7.59  .0     .00      .00      7.59      7.59     .00
            D, 50 HP   L LABOR   day     .16  .0     .00      .00       .16       .16     .00
            MATERIAL   M MATERIAL        .00  .0     .00      .00       .00       .00     .00
                                      -----------  --------  --------  ---------  --------  --------
            TOTAL :                    18.98  .0     .00      .00     18.98     18.98     .00

     800  Doors & Windows
          RD   4 ES  7FEB97  EF 12FEB97  LS 12FEB97  LF 17FEB97  TF   3

            CARPENTR   L LABOR   day    7.13  .0     .00      .00      7.13      7.13     .00
            CRPN HLP   L LABOR   day    7.13  .0     .00      .00      7.13      7.13     .00
            MATERIAL   M MATERIAL        .00  .0     .00      .00       .00       .00     .00
            SUB        S SUBCNT'R        .00  .0     .00      .00       .00       .00     .00
                                      -----------  --------  --------  ---------  --------  --------
            TOTAL :                    14.26  .0     .00      .00     14.26     14.26     .00

     900  Ext Wall Board
          RD   2 ES  5FEB97  EF  6FEB97  LS  5FEB97  LF  6FEB97  TF   0

            CARPENTR   L LABOR   day     .82  .0     .00      .00       .82       .82     .00
            CRPN FOR   L LABOR   day     .41  .0     .00      .00       .41       .41     .00
            CRPN HLP   L LABOR   day     .82  .0     .00      .00       .82       .82     .00
            MATERIAL   M MATERIAL        .00  .0     .00      .00       .00       .00     .00
                                      -----------  --------  --------  ---------  --------  --------
            TOTAL :                     2.05  .0     .00      .00      2.05      2.05     .00

    1000  Ext Wall Insulation
          RD   1 ES 18FEB97  EF 18FEB97  LS 18FEB97  LF 18FEB97  TF   0

            SUB        S SUBCNT'R        .00  .0     .00      .00       .00       .00     .00
                                      -----------  --------  --------  ---------  --------  --------
            TOTAL :                      .00  .0     .00      .00       .00       .00     .00

    1100  Rough Plumbing
          RD   4 ES  7FEB97  EF 12FEB97  LS  7FEB97  LF 12FEB97  TF   0

            SUB        S SUBCNT'R        .00  .0     .00      .00       .00       .00     .00
                                      -----------  --------  --------  ---------  --------  --------
            TOTAL :                      .00  .0     .00      .00       .00       .00     .00

    1200  Rough HVAC
          RD   3 ES  7FEB97  EF 11FEB97  LS 10FEB97  LF 12FEB97  TF   1

            SUB        S SUBCNT'R        .00  .0     .00      .00       .00       .00     .00
                                      -----------  --------  --------  ---------  --------  --------
            TOTAL :                      .00  .0     .00      .00       .00       .00     .00

    1300  Rough Elect
          RD   3 ES 12FEB97  EF 14FEB97  LS 13FEB97  LF 17FEB97  TF   1

            SUB        S SUBCNT'R        .00  .0     .00      .00       .00       .00     .00
                                      -----------  --------  --------  ---------  --------  --------
            TOTAL :                      .00  .0     .00      .00       .00       .00     .00

    1400  Shingles
          RD   3 ES 13FEB97  EF 17FEB97  LS 13FEB97  LF 17FEB97  TF   0

            MATERIAL   M MATERIAL        .00  .0     .00      .00       .00       .00     .00
            SUB        S SUBCNT'R        .00  .0     .00      .00       .00       .00     .00
                                      -----------  --------  --------  ---------  --------  --------
            TOTAL :                      .00  .0     .00      .00       .00       .00     .00

    1500  Ext Siding
          RD   3 ES 20FEB97  EF 24FEB97  LS  3MAR97  LF  5MAR97  TF   7

            CARPENTR   L LABOR   day     .36  .0     .00      .00       .36       .36     .00
            CRPN FOR   L LABOR   day     .36  .0     .00      .00       .36       .36     .00
            MATERIAL   M MATERIAL        .00  .0     .00      .00       .00       .00     .00
            SUB        S SUBCNT'R        .00  .0     .00      .00       .00       .00     .00
                                      -----------  --------  --------  ---------  --------  --------
            TOTAL :                      .72  .0     .00      .00       .72       .72     .00
```

Figure 9-57b Resource Control Activity Report Print—Wood Frame House, page 2

```
------------------------------------------------------------------------------------------------------------------
Student Constructors                      PRIMAVERA PROJECT PLANNER              Sample Project #1

REPORT DATE  30DEC96  RUN NO.   32        RESOURCE CONTROL ACTIVITY REPORT       START DATE  5JAN97  FIN DATE 20MAR97
             6:28
Resource Control - Detail By Activity                                           DATA DATE  5JAN97   PAGE NO.    3
------------------------------------------------------------------------------------------------------------------
```

ACTIVITY ID	RESOURCE	COST ACCOUNT	ACCOUNT CATEGORY	UNIT MEAS	BUDGET	PCT CMP	ACTUAL TO DATE	ACTUAL THIS PERIOD	ESTIMATE TO COMPLETE	FORECAST	VARIANCE

```
      1600  Ext Finish Carpentry
            RD   2 ES 18FEB97  EF 19FEB97  LS 27FEB97  LF 28FEB97  TF    7

            CARPENTR            L LABOR     day     .04    .0       .00       .00       .04      .04      .00
            CRPN FOR            L LABOR     day     .04    .0       .00       .00       .04      .04      .00
            MATERIAL            M MATERIAL          .00    .0       .00       .00       .00      .00      .00
            SUB                 S SUBCNT'R          .00    .0       .00       .00       .00      .00      .00
                                               ------------ ----- ------------ ------------ ------------ ------------ ------------
            TOTAL :                                .08    .0       .00       .00       .08      .08      .00

      1700  Hang Drywall
            RD   4 ES 19FEB97  EF 24FEB97  LS 19FEB97  LF 24FEB97  TF    0

            LABOR               L LABOR             .00    .0       .00       .00       .00      .00      .00
            MATERIAL            M MATERIAL          .00    .0       .00       .00       .00      .00      .00
            SUB                 S SUBCNT'R          .00    .0       .00       .00       .00      .00      .00
                                               ------------ ----- ------------ ------------ ------------ ------------ ------------
            TOTAL :                                .00    .0       .00       .00       .00      .00      .00

      1800  Finish Drywall
            RD   4 ES 25FEB97  EF 28FEB97  LS 25FEB97  LF 28FEB97  TF    0

            LABOR               L LABOR             .00    .0       .00       .00       .00      .00      .00
            MATERIAL            M MATERIAL          .00    .0       .00       .00       .00      .00      .00
            SUB                 S SUBCNT'R          .00    .0       .00       .00       .00      .00      .00
                                               ------------ ----- ------------ ------------ ------------ ------------ ------------
            TOTAL :                                .00    .0       .00       .00       .00      .00      .00

      1900  Cabinets
            RD   2 ES  3MAR97  EF  4MAR97  LS  3MAR97  LF  4MAR97  TF    0

            SUB                 S SUBCNT'R          .00    .0       .00       .00       .00      .00      .00
                                               ------------ ----- ------------ ------------ ------------ ------------ ------------
            TOTAL :                                .00    .0       .00       .00       .00      .00      .00

      2000  Ext Paint
            RD   3 ES 25FEB97  EF 27FEB97  LS  6MAR97  LF 10MAR97  TF    7

            SUB                 S SUBCNT'R          .00    .0       .00       .00       .00      .00      .00
                                               ------------ ----- ------------ ------------ ------------ ------------ ------------
            TOTAL :                                .00    .0       .00       .00       .00      .00      .00

      2100  Int Finish Carpentry
            RD   4 ES  5MAR97  EF 10MAR97  LS  5MAR97  LF 10MAR97  TF    0

            CARPENTR            L LABOR     day    2.89    .0       .00       .00      2.89     2.89      .00
            CRPN HLP            L LABOR     day    2.89    .0       .00       .00      2.89     2.89      .00
            MATERIAL            M MATERIAL          .00    .0       .00       .00       .00      .00      .00
                                               ------------ ----- ------------ ------------ ------------ ------------ ------------
            TOTAL :                               5.78    .0       .00       .00      5.78     5.78      .00

      2200  Int Paint
            RD   3 ES 11MAR97  EF 13MAR97  LS 11MAR97  LF 13MAR97  TF    0

            SUB                 S SUBCNT'R          .00    .0       .00       .00       .00      .00      .00
                                               ------------ ----- ------------ ------------ ------------ ------------ ------------
            TOTAL :                                .00    .0       .00       .00       .00      .00      .00

      2300  Finish Plumbing
            RD   2 ES  5MAR97  EF  6MAR97  LS  7MAR97  LF 10MAR97  TF    2

            SUB                 S SUBCNT'R          .00    .0       .00       .00       .00      .00      .00
                                               ------------ ----- ------------ ------------ ------------ ------------ ------------
            TOTAL :                                .00    .0       .00       .00       .00      .00      .00
```

Figure 9-57c Resource Control Activity Report Print—Wood Frame House, page 3

```
-----------------------------------------------------------------------------------------------------
Student Constructors          ·           PRIMAVERA PROJECT PLANNER        Sample Project #1
REPORT DATE  30DEC96  RUN NO.   32          RESOURCE CONTROL ACTIVITY REPORT    START DATE  5JAN97  FIN DATE 20MAR97
            6:28
Resource Control - Detail By Activity                                          DATA DATE   5JAN97   PAGE NO.    4
-----------------------------------------------------------------------------------------------------

                         COST    ACCOUNT  UNIT              PCT    ACTUAL      ACTUAL    ESTIMATE TO
ACTIVITY ID RESOURCE   ACCOUNT  CATEGORY  MEAS   BUDGET     CMP   TO DATE   THIS PERIOD   COMPLETE    FORECAST    VARIANCE
----------- --------   -------  --------  ----  ---------- -----  --------  ----------- ------------ ----------- -----------

     2400  Finish HVAC
           RD    3 ES  3MAR97  EF  5MAR97  LS  6MAR97  LF 10MAR97  TF    3

           SUB              S SUBCNT'R           .00    .0       .00        .00         .00        .00        .00
                                           ------------ ----- ------------ ----------- ------------ ----------- ------------
           TOTAL :                                .00    .0       .00        .00         .00        .00        .00

     2500  Finish Elect
           RD    2 ES  3MAR97  EF  4MAR97  LS  7MAR97  LF 10MAR97  TF    4

           SUB              S SUBCNT'R           .00    .0       .00        .00         .00        .00        .00
                                           ------------ ----- ------------ ----------- ------------ ----------- ------------
           TOTAL :                                .00    .0       .00        .00         .00        .00        .00

     2600  Flooring
           RD    3 ES 14MAR97  EF 18MAR97  LS 14MAR97  LF 18MAR97  TF    0

           SUB              S SUBCNT'R           .00    .0       .00        .00         .00        .00        .00
                                           ------------ ----- ------------ ----------- ------------ ----------- ------------
           TOTAL :                                .00    .0       .00        .00         .00        .00        .00

     2700  Grading & Landscaping
           RD    4 ES 18FEB97  EF 21FEB97  LS 13MAR97  LF 18MAR97  TF   17

           SUB              S SUBCNT'R           .00    .0       .00        .00         .00        .00        .00
                                           ------------ ----- ------------ ----------- ------------ ----------- ------------
           TOTAL :                                .00    .0       .00        .00         .00        .00        .00

     2800  Punch List
           RD    2 ES 19MAR97  EF 20MAR97  LS 19MAR97  LF 20MAR97  TF    0

           TOTAL :                                .00    .0       .00        .00         .00        .00        .00

                                           ------------ ----- ------------ ----------- ------------ ----------- ------------
                     REPORT TOTALS             86.33    .0       .00        .00       86.33      86.33        .00
=====================================================================================================
```

Figure 9-57d　Resource Control Activity Report Print—Wood Frame House, page 4

```
------------------------------------------------------------------------------------------------------------
Student Constructors                    PRIMAVERA PROJECT PLANNER          Sample Project #1

REPORT DATE  30DEC96  RUN NO.   35      COST CONTROL ACTIVITY REPORT       START DATE  5JAN97  FIN DATE 20MAR97
                      6:36
Cost Control - Detailed by Activity                                       DATA DATE  5JAN97   PAGE NO.    1
------------------------------------------------------------------------------------------------------------
```

ACTIVITY ID	RESOURCE	COST ACCOUNT	ACCOUNT CATEGORY	UNIT MEAS	BUDGET	PCT CMP	ACTUAL TO DATE	ACTUAL THIS PERIOD	ESTIMATE TO COMPLETE	FORECAST	VARIANCE
100	Clear Site										
	RD 2 ES 6JAN97 EF 7JAN97 LS 6JAN97 LF 7JAN97 TF 0										
	SUB		S SUBCNT'R		1280.00	.0	.00	.00	1280.00	1280.00	.00
	TOTAL :				1280.00	.0	.00	.00	1280.00	1280.00	.00
200	Building Layout										
	RD 1 ES 8JAN97 EF 8JAN97 LS 8JAN97 LF 8JAN97 TF 0										
	CARPENTR		L LABOR	day	9.96	.0	.00	.00	9.96	9.96	.00
	CRPN FOR		L LABOR	day	10.76	.0	.00	.00	10.76	10.76	.00
	LAB CL 1		L LABOR	day	12.75	.0	.00	.00	12.75	12.75	.00
	LAB CL 2		L LABOR	day	11.15	.0	.00	.00	11.15	11.15	.00
	MATERIAL		M MATERIAL		70.87	.0	.00	.00	70.87	70.87	.00
	SUB		S SUBCNT'R		15.00	.0	.00	.00	15.00	15.00	.00
	TOTAL :				130.49	.0	.00	.00	130.49	130.49	.00
300	Form/Pour Footings										
	RD 3 ES 9JAN97 EF 13JAN97 LS 9JAN97 LF 13JAN97 TF 0										
	CARPENTR		L LABOR	day	89.62	.0	.00	.00	89.62	89.62	.00
	CRPN FOR		L LABOR	day	96.80	.0	.00	.00	96.80	96.80	.00
	LAB CL 1		L LABOR	day	114.72	.0	.00	.00	114.72	114.72	.00
	LAB CL 2		L LABOR	day	100.38	.0	.00	.00	100.38	100.38	.00
	MATERIAL		M MATERIAL		637.80	.0	.00	.00	637.80	637.80	.00
	SUB		S SUBCNT'R		135.00	.0	.00	.00	135.00	135.00	.00
	TOTAL :				1174.32	.0	.00	.00	1174.32	1174.32	.00
400	Pier Masonry										
	RD 2 ES 14JAN97 EF 15JAN97 LS 14JAN97 LF 15JAN97 TF 0										
	CARPENTR		L LABOR	day	51.25	.0	.00	.00	51.25	51.25	.00
	CRPN FOR		L LABOR	day	27.68	.0	.00	.00	27.68	27.68	.00
	CRPN HLP		L LABOR	day	36.90	.0	.00	.00	36.90	36.90	.00
	LAB CL 1		L LABOR	day	65.77	.0	.00	.00	65.77	65.77	.00
	LAB CL 2		L LABOR	day	57.55	.0	.00	.00	57.55	57.55	.00
	MASON		L LABOR	day	180.87	.0	.00	.00	180.87	180.87	.00
	MATERIAL		M MATERIAL		547.27	.0	.00	.00	547.27	547.27	.00
	TOTAL :				967.29	.0	.00	.00	967.29	967.29	.00
500	Wood Floor System										
	RD 4 ES 16JAN97 EF 21JAN97 LS 16JAN97 LF 21JAN97 TF 0										
	CARPENTR		L LABOR	day	667.61	.0	.00	.00	667.61	667.61	.00
	CRPN FOR		L LABOR	day	360.51	.0	.00	.00	360.51	360.51	.00
	CRPN HLP		L LABOR	day	480.67	.0	.00	.00	480.67	480.67	.00
	MATERIAL		M MATERIAL		2672.25	.0	.00	.00	2672.25	2672.25	.00
	TOTAL :				4181.04	.0	.00	.00	4181.04	4181.04	.00
600	Rough Framing Walls										
	RD 6 ES 22JAN97 EF 29JAN97 LS 22JAN97 LF 29JAN97 TF 0										
	CARPENTR		L LABOR	day	655.03	.0	.00	.00	655.03	655.03	.00
	CRPN FOR		L LABOR	day	353.72	.0	.00	.00	353.72	353.72	.00
	CRPN HLP		L LABOR	day	471.63	.0	.00	.00	471.63	471.63	.00
	MATERIAL		M MATERIAL		1843.61	.0	.00	.00	1843.61	1843.61	.00
	TOTAL :				3323.99	.0	.00	.00	3323.99	3323.99	.00
700	Rough Framing Roof										
	RD 4 ES 30JAN97 EF 4FEB97 LS 30JAN97 LF 4FEB97 TF 0										
	CARPENTR		L LABOR	day	751.15	.0	.00	.00	751.15	751.15	.00
	CRPN FOR		L LABOR	day	401.29	.0	.00	.00	401.29	401.29	.00

Figure 9-58a Cost Control Activity Report Print—Wood Frame House, page 1

```
--------------------------------------------------------------------------------------------------------
Student Constructors                    PRIMAVERA PROJECT PLANNER          Sample Project #1

REPORT DATE  30DEC96  RUN NO.   35         COST CONTROL ACTIVITY REPORT       START DATE  5JAN97  FIN DATE 20MAR97
               6:36
Cost Control - Detailed by Activity                                          DATA DATE   5JAN97   PAGE NO.    2
--------------------------------------------------------------------------------------------------------
                          COST      ACCOUNT  UNIT              PCT    ACTUAL     ACTUAL     ESTIMATE TO
ACTIVITY ID RESOURCE    ACCOUNT    CATEGORY  MEAS    BUDGET    CMP    TO DATE  THIS PERIOD   COMPLETE    FORECAST    VARIANCE
----------- --------   --------    --------  ----  ---------- -----  -------- ------------ ----------- ----------- -----------
            CRPN HLP               L LABOR   day     546.57    .0      .00        .00        546.57      546.57        .00
            D, 50 HP              L LABOR   day      76.80    .0      .00        .00         76.80       76.80        .00
            MATERIAL             M MATERIAL        1692.59    .0      .00        .00       1692.59     1692.59        .00
                                                   ----------        -------- ------------ ----------- -----------  ----------
            TOTAL :                               3468.40    .0      .00        .00       3468.40     3468.40        .00

       800  Doors & Windows
            RD   4 ES  7FEB97  EF 12FEB97  LS 12FEB97  LF 17FEB97  TF   3

            CARPENTR             L LABOR   day     712.50    .0      .00        .00        712.50      712.50        .00
            CRPN HLP             L LABOR   day     513.00    .0      .00        .00        513.00      513.00        .00
            MATERIAL             M MATERIAL       2522.90    .0      .00        .00       2522.90     2522.90        .00
            SUB                  S SUBCNT'R        247.00    .0      .00        .00        247.00      247.00        .00
                                                   ----------        -------- ------------ ----------- -----------  ----------
            TOTAL :                               3995.40    .0      .00        .00       3995.40     3995.40        .00

       900  Ext Wall Board
            RD   2 ES  5FEB97  EF  6FEB97  LS  5FEB97  LF  6FEB97  TF   0

            CARPENTR             L LABOR   day      82.08    .0      .00        .00         82.08       82.08        .00
            CRPN FOR             L LABOR   day      44.33    .0      .00        .00         44.33       44.33        .00
            CRPN HLP             L LABOR   day      59.10    .0      .00        .00         59.10       59.10        .00
            MATERIAL             M MATERIAL        224.46    .0      .00        .00        224.46      224.46        .00
                                                   ----------        -------- ------------ ----------- -----------  ----------
            TOTAL :                                409.97    .0      .00        .00        409.97      409.97        .00

      1000  Ext Wall Insulation
            RD   1 ES 18FEB97  EF 18FEB97  LS 18FEB97  LF 18FEB97  TF   0

            SUB                  S SUBCNT'R        385.32    .0      .00        .00        385.32      385.32        .00
                                                   ----------        -------- ------------ ----------- -----------  ----------
            TOTAL :                                385.32    .0      .00        .00        385.32      385.32        .00

      1100  Rough Plumbing
            RD   4 ES  7FEB97  EF 12FEB97  LS  7FEB97  LF 12FEB97  TF   0

            SUB                  S SUBCNT'R        750.00    .0      .00        .00        750.00      750.00        .00
                                                   ----------        -------- ------------ ----------- -----------  ----------
            TOTAL :                                750.00    .0      .00        .00        750.00      750.00        .00

      1200  Rough HVAC
            RD   3 ES  7FEB97  EF 11FEB97  LS 10FEB97  LF 12FEB97  TF   1

            SUB                  S SUBCNT'R       1168.75    .0      .00        .00       1168.75     1168.75        .00
                                                   ----------        -------- ------------ ----------- -----------  ----------
            TOTAL :                               1168.75    .0      .00        .00       1168.75     1168.75        .00

      1300  Rough Elect
            RD   3 ES 12FEB97  EF 14FEB97  LS 13FEB97  LF 17FEB97  TF   1

            SUB                  S SUBCNT'R        940.00    .0      .00        .00        940.00      940.00        .00
                                                   ----------        -------- ------------ ----------- -----------  ----------
            TOTAL :                                940.00    .0      .00        .00        940.00      940.00        .00

      1400  Shingles
            RD   3 ES 13FEB97  EF 17FEB97  LS 13FEB97  LF 17FEB97  TF   0

            MATERIAL             M MATERIAL        834.52    .0      .00        .00        834.52      834.52        .00
            SUB                  S SUBCNT'R        256.77    .0      .00        .00        256.77      256.77        .00
                                                   ----------        -------- ------------ ----------- -----------  ----------
            TOTAL :                               1091.29    .0      .00        .00       1091.29     1091.29        .00

      1500  Ext Siding
            RD   3 ES 20FEB97  EF 24FEB97  LS  3MAR97  LF  5MAR97  TF   7

            CARPENTR             L LABOR   day      36.00    .0      .00        .00         36.00       36.00        .00
            CRPN FOR             L LABOR   day      38.88    .0      .00        .00         38.88       38.88        .00
            MATERIAL             M MATERIAL         36.86    .0      .00        .00         36.86       36.86        .00
            SUB                  S SUBCNT'R       1598.80    .0      .00        .00       1598.80     1598.80        .00
                                                   ----------        -------- ------------ ----------- -----------  ----------
            TOTAL :                               1710.54    .0      .00        .00       1710.54     1710.54        .00
```

Figure 9-58b   Cost Control Activity Report Print—Wood Frame House, page 2

```
--------------------------------------------------------------------------------------------------------
Student Constructors                    PRIMAVERA PROJECT PLANNER          Sample Project #1

REPORT DATE  30DEC96  RUN NO.   35           COST CONTROL ACTIVITY REPORT      START DATE  5JAN97  FIN DATE 20MAR97
             6:36
Cost Control - Detailed by Activity                                           DATA DATE  5JAN97   PAGE NO.    3
--------------------------------------------------------------------------------------------------------

                          COST      ACCOUNT UNIT           PCT    ACTUAL    ACTUAL   ESTIMATE TO
ACTIVITY ID RESOURCE     ACCOUNT   CATEGORY MEAS   BUDGET   CMP   TO DATE  THIS PERIOD  COMPLETE   FORECAST    VARIANCE
---------- --------     -------- -------- ----   ------   ---   -------  -----------  --------   --------    --------

   1600  Ext Finish Carpentry
         RD    2 ES 18FEB97  EF 19FEB97  LS 27FEB97  LF 28FEB97  TF    7

         CARPENTR        L LABOR   day     4.00    .0     .00      .00        4.00       4.00        .00
         CRPN FOR        L LABOR   day     4.32    .0     .00      .00        4.32       4.32        .00
         MATERIAL        M MATERIAL        4.10    .0     .00      .00        4.10       4.10        .00
         SUB             S SUBCNT'R      177.64    .0     .00      .00      177.64     177.64        .00
                                      ----------- -----  -------- -----------  --------   --------    --------
         TOTAL :                        190.06    .0     .00      .00      190.06     190.06        .00

   1700  Hang Drywall
         RD    4 ES 19FEB97  EF 24FEB97  LS 19FEB97  LF 24FEB97  TF    0

         LABOR           L LABOR         101.92    .0     .00      .00      101.92     101.92        .00
         MATERIAL        M MATERIAL      477.25    .0     .00      .00      477.25     477.25        .00
         SUB             S SUBCNT'R     1265.43    .0     .00      .00     1265.43    1265.43        .00
                                      ----------- -----  -------- -----------  --------   --------    --------
         TOTAL :                       1844.60    .0     .00      .00     1844.60    1844.60        .00

   1800  Finish Drywall
         RD    4 ES 25FEB97  EF 28FEB97  LS 25FEB97  LF 28FEB97  TF    0

         LABOR           L LABOR          43.68    .0     .00      .00       43.68      43.68        .00
         MATERIAL        M MATERIAL      204.53    .0     .00      .00      204.53     204.53        .00
         SUB             S SUBCNT'R      542.33    .0     .00      .00      542.33     542.33        .00
                                      ----------- -----  -------- -----------  --------   --------    --------
         TOTAL :                        790.54    .0     .00      .00      790.54     790.54        .00

   1900  Cabinets
         RD    2 ES  3MAR97  EF  4MAR97  LS  3MAR97  LF  4MAR97  TF    0

         SUB             S SUBCNT'R     1618.00    .0     .00      .00     1618.00    1618.00        .00
                                      ----------- -----  -------- -----------  --------   --------    --------
         TOTAL :                       1618.00    .0     .00      .00     1618.00    1618.00        .00

   2000  Ext Paint
         RD    3 ES 25FEB97  EF 27FEB97  LS  6MAR97  LF 10MAR97  TF    7

         SUB             S SUBCNT'R      525.00    .0     .00      .00      525.00     525.00        .00
                                      ----------- -----  -------- -----------  --------   --------    --------
         TOTAL :                        525.00    .0     .00      .00      525.00     525.00        .00

   2100  Int Finish Carpentry
         RD    4 ES  5MAR97  EF 10MAR97  LS  5MAR97  LF 10MAR97  TF    0

         CARPENTR        L LABOR   day   289.28    .0     .00      .00      289.28     289.28        .00
         CRPN HLP        L LABOR   day   208.29    .0     .00      .00      208.29     208.29        .00
         MATERIAL        M MATERIAL      705.58    .0     .00      .00      705.58     705.58        .00
                                      ----------- -----  -------- -----------  --------   --------    --------
         TOTAL :                       1203.15    .0     .00      .00     1203.15    1203.15        .00

   2200  Int Paint
         RD    3 ES 11MAR97  EF 13MAR97  LS 11MAR97  LF 13MAR97  TF    0

         SUB             S SUBCNT'R     4725.00    .0     .00      .00     4725.00    4725.00        .00
                                      ----------- -----  -------- -----------  --------   --------    --------
         TOTAL :                       4725.00    .0     .00      .00     4725.00    4725.00        .00

   2300  Finish Plumbing
         RD    2 ES  5MAR97  EF  6MAR97  LS  7MAR97  LF 10MAR97  TF    2

         SUB             S SUBCNT'R     3000.00    .0     .00      .00     3000.00    3000.00        .00
                                      ----------- -----  -------- -----------  --------   --------    --------
         TOTAL :                       3000.00    .0     .00      .00     3000.00    3000.00        .00
```

Figure 9-58c Cost Control Activity Report Print—Wood Frame House, page 3

```
-------------------------------------------------------------------------------------------------
Student Constructors                        PRIMAVERA PROJECT PLANNER          Sample Project #1

REPORT DATE  30DEC96  RUN NO.   35          COST CONTROL ACTIVITY REPORT       START DATE  5JAN97  FIN DATE 20MAR97
             6:36
Cost Control - Detailed by Activity                                           DATA DATE   5JAN97  PAGE NO.    4

-------------------------------------------------------------------------------------------------

                          COST     ACCOUNT  UNIT              PCT   ACTUAL    ACTUAL     ESTIMATE TO
ACTIVITY ID RESOURCE      ACCOUNT  CATEGORY MEAS    BUDGET    CMP   TO DATE   THIS PERIOD COMPLETE   FORECAST    VARIANCE
----------- --------      -------- -------- ----  ---------- ----- --------- ----------- ----------- ---------- -----------

      2400  Finish HVAC
            RD    3 ES  3MAR97  EF  5MAR97  LS  6MAR97  LF 10MAR97  TF    3

            SUB               S SUBCNT'R       3506.25   .0        .00          .00      3506.25    3506.25         .00
                                              ----------- -----  --------- ----------- ----------- ---------- -----------
            TOTAL :                            3506.25   .0        .00          .00      3506.25    3506.25         .00

      2500  Finish Elect
            RD    2 ES  3MAR97  EF  4MAR97  LS  7MAR97  LF 10MAR97  TF    4

            SUB               S SUBCNT'R       1410.00   .0        .00          .00      1410.00    1410.00         .00
                                              ----------- -----  --------- ----------- ----------- ---------- -----------
            TOTAL :                            1410.00   .0        .00          .00      1410.00    1410.00         .00

      2600  Flooring
            RD    3 ES 14MAR97  EF 18MAR97  LS 14MAR97  LF 18MAR97  TF    0

            SUB               S SUBCNT'R       1583.34   .0        .00          .00      1583.34    1583.34         .00
                                              ----------- -----  --------- ----------- ----------- ---------- -----------
            TOTAL :                            1583.34   .0        .00          .00      1583.34    1583.34         .00

      2700  Grading & Landscaping
            RD    4 ES 18FEB97  EF 21FEB97  LS 13MAR97  LF 18MAR97  TF   17

            SUB               S SUBCNT'R        600.00   .0        .00          .00       600.00     600.00         .00
                                              ----------- -----  --------- ----------- ----------- ---------- -----------
            TOTAL :                             600.00   .0        .00          .00       600.00     600.00         .00

      2800  Punch List
            RD    2 ES 19MAR97  EF 20MAR97  LS 19MAR97  LF 20MAR97  TF    0

            TOTAL :                                .00   .0        .00          .00          .00        .00         .00

                                              ----------- -----  --------- ----------- ----------- ---------- -----------
            REPORT TOTALS                     45972.74   .0        .00          .00     45972.74   45972.74         .00
=================================================================================================
```

Figure 9-58d   Cost Control Activity Report Print—Wood Frame House, page 4

EXERCISES

1. Small Commercial Concrete Block Building—Tabular Reports

Prepare tabular reports for the small commercial concrete block building located in the appendix. The following tabular reports should be prepared:

1. Classic Schedule Report—Sort by ES, TF
2. Resource Control Report—Detailed by Activity
3. Cost Control Report—Detailed by Activity

2. Large Commercial Building—Logic Diagram Prints

Prepare tabular reports for the large commercial building located in the appendix. The following tabular reports should be prepared:

1. Classic Schedule Report—Sort by ES, TF
2. Resource Control Report—Detailed by Activity
3. Cost Control Report—Detailed by Activity

Section 3

Controlling

10 | Updating the Schedule Using *P3*

Objectives

Upon completion of this chapter, the reader should be able to:

- Copy a *P3* schedule
- Establish a target schedule
- Record progress
- Establish a new data date
- Use the activity log
- Input duration changes
- Input logic changes
- Input added activities
- Produce updated reports

NECESSITY FOR RESOURCE OPTIMIZATION

Keeping Current

Once a project is under way, it is necessary to modify the schedule to keep it current. Very few projects ever go exactly according to the original schedule. Among the influences likely to change the original plan are weather, acts of God, better or worse productivity than anticipated, delivery problems, labor problems, changes in the scope of the project, interferences between crafts or subcontractors, and mismanagement in the flow of work.

Changes   For a schedule to remain viable, the project changes must continually be incorporated in the current schedule. The target (original) schedule is copied and then modified to reflect changes that make it the current schedule. Generally a project is updated at the same time each month. The data date for recording progress is usually at the beginning of the new month. The current schedule is thus the current plan for project completion.

Monitoring Progress

Tracking Physical Progress   There are two phases to keeping the schedule current. The first is monitoring all activities to determine their status. Each activity is either complete or partially complete or no work has been accomplished. The activities for which there is no physical progress keep the original activity relationships and durations. Where there is partial physical progress, the original activity relationships are kept and the remaining duration is calculated either as number of days or percent complete. After all progress to date is recorded for each activity, the schedule is recalculated. Individual activities and the ensemble are determined to be either on, ahead, or behind schedule.

Documenting Changes   The second phase to keeping the schedule current is documenting changes. Very seldom is a project built in exactly the sequence planned. Either more detail is needed or, with more time for analysis, better plans are developed. When project scope changes, change orders must be incorporated. As the approach to building the project changes, the constructor needs to modify the schedule to show the changes.

GETTING STARTED

Revising the Schedule

P3 offers tools to aid the scheduler in keeping the schedule current. The first step is to copy the previous schedule.

Copy Dialog Box   Monitoring progress requires comparing all activities with the target schedule—the baseline or "original" schedule. It is convenient to record changes on a copy of the target schedule (Figure 10-1). To copy the schedule with *P3*, click on the **Tools** main pull-down menu. Select **Project Utilities**, then **Copy...** (Figure 10-2). The **Copy** dialog box will appear (Figure 10-3).

From   The **Copy** dialog box is divided into two parts. The top half, or the **From** portion, is used to define the schedule that will be copied. The **Projects:** window provides a listing of all projects in the *P3* default directory. Use the **Dir...** button to change directories if the desired project is located in a directory other than the default directory. Once the desired project, in this case SA10, has been selected from the **Projects:** window, the name automatically appears in the **Project group/Project name:** (under **From**). As in Figure 10-3, if you have already opened the file for the project to be copied, that project name will be in the **Projects:** window of the **Copy** dialog box. Since

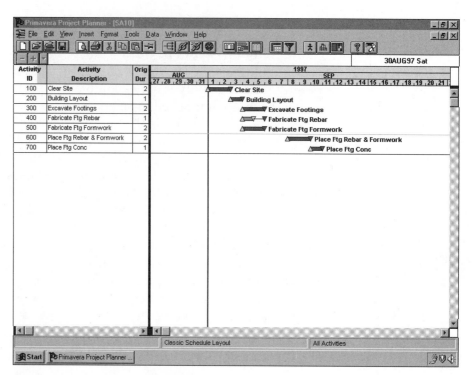

Figure 10-1   Sample Schedule

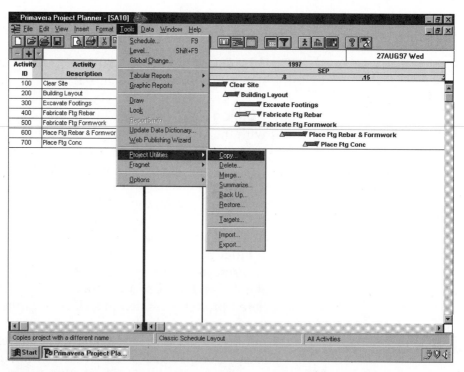

Figure 10-2   Tools Main Pull-Down Menu—Project Utilities and Copy
Options

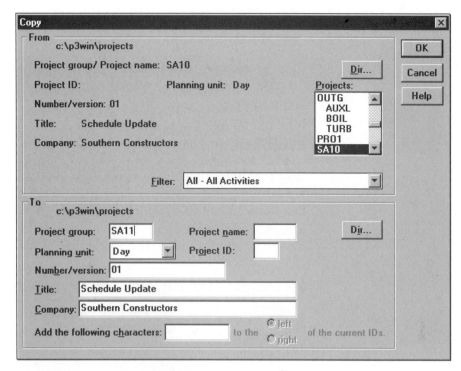

Figure 10-3   Copy Dialog Box

an existing project is being copied, the **Project group/Project name:,
Project ID:, Number/version:, Title:,** and **Company:** cannot be modified. They are provided for information purposes.

Filter: The only field available for possible modification in the **From**
portion is the **Filter:** field. If only certain activities based on some criteria are to be selected, the **Filter:** field can be used. The filter criteria
must be defined before the schedule copy is made. From the **Format**
pull-down menu select **Filter....** The **Filter** dialog box will appear,
and the filter criteria can be defined.

To   The bottom half of the **Copy** dialog box is the **To** portion. It is used
to define the new schedule to be created from the copied schedule.
The new identification fields are:

- **Project group:**
- **Planning unit:**
- **Number/version:**
- **Title:**
- **Company:**
- **Project name:**
- **Project ID:**

These fields were discussed in Chapter 4 for the **Add a New Project** dialog box.

Add the following characters: The **Add the following characters:** field is used to add a prefix or suffix to activity IDs of the copied schedule for identification purposes.

Establishing New Targets

Targets Dialog Box   To establish the actual target relationship between the original and copied schedules, select **Project Utilities** from the **Tools** main pull-down menu (Figure 10-4). Then select the **Targets...** option. The **Targets** dialog box will appear (Figure 10-5). Notice that the **Project group/Project name:** appearing in the **Targets** dialog box is the

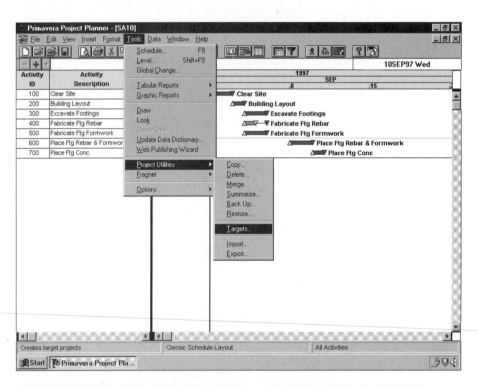

Figure 10-4   Tools Main Pull-Down Menu—Project Utilities and Targets Options

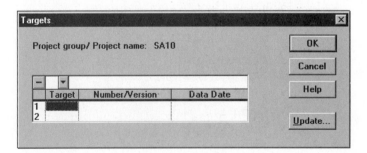

Figure 10-5   Targets Dialog Box

current project presently on-screen in bar chart format (SA10). The current project is the schedule that will be updated. The target schedule is the original or baseline schedule that is kept for comparison purposes. Click on the down arrow, and the project name listing will appear. The copied schedule to be used as the target was named SA11 (Figure 10-6). Click on the **OK** button of the **Targets** dialog box, and the relationship between the master project and the target schedule is established. Note: You don't actually need to copy a project before making it a target. If there are no projects by the input target name, a copy will be made automatically when you input the target name.

Establishing a New Data Date

Schedule Dialog Box   To recalculate the schedule and change the data date, select the **Tools** main pull-down menu and the **Schedule...** option. The **Schedule** dialog box will appear (Figure 10-7).

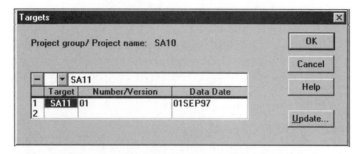

Figure 10-6   Targets Dialog Box—Completed

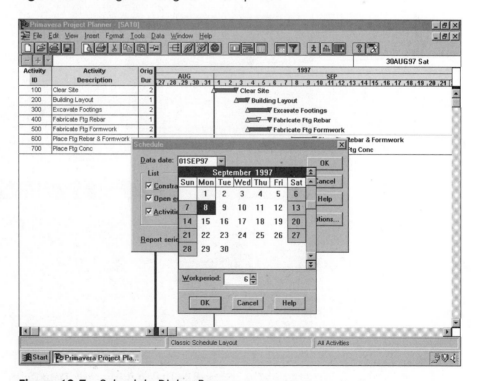

Figure 10-7   Schedule Dialog Box

Data date: Notice that in Figure 10-8, the **Data date:** is 01SEP97. This data date was established when the schedule was originally created. Since actual progress is being recorded, it is time to change the data date. Click on the down arrow and use the pop-up calendar, or simply type in the new date. In Figure 10-9, the new data date of 08SEP97 has been selected. To complete the rescheduling process, select the **OK** button.

Figure 10-10 is the on-screen bar chart of the updated schedule with physical progress as of September 8, 1997. Compare Figure 10-10 to Figure 10-1 (p. 262), the schedule before updating.

Recording Progress

Datometer   The next step is to record physical progress. Select the first activity for recording progress. Choose Activity 100, Clear Site. To record the actual start, place the mouse arrow at the triangle at the beginning or left of the activity bar. The mouse arrow becomes a horizontal line with arrowheads on both ends. When the mouse arrow changes in appearance, hold down the Shift key on the keyboard, and the two-headed arrow will change in appearance again. The appear-

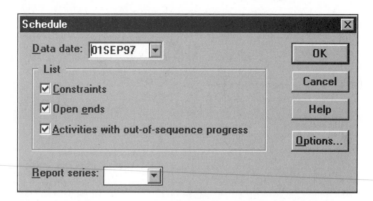

Figure 10-8   Schedule Dialog Box—Original Data Date

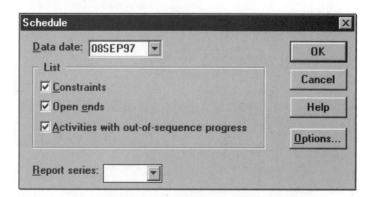

Figure 10-9   Schedule Dialog Box—New Data Date

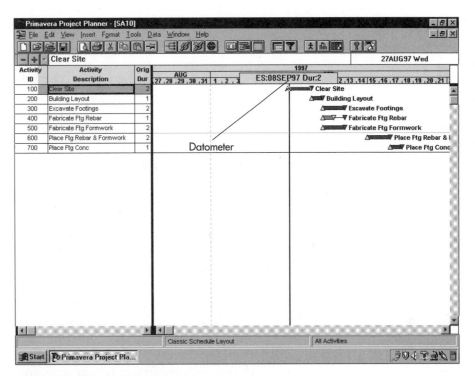

Figure 10-10   Datometer

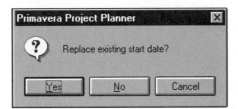

Figure 10-11   Replace Existing Start Date? Caution Box

ance now will be a two-headed arrow with a ball in the center and a large **A** above and to the left. The **A** stands for actual durations. Depress and hold down the left mouse button, and the Datometer will appear (Figure 10-10). Moving the mouse to the right or left causes the Datometer date to change.

Another way to input physical progress is to use the **Activity** form from the **View** main pull-down menu. When an activity is selected, select the **ES** (early start) check box and it becomes **AS** or actual start. Use the **Pct** (percent complete) and **RD** (remaining duration) fields to define activities that are partially complete.

Replace Existing Start Date? Caution Box   Notice in Figure 10-10 that the activity bar for Activity 100 now has three triangles. The center triangle shows the location to where the mouse has been moved to establish the actual start of Activity 100. Release the left mouse button, and the **Replace existing start date?** caution box will appear (Figure 10-11).

Select the **No** button, and the **Progress** dialog box will appear (Figure 10-12).

Progress Dialog Box As you click on the **ES** (early start date) check box, it becomes **AS** (actual start date) of Activity 100. The actual early start of Activity 100 is 01SEP97. The original early start was 01SEP97, before entering actual progress. Now that the actual start has been established, progress can be determined in one of three ways: by **EF**, or early finish, by **Remaining duration:**, or by **Percent complete:**.

*Percent **complete:*** Inputting percent complete can be accomplished in one of three ways: (1) click in the field itself, and type in the number up to 100 percent complete; (2) click on the up/down arrow boxes beside the **Percent complete:** field to adjust the number; (3) use the slide button bar above the AS and and EF fields. Notice that the button on the slide bar moves as the number in the **Percent complete:** field changes. The **Percent complete:** field is calibrated with the slide bar. Also, the activity bar now has two horizontal components. As the button is moved across the slide bar, the darker component shows percent complete, and the lighter component shows remaining duration. For the example schedule, the slide button was moved all the way to the right, or 100 percent complete, for Activity 100.

OK Click on the **OK** button of the **Progress** dialog box to accept the updated physical progress for an activity. The **Progress** dialog box disappears, and the on-screen bar chart is modified (Figure 10-13).

Changing Bar Color Notice that the activity bar color of Activity 100 has changed to show progress. The on-screen color of Activity 100 is blue (default *P3* color configuration). The default color can be changed. Click the **Format** main pull-down menu. Choose the **Bars...** option. Within the **Display progress on bars based on** section, click on the **Color:** button. From the color chart that appears, choose the desired color for the bars that show progress.

Pop-up Calendar Now that Activity 100 has been updated, select the next activity to be updated. Click on Activity 200 (Figure 10-14), then click on the **ES** (early start) check box and it becomes **AS** (actual

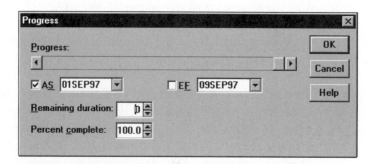

Figure 10-12 Progress Dialog Box—Activity 100

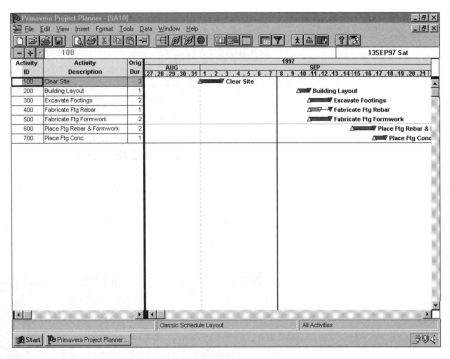

Figure 10-13   Progress—Activity 100

start). Also, by clicking on the **EF** (early finish) check box, it becomes **AF** (actual finish). Changing the **Percent complete** field to 100 percent will also cause the **EF** check box to show **AF**.

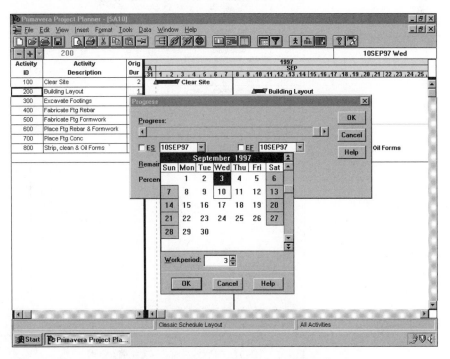

Figure 10-14   Progress Dialog Box—Pull-Down Calendar

Click on the down arrow, and the pop-up calendar appears. For example, from recording physical progress at the job site, you have determined that Activity 200 was finished on September 3, 1997. The activity had an actual start of the beginning of the day on September 3. Since the default of days was chosen when the sample project was originally opened in *P3*, a part of a day cannot be chosen. So, it started at the beginning of the day and was completed the same day. Its actual finish was the end of the day on September 3. Even though the date 03SEP97 appears in both the **A<u>S</u>** and the **A<u>F</u>** boxes, one day has elapsed (Figure 10-15). Click the **OK** button to accept this progress, and the **Progress** dialog box disappears. Compare the schedule shown in Figure 10-16 with the previous schedule in Figure 10-13 (p. 269). The actual start and finish dates for Activity 200 are shown, and the activity bar has changed color to show progress.

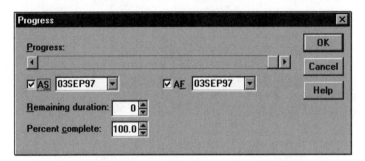

Figure 10-15 Progress Dialog Box—Activity 200

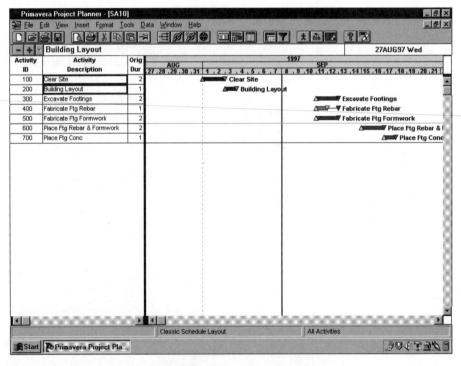

Figure 10-16 Progress—Activity 200

Percent Complete   For the example schedule, physical progress is input on three more activities (300, 400, and 500). No work has been accomplished on Activities 600 and 700. Activity 300 has an actual start of September 4 and is 50 percent complete (see Figures 10-17 and 10-18). Actual start for Activity 400 is input 04SEP97, and the activity is 100 percent complete. Actual start for Activity 500 is input 04SEP97, and the activity is 50 percent complete.

When the progress is input through Activity 500 and the schedule is recalculated (using the **Schedule...** option from the **Tools** main pull-down menu), Figure 10-19 is the result. These progress changes have been made on the master schedule and can be compared back to the original, or target, schedule.

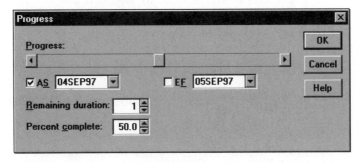

Figure 10-17   Progress Dialog Box—Activity 300

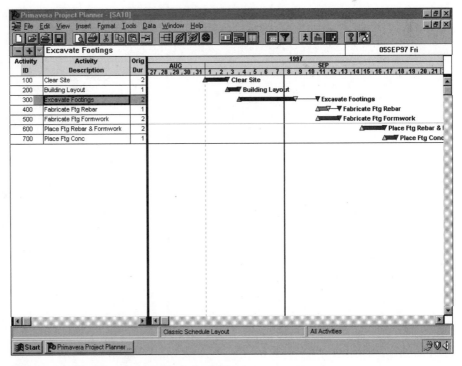

Figure 10-18   Progress—Activity 300

Inputting Progress with Activity Table

Another convenient way to update physical progress is to use the activity table. Place the mouse arrow on the heavy dark vertical line between the activity table and the bar chart. The mouse arrow becomes two short vertical lines. Hold the left mouse button down to adjust the screen to expose more of the activity table (Figure 10-20). Compare Figure 10-19 to Figure 10-20. Three more columns of the activity table have been exposed. The **Rem Dur** (remaining duration) and the **%** (percentage complete) columns can be used for inputting physical progress. Notice that in Figure 10-20, the early start for Activity 100 is listed as 01SEP97A. The A at the end of the date denotes that the date given was the actual start of Activity 100. Note that work on Activity 600 has not commenced, so there is no A by the early start date. To change the **Early Start** date column in the activity table, you must still employ the Datometer using the bar on the bar chart side of the screen, or you can use the **Activity** form. The activity table is another way to input data the **Progress** dialog box.

Documentation Using Activity Log

As observed before, the number one reason that contractors lose schedule-related claims in courts of law is a lack of documentation. Hence documentation of how the project has progressed is worthwhile. The delay of an activity for whatever reason may impact other activities and

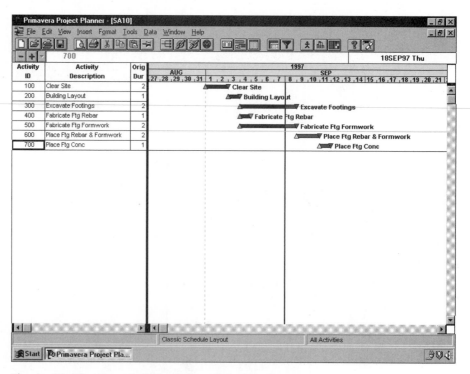

Figure 10-19   Sample Schedule—Calculated

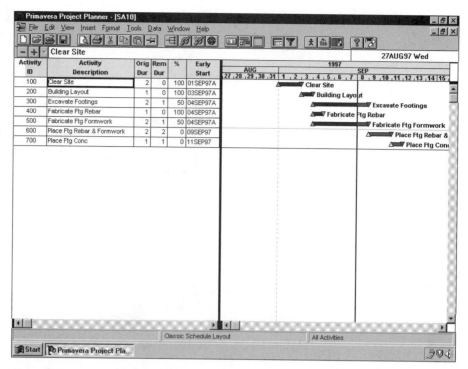

Figure 10-20   Sample Schedule—Activity Table Exposed

possibly project duration. *P3* provides a convenient activity log for documentation purposes.

Log Dialog Box   With Activity 100, Clear Site, highlighted, click the right mouse button, and the **Activity Detail** selection menu will appear (Figure 10-21). Select **Log**, and the **Log** dialog box will appear (Figure 10-22). If the actual start of Activity 100 were delayed, the log is an ideal way to document it. The documentation was entered under line **1** for Activity 100 (Figure 10-22). The **Mask** field can be used to keep a particular log entry from appearing on the on-screen bar chart or hard copy print when the activity logs are requested.

Modify Bar Definition Dialog Box   For the log entries to appear on the on-screen bar chart, the activity bar information must be reconfigured. Select **Bars...** from the **Format** main pull-down menu (Figure 10-23). The **Bars** dialog box will appear (Figure 10-24). Make sure the **Bar Description** item that is to be modified is selected. The **Early Bar** option in Figure 10-24 was selected. Select the **Modify...** button, and the **Modify Bar Definition** dialog box will appear (Figure 10-25). The lower half of the dialog box defines the information that will appear with the on-screen activity bar.

Text: Click under the **Position** field under **Text:** to add a new entry. Click the down arrow to bring up the **Position** selection box (Figure 10-26).

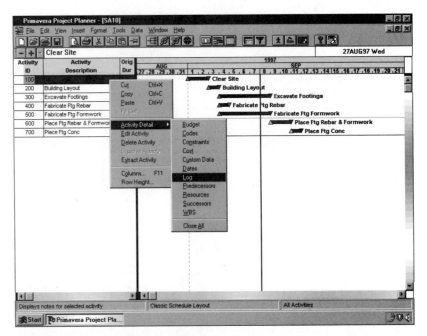

Figure 10-21   Activity Detail Selection Menu—Log Option

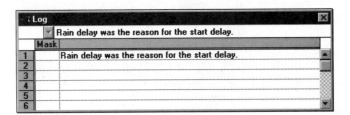

Figure 10-22   Log Dialog Box

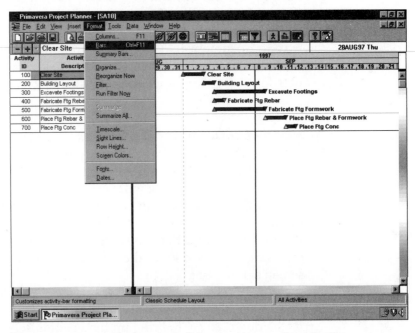

Figure 10-23   Format Main Pull-Down Menu—Bars Option

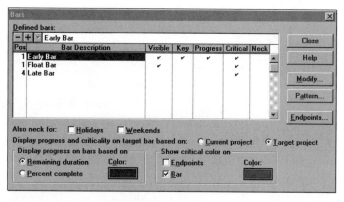

Figure 10-24   Bars Dialog Box

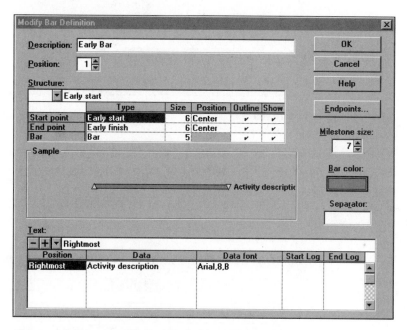

Figure 10-25   Modify Bar Definition Dialog Box

Options are **Bottom, Bottom-left, Bottom-right, Left, Leftmost, Right, Rightmost, Top, Top-left,** and **Top-right**.

For this example, select **Bottom-right**, then click on the **Data** field, click the down arrow, and select **Log record** (Figure 10-27). Click on the **OK** button of the **Modify Bar Definition** dialog box to save the reconfiguration of the on-screen activity bar. The on-screen bar chart

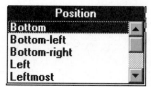

Figure 10-26   Modify Bar Definition Dialog Box—Position Selection Box

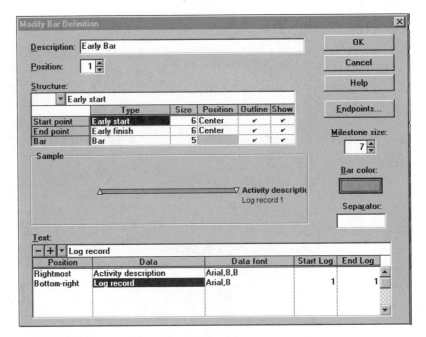

Figure 10-27   Modify Bar Definition Dialog Box—Data Field

now includes an activity log entry for Activity 100 (Figure 10-28). Compare Figure 10-28 to Figure 10-20 (p. 275). An extra line after each activity is provided automatically by *P3* for activity log infor-

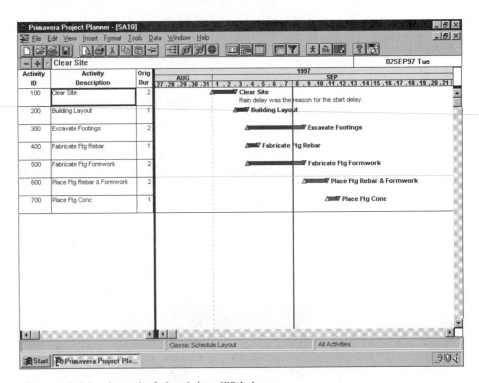

Figure 10-28   Sample Schedule—With Log

mation. The activity log information can also be printed out on various reports.

Producing Reports

SR-01, Classic Schedule Report - Sort by ES, TF   Target comparison reports are very helpful in determining progress against the original or baseline schedule. Look first at a tabular report without a target comparison. From the **Tools** main pull-down menu, select the **Tabular Reports** option. Then pick the **Schedule...** option, and the **Schedule Reports** title selection box will appear (Figure 10-29). The report structure in Figure 10-29 will appear this way only if no modifications have been made to the original list of reports. Figure 10-30 is a printout of the updated schedule run in report format SR-01, Classic Schedule Report - Sort by ES, TF. Compare this report to Figure 9-3 (p. 217), the same report for the target schedule. It would be difficult to determine progress by looking at just Figure 10-30.

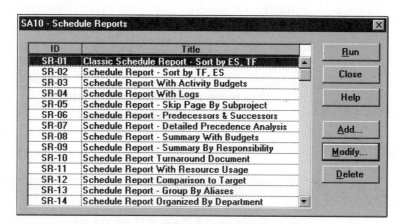

Figure 10-29   Tabular Schedule Reports Title Selection Box—Classic Schedule Report

```
-----------------------------------------------------------------------------------------------------------
Southern Constructors              PRIMAVERA PROJECT PLANNER              Schedule Update

REPORT DATE  2JAN97  RUN NO.   57                            START DATE  1SEP97  FIN DATE 11SEP97
             13:06
Classic Schedule Report - Sort by ES, TF                    DATA DATE   8SEP97  PAGE NO.    1
```

ACTIVITY ID	ORIG DUR	REM DUR	%	CODE	ACTIVITY DESCRIPTION	EARLY START	EARLY FINISH	LATE START	LATE FINISH	TOTAL FLOAT
100	2	0	100		Clear Site	1SEP97A				
200	1	0	100		Building Layout	3SEP97A				
400	1	0	100		Fabricate Ftg Rebar	4SEP97A	4SEP97A			
300	2	1	50		Excavate Footings	4SEP97A	8SEP97		8SEP97	0
500	2	1	50		Fabricate Ftg Formwork	4SEP97A	8SEP97		8SEP97	0
600	2	2	0		Place Ftg Rebar & Formwork	9SEP97	10SEP97	9SEP97	10SEP97	0
700	1	1	0		Place Ftg Conc	11SEP97	11SEP97	11SEP97	11SEP97	0

Figure 10-30   Classic Schedule Report Print—Sort by ES, TF

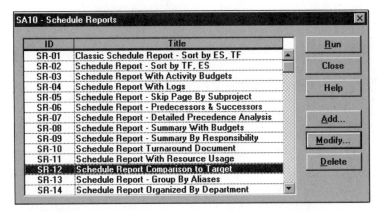

Figure 10-31   Schedule Reports Title Selection Box—Schedule Report Comparison to Target

SR-12, Schedule Report Comparison to Target   Select   SR-12, **Schedule Report Comparison to Target** (Figure 10-31), to run a comparison report. Select the **Modify...** button and the **Schedule Reports** dialog box appears. On the **Format** screen (Figure 10-32), under **Target comparison**, click the check box **Target 1**. When you click on the **Run** button, Figure 10-33 is the resulting report. It gives target versus current duration. Percent complete is provided. It also gives current versus target early start and current early finish versus target early finish. A variance column compares how far each activity that is yet to be completed is either ahead of or behind schedule. Activities 300, 500, 600, and 700 have a variance of –1. This means that each activity's current early start

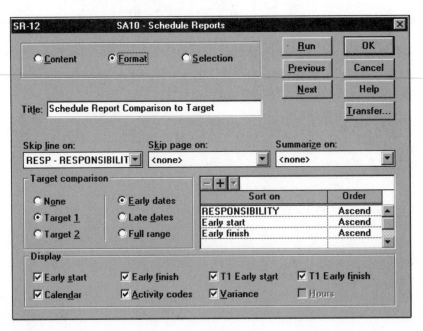

Figure 10-32   Schedule Reports Dialog Box—Schedule Report Comparison to Target

```
--------------------------------------------------------------------------------------------------
Southern Constructors               PRIMAVERA PROJECT PLANNER          Schedule Update

REPORT DATE 2JAN97  RUN NO.   60                                START DATE 1SEP97 FIN DATE 11SEP97
            13:14
Schedule Report Comparison to Target                            DATA DATE  8SEP97 PAGE NO.    1


----- -----   ----  ----  -  ---  ----------  -------------------------------  -------- -------- -------- -------- -----
ACTIVITY    TAR  CUR              ACTIVITY DESCRIPTION             CURRENT  EARLY    TARGET   EARLY
   ID       DUR  DUR   %   CODE                                    START    FINISH   START    FINISH   VAR.
----- -----   ----  ----  -  ---  ----------  -------------------------------  -------- -------- -------- -------- -----

       100    2    5   100        Clear Site                       1SEP97A           1SEP97   2SEP97
       200    1    3   100        Building Layout                  3SEP97A           3SEP97   3SEP97
       400    1    1   100        Fabricate Ftg Rebar              4SEP97A  4SEP97A  4SEP97   4SEP97     0
       300    2    1    50        Excavate Footings                4SEP97A  8SEP97   4SEP97   5SEP97    -1
       500    2    1    50        Fabricate Ftg Formwork           4SEP97A  8SEP97   4SEP97   5SEP97    -1
       600    2    2     0        Place Ftg Rebar & Formwork       9SEP97   10SEP97  8SEP97   9SEP97    -1
       700    1    1     0        Place Ftg Conc                   11SEP97  11SEP97  10SEP97  10SEP97   -1
```

Figure 10-33   Schedule Report Comparison to Target Print

is one day behind the target early start for that activity. This project is anticipated to finish a day late compared to the target schedule.

Bar Chart Comparison Report   Another good example of a comparison report is the bar chart comparison report. From the **Tools** main pull-down menu, select the **Graphic Reports** and then the **Bar...** option. Choose BC-23, click **Modify...**, then select the **Content** screen (Figure 10-34). Under the **Dates** portion, check the **Current schedule and Target schedule**. Select **Target 1** and then click on the **Run** button. Figure 10-35 is the hard copy run with the comparison configuration. The thicker activity bar lines represent the present or master schedule, while the thinner activity bar lines represent the original or target schedule.

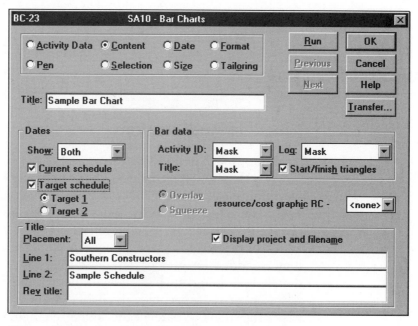

Figure 10-34   Bar Charts Dialog Box—Content Screen

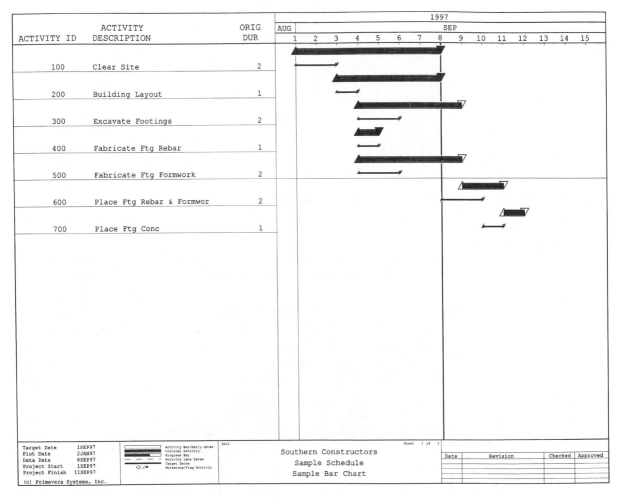

Figure 10-35 Sample Schedule Bar Chart Print

DOCUMENTING CHANGES

Schedule Modifications

As the approach to building the project changes, the schedule should be modified to show the changes. There are many kinds of changes. Activity durations may change if crew sizes are increased to make up for lost time. The logic of activity sequences and interrelationships may be changed to more accurately reflect interferences. Activities may be added or deleted to reflect changes in the scope of the project. Activities may be added for greater detail. As the project proceeds, usually more detailed planning, such as the so-called 10-day look-ahead, is required. Also, better data becomes available with actual activity on which to base projections. Whatever the source, changes need to be incorporated into the schedule.

Duration Changes

A change in duration may occur because similar activities are performed repeatedly within the schedule. An example is pouring a multistory concrete building, with each floor divided into several pours. By the time several floors have been placed, very good empirical data on time required for each pour is available. This information is used to adjust projections of activities remaining to be completed.

The original duration of Activity 600, Place Ftg Rebar & Formwork, was 2 days (Figure 10-36). In Figure 10-37, it was changed to 3 days. This modification was made by changing the number in the **Orig Dur** field of the activity table. Notice that since the schedule has not been calculated since the change, Activity 700 (a successor to Activity 600) has not been affected. In Figure 10-38, the schedule has been calculated since the change in duration to Activity 600. The project duration has been extended from the end of September 11 to September 12.

Logic Changes

The original logic of the relationships between activities may change for any number of reasons. The following is an example of a logic change and how to incorporate it into the schedule. It is necessary for inspection purposes to include one day of lag between Activity 600 and 700.

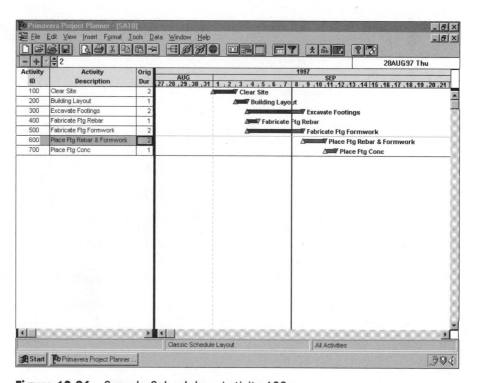

Figure 10-36   Sample Schedule—Activity 600

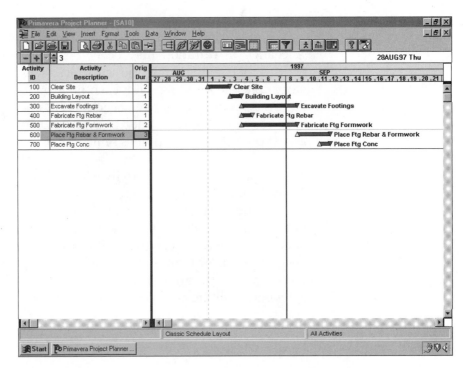

Figure 10-37 Sample Schedule—Forecast for Activity 600

Select Activity 600 by clicking the mouse arrow on any of the activity table fields, then depress the right mouse button and the **Activity Detail** selection menu (Figure 10-38) will appear. Select **Successors**, and

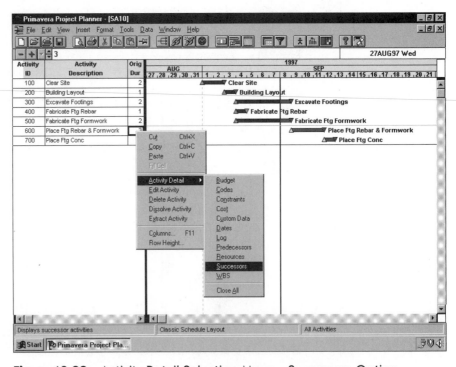

Figure 10-38 Activity Detail Selection Menu—Successors Option

the **Successors** dialog box will appear (Figure 10-39). Activity 700 is a successor to Activity 600. Select Activity 700 for changes, then click on the **Lag** field and enter 1. Compare the calculated schedule with this change incorporated (Figure 10-40) to the schedule before the change (Figure 10-38, p. 282). In the schedule before the change, Activity 600 ends at the end of the day of Thursday, September 11. Activity 700 ends at the end of the day of Friday, September 12. With the addition of the lag day in Activity 600, Friday, September 12, becomes the lag day. Saturday (the 13th) and Sunday September 14 are nonworkdays according to the project calendar configuration, so Activity 700 now begins and ends on Monday, September 15.

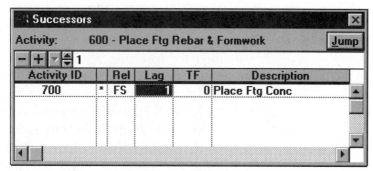

Figure 10-39   Successors Dialog Box—Activity 600

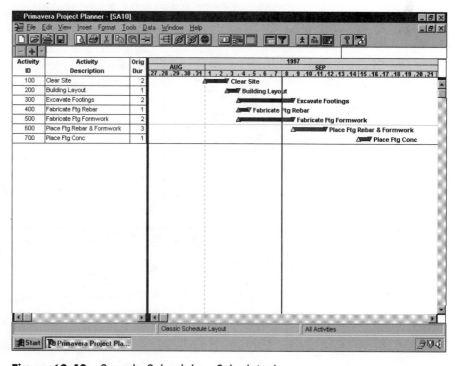

Figure 10-40   Sample Schedule—Calculated

Added Activities

Activity Table Sometimes in documenting changes, it is necessary either to add or to delete activities from the schedule. Adding activities to the schedule within *P3* is not difficult. Place the mouse cursor on the activity after which the new activity will be placed and double-click on the left mouse button. Then click in the empty space after the last activity. Another way is to click on the activity after which the new activity is to be placed, highlight it, and then click in the empty space in the activity table after the last activity. The new activity will be added. To input the rest of the activity information, click on the **View** main pull-down menu and select **Activity Form**. The **Activity** form will appear at the bottom of the screen (Figure 10-41). The new activity will appear after the last activity, 700, Place Ftg Conc. In Figure 10-42, activity 800, Strip, clean & Oil Forms, has been added to the **Activity** form. An original duration (**OD**) of 1 is accepted.

Another method for inserting an activity is through the **Insert** main pull-down menu. Select **Activity Ins**.

Relationships Next, relationships need to be defined. Since Activity 800 will be the last activity of the schedule and has no successors, predecessors need to be established. Click on the **Pred** button of the **Activity** form, and the **Predecessors** dialog box will appear (Figure 10-43). Activity 700, Place Ftg Conc, is made a predecessor to Activity 800,

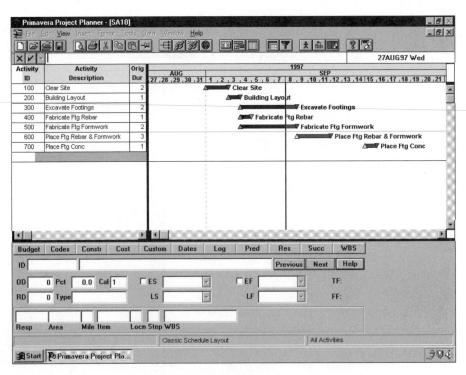

Figure 10-41 Activity Form—New Activity

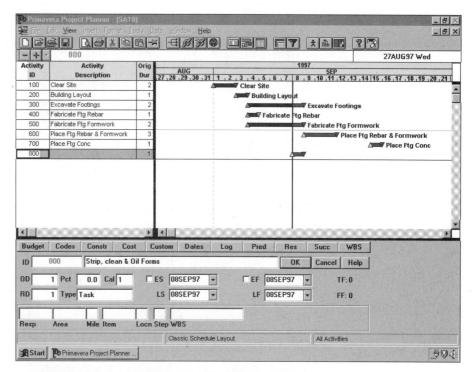

Figure 10-42   Activity Form—Activity 800

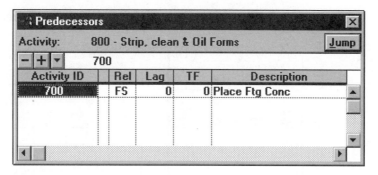

Figure 10-43   Predecessors Dialog Box—Activity 800

Strip, clean & Oil Forms. The default relationship (**Rel**) of finish to start (**FS**) is accepted, and no lag time is entered. When the **Predecessors** dialog box is closed, click the **OK** button on the **Activity** form to accept the new activity, and the resulting screen will look like Figure 10-44. To incorporate the new relationships into the schedule requires schedule calculations. Figure 10-45 shows the schedule calculated with the new activity.

Timescale Definition Dialog Box   The entire activity description of the activity bar for Activity 800 is not on the screen. Obviously, for a larger schedule, all of the information will not fit on a single screen. The purpose of the slide button bars located to the right and at the bottom of the screen is to move to the desired part of the schedule. As discussed in

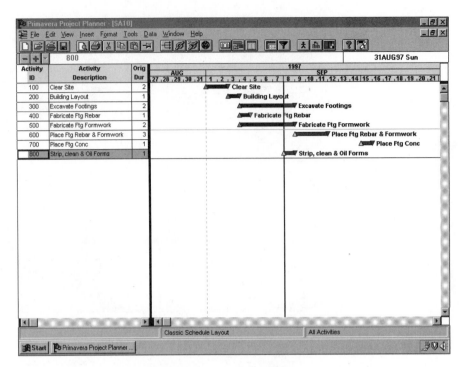

Figure 10-44 Sample Schedule—Activity 800

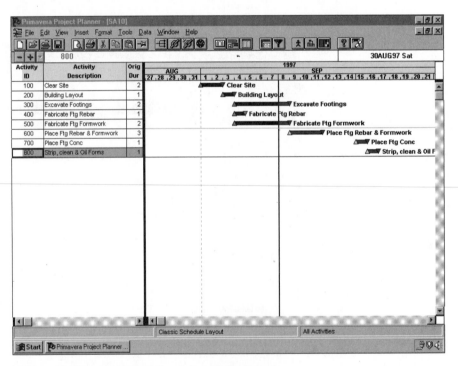

Figure 10-45 Sample Schedule—Calculated

Chapter 5, the **Timescale Definition** dialog box (Figure 10-46) can be used to calibrate the timescale to fit more information on a single screen. Access the **Timescale Definition** dialog box by selecting **Timescale...** from the **Format** main pull-down menu. In the sample schedule

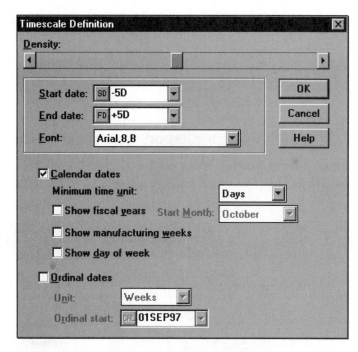

Figure 10-46　Timescale Definition Dialog Box

in Figure 10-45 (p. 286), the space between August 27 and August 31 is wasted. Click on the down arrow under **Start date:** and the pop-up calendar will appear (Figure 10-47). This start date defines the start of the on-screen bar chart and is not associated with the start date of the

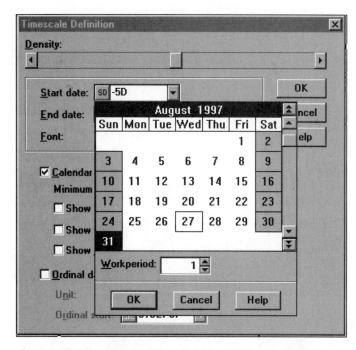

Figure 10-47　Timescale Definition Dialog Box—Pop-up Calendar

schedule for calculation purposes. Select August 31, 1997, as the start date, and click the **OK** button to accept the change. Figure 10-48 is the result of the change in start date.

Producing Reports

SR-12, Schedule Report Comparison to Target

Figure 10-49 is the SR-12, Schedule Report Comparison to Target, report run with the documentation changes incorporated. Compare it to Figure 10-33 (p. 279).

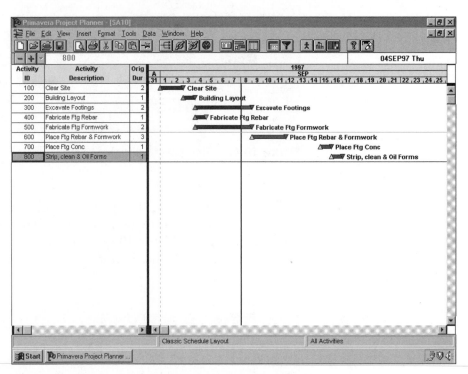

Figure 10-48   Sample Schedule

```
------------------------------------------------------------------------------------------------------------
Southern Constructors                    PRIMAVERA PROJECT PLANNER              Schedule Update

REPORT DATE   2JAN97  RUN NO.   66                                   START DATE  1SEP97  FIN DATE 16SEP97
              16:12
Schedule Report Comparison to Target                                 DATA DATE   8SEP97  PAGE NO.    1


----- -----   --- ---- - ---  --------- ------------------------------   -------- ------- -------- -------- -----
ACTIVITY      TAR CUR                        ACTIVITY DESCRIPTION         CURRENT  EARLY   TARGET   EARLY
   ID         DUR DUR    %  CODE                                          START    FINISH  START    FINISH   VAR.
----- -----   --- ---- - ---  --------- ------------------------------   -------- ------- -------- -------- -----

       100     2   5   100         Clear Site                            1SEP97A           1SEP97   2SEP97
       200     1   3   100         Building Layout                       3SEP97A           3SEP97   3SEP97
       400     1   1   100         Fabricate Ftg Rebar                   4SEP97A  4SEP97A  4SEP97   4SEP97      0
       300     2   1    50         Excavate Footings                     4SEP97A  8SEP97   4SEP97   5SEP97     -1
       500     2   1    50         Fabricate Ftg Formwork                4SEP97A  8SEP97   4SEP97   5SEP97     -1
       600     2   3     0         Place Ftg Rebar & Formwork            9SEP97   11SEP97  8SEP97   9SEP97     -2
       700     1   1     0         Place Ftg Conc                        15SEP97  15SEP97  10SEP97  10SEP97    -3
       800     0   1     0         Strip, clean & Oil Forms              16SEP97  16SEP97
```

Figure 10-49   Schedule Report Comparison to Target Print

Notice that the current start (projected) for Activity 700 is three work-days (five calendar days) behind the target start. Note, there is no target start, early finish, or variance for Activity 800, since it was not included in the target schedule.

Bar Chart Comparison Report   Figure 10-50 is the bar chart comparison report run with the documentation changes incorporated. Compare it to Figure 10-35 (p. 280). There is no target bar for Activity 800, since it was not included in the target schedule.

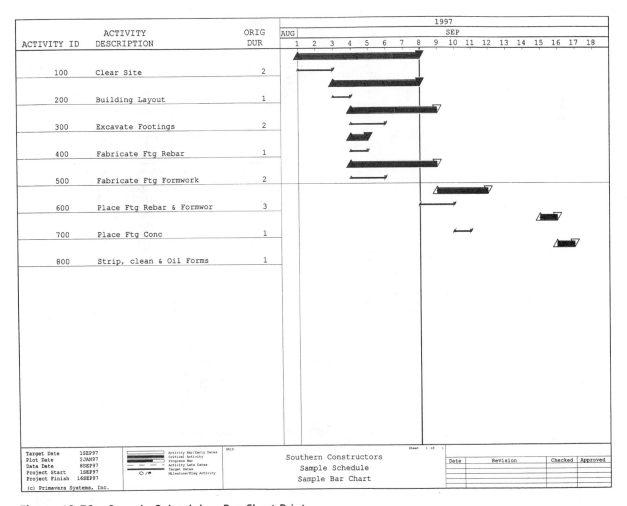

Figure 10-50   Sample Schedule—Bar Chart Print

EXAMPLE PROBLEM: Updating

Figures 10-51 and 10-52 are hard copy prints of updated reports for a house put together as an example for student use (see the wood frame house drawings in the Appendix.)

Figure 10-51 is a copy of the schedule report comparison to target in tabular format. Figures 10-52a and b are hard copy prints of the updated on-screen bar chart using the **Print...** option from the **File** main pull-down menu.

```
-------------------------------------------------------------------------------------------------------------------
Student Constructors                      PRIMAVERA PROJECT PLANNER           Sample Project #1

REPORT DATE  4JAN97  RUN NO.   30                                             START DATE  5JAN97  FIN DATE 14MAR97
             16:47
Schedule Report Comparison to Target                                         DATA DATE   3FEB97  PAGE NO.    1
```

ACTIVITY ID	TAR DUR	CUR DUR	%	ACTIVITY DESCRIPTION	CURRENT START	EARLY FINISH	TARGET START	EARLY FINISH	VAR.
100	2	3	100	Clear Site	6JAN97A	8JAN97A	6JAN97	7JAN97	-1
200	1	1	100	Building Layout	8JAN97A	8JAN97A	8JAN97	8JAN97	0
300	3	2	100	Form/Pour Footings	9JAN97A	10JAN97A	9JAN97	13JAN97	1
400	2	2	100	Pier Masonry	15JAN97A	16JAN97A	14JAN97	15JAN97	-1
500	4	4	100	Wood Floor System	17JAN97A	22JAN97A	16JAN97	21JAN97	-1
600	6	5	100	Rough Framing Walls	23JAN97A	29JAN97A	22JAN97	29JAN97	0
700	4	3	25	Rough Framing Roof	28JAN97A	5FEB97	30JAN97	4FEB97	-1
800	4	4	0	Doors & Windows	10FEB97	13FEB97	7FEB97	12FEB97	-1
900	2	2	0	Ext Wall Board	6FEB97	7FEB97	5FEB97	6FEB97	-1
1000	1	1	0	Ext Wall Insulation	14FEB97	14FEB97	13FEB97	13FEB97	-1
1100	4	4	0	Rough Plumbing	10FEB97	13FEB97	7FEB97	12FEB97	-1
1200	3	3	0	Rough HVAC	6FEB97	10FEB97	5FEB97	7FEB97	-1
1300	3	3	0	Rough Elect	11FEB97	13FEB97	10FEB97	12FEB97	-1
1400	3	3	0	Shingles	14FEB97	18FEB97	13FEB97	17FEB97	-1
1500	3	3	0	Ext Siding	18FEB97	20FEB97	17FEB97	19FEB97	-1
1600	2	2	0	Ext Finish Carpentry	14FEB97	17FEB97	13FEB97	14FEB97	-1
1700	4	4	0	Hang Drywall	17FEB97	20FEB97	14FEB97	19FEB97	-1
1800	4	4	0	Finish Drywall	21FEB97	26FEB97	20FEB97	25FEB97	-1
1900	2	2	0	Cabinets	27FEB97	28FEB97	26FEB97	27FEB97	-1
2000	3	3	0	Ext Paint	21FEB97	25FEB97	20FEB97	24FEB97	-1
2100	4	4	0	Int Finish Carpentry	3MAR97	6MAR97	28FEB97	5MAR97	-1
2200	3	3	0	Int Paint	5MAR97	7MAR97	4MAR97	6MAR97	-1
2300	2	2	0	Finish Plumbing	3MAR97	4MAR97	28FEB97	3MAR97	-1
2400	3	3	0	Finish HVAC	27FEB97	3MAR97	26FEB97	28FEB97	-1
2500	2	2	0	Finish Elect	27FEB97	28FEB97	26FEB97	27FEB97	-1
2600	3	3	0	Flooring	10MAR97	12MAR97	7MAR97	11MAR97	-1
2700	4	4	0	Grading & Landscaping	5MAR97	10MAR97	4MAR97	7MAR97	-1
2800	2	2	0	Punch List	13MAR97	14MAR97	12MAR97	13MAR97	-1

Figure 10-51   Schedule Report Comparison to Target Print—Wood Frame House

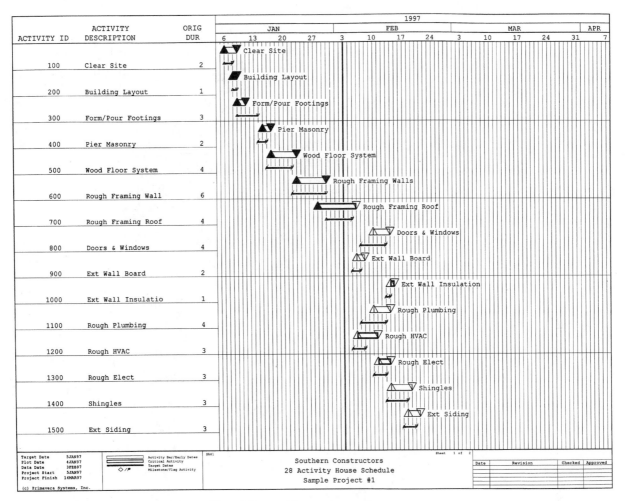

Figure 10-52a   Bar Chart Print—Wood Frame House, page 1

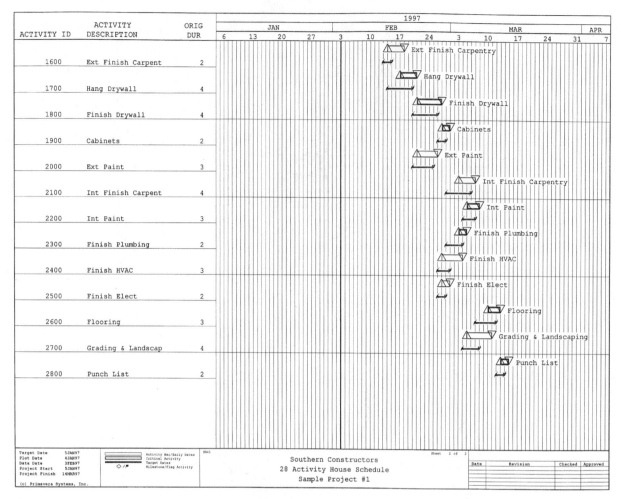

Figure 10-53b   Bar Chart Print—Wood Frame House, page 2

EXERCISES

1. Small Commercial Concrete Block Building—Updating

Prepare the following reports for the small commercial concrete block building located in the Appendix.

1. Updated bar chart
2. Schedule report comparison to target in tabular format

2. Large Commercial Building—Updating

Prepare the following reports for the large commercial building located in the Appendix.

1. Updated bar chart
2. Schedule report comparison to target in tabular format

Schedule Optimization Using *P3*

Objectives

Upon completion of this chapter, the reader should be able to:

- Analyze resources
- Optimize resources
- Level resources
- Compress or decompress the schedule
- Update the schedule using *P3*

EFFICIENCY ANALYSIS

Attention to efficiency in the use of resources is a critical part of construction management. The efficient layout, organization, and use of assets is the very essence of good planning, scheduling, and project management.

Labor

On most construction projects, labor requirements follow the shape of a bell curve. The project begins with mobilization activities and some site work with little labor required. The curve rises slowly as more crafts are required (Figure 11-1). The middle of project duration is usually the peak of labor requirements. At this point many things are going on at the project at one time. Many crafts are working and a lot of activity is taking place. At the end of the project, the labor curve flattens out. The job is being demobilized; people are moving on to other projects.

A contractor needs to make this curve as smooth as possible. No contractor wants to have to hire five carpenters one week, fifteen the next, two the next, then ten the next. It would be much more efficient to flatten the labor curve by hiring eight carpenters for the entire four weeks. The same principle applies to subcontractors as well.

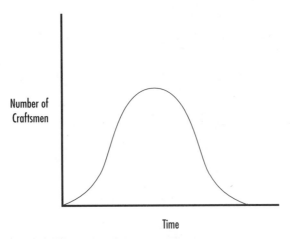

Figure 11-1   Manpower Bell Curve

Subcontractors

The way a subcontractor makes money is to come to the job site, get the work done as efficiently as possible, and then leave. Unnecessary starts and stops rob productivity and efficiency.

To achieve a continuous smooth flow of work the subcontractor wants to get the crew to the job site when the job site is ready and not before, and to let them work uninterrupted as long as possible. Productivity is lost when a sub has to constantly mobilize and demobilize. Obviously this assumes there are no interferences from activities of other crafts.

Materials

Smoothing demand also applies to improving efficiency of materials use. An ideal example is formwork. The way to stretch funds spent on formwork is to reuse it as many times as possible by carefully sequencing the pours.

RESOURCE ANALYSIS

Target Schedule

Figure 11-2 is a slight modification of the target schedule originally put together for the project with the resources input in Chapter 5. The changes made in Chapter 10 to update durations and actual progress will not be used for the discussions in this chapter. Schedule optimization can be used at any time during the life of the project, so for simplicity's sake the original schedule will be used.

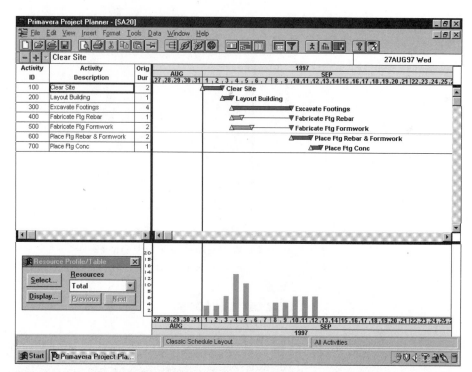

Figure 11-2 Sample Schedule—Target Schedule (Changed for this Chapter for Demonstration Purposes)

Labor Requirements

The labor profile (as discussed in Chapter 5), or labor curve, is used in craft or labor analysis. Table 11-1 shows the craft requirements by day and by activity for the target schedule. The requirements are 3 craftspersons on September 1; 6 on September 3; 13 on September 4; 10 on September 5; 4 on September 8; and 6 on September 10—not a very efficient use of resources. There is room for improving this situation.

Figures 11-3 through 11-9 detail the target (original) labor resource requirements for each activity. Table 11-1 summarizes these figures, showing the

	1	2	3	4	5	6	7	8	9	10	11	12
CARP 1	100 - 3	100 - 3	200 - 2	300 - 2 500 - 2	300 - 2 500 - 2			300 - 2	300 - 2	600 - 2	600 - 2	700 - 2
CARP 2			200 - 2	500 - 2	500 - 2							
EQUIP OP				300 - 1	300 - 1			300 - 1	300 - 1	600 - 2	600 - 2	700 - 2
LAB 1			200 - 2	300 - 1 400 - 1	300 - 1			300 - 1	300 - 1			700 - 2
LAB 2				400 - 2 500 - 2	500 - 2					600 - 2	600 - 2	
Total	3	3	6	13	10			4	4	6	6	6

Table 11-1 Craft Requirements by Activity by Day—Target Schedule

Resources — CARP 1

Resource	CARP 1
Cost Acct/Category	
Driving	
Curve	
Units per day	3.00
Budgeted quantity	6.00
Res Lag/Duration	0
Percent complete	
Actual this period	0.00
Actual to date	0.00
To complete	6.00
At completion	6.00
Variance (units)	0.00
Early start	01SEP97
Early finish	02SEP97
Late start	01SEP97
Late finish	02SEP97

Figure 11-3   Resources Dialog Box—Activity 100

Resources — CARP 2

Resource	CARP 2	LAB 1	CARP 1
Cost Acct/Category			
Driving			
Curve			
Units per day	2.00	2.00	2.00
Budgeted quantity	2.00	2.00	2.00
Res Lag/Duration	0	0	0
Percent complete			
Actual this period	0.00	0.00	0.00
Actual to date	0.00	0.00	0.00
To complete	2.00	2.00	2.00
At completion	2.00	2.00	2.00
Variance (units)	0.00	0.00	0.00
Early start	03SEP97	03SEP97	03SEP97
Early finish	03SEP97	03SEP97	03SEP97
Late start	03SEP97	03SEP97	03SEP97
Late finish	03SEP97	03SEP97	03SEP97

Figure 11-4   Resources Dialog Box—Activity 200

Resources — LAB 1

Resource	LAB 1	EQUIP OP	CARP 1
Cost Acct/Category			
Driving			
Curve			
Units per day	1.00	1.00	2.00
Budgeted quantity	4.00	4.00	8.00
Res Lag/Duration	0	0	0
Percent complete			
Actual this period	0.00	0.00	0.00
Actual to date	0.00	0.00	0.00
To complete	4.00	4.00	8.00
At completion	4.00	4.00	8.00
Variance (units)	0.00	0.00	0.00
Early start	04SEP97	04SEP97	04SEP97
Early finish	09SEP97	09SEP97	09SEP97
Late start	04SEP97	04SEP97	04SEP97
Late finish	09SEP97	09SEP97	09SEP97

Figure 11-5   Resources Dialog Box—Activity 300

Resources — LAB 1	LAB 1	LAB 2	
Resource	LAB 1	LAB 2	
Cost Acct/Category			
Driving	✔		
Curve			
Units per day	1.00	2.00	
Budgeted quantity	1.00	2.00	
Res Lag/Duration	0 1	0	
Percent complete			
Actual this period	0.00	0.00	
Actual to date	0.00	0.00	
To complete	1.00	2.00	
At completion	1.00	2.00	
Variance (units)	0.00	0.00	
Early start	04SEP97	04SEP97	
Early finish	04SEP97	04SEP97	
Late start	09SEP97	09SEP97	
Late finish	09SEP97	09SEP97	

Figure 11-6   Resources Dialog Box—Activity 400

Resources — CARP 2	CARP 2	LAB 2	CARP 1
Resource	CARP 2	LAB 2	CARP 1
Cost Acct/Category			
Driving			
Curve			
Units per day	2.00	2.00	2.00
Budgeted quantity	4.00	4.00	4.00
Res Lag/Duration	0	0	0
Percent complete			
Actual this period	0.00	0.00	0.00
Actual to date	0.00	0.00	0.00
To complete	4.00	4.00	4.00
At completion	4.00	4.00	4.00
Variance (units)	0.00	0.00	0.00
Early start	04SEP97	04SEP97	04SEP97
Early finish	05SEP97	05SEP97	05SEP97
Late start	08SEP97	08SEP97	08SEP97
Late finish	09SEP97	09SEP97	09SEP97

Figure 11-7   Resources Dialog Box—Activity 500

Resources — EQUIP OP	EQUIP OP	LAB 2	CARP 1
Resource	EQUIP OP	LAB 2	CARP 1
Cost Acct/Category			
Driving			
Curve			
Units per day	2.00	2.00	2.00
Budgeted quantity	4.00	4.00	4.00
Res Lag/Duration	0	0	0
Percent complete			
Actual this period	0.00	0.00	0.00
Actual to date	0.00	0.00	0.00
To complete	4.00	4.00	4.00
At completion	4.00	4.00	4.00
Variance (units)	0.00	0.00	0.00
Early start	10SEP97	10SEP97	10SEP97
Early finish	11SEP97	11SEP97	11SEP97
Late start	10SEP97	10SEP97	10SEP97
Late finish	11SEP97	11SEP97	11SEP97

Figure 11-8   Resources Dialog Box—Activity 600

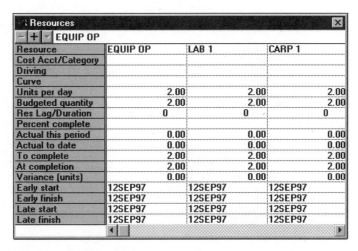

Figure 11-9   Resources Dialog Box—Activity 700

labor resource requirements by activity, the totals by craft by day, and the total labor requirements for each day for the target schedule. The labor requirements for the activities were derived from the estimate, which specified the quantity of work, productivity rate, crew size, and, therefore, the duration of each activity. Earlier discussions did not consider the relationship between total labor needs and the totals by craft. This caused the fluctuations in labor requirements discussed earlier.

RESOURCE AVAILABILITY

The resource availability for five resources will be covered in this section. The five resources are CARP 1, CARP 2, EQUIP OP, LAB 1, and LAB 2.

CARP 1

Resource Dictionary Dialog Box   One way to "flatten" or smooth project resource requirements is the float. Click on the **Data** main pulldown menu. Select **Resources...**, and the **Resource Dictionary** dialog box will appear (Figure 11-10). Under **Resources:**, select **CARP 1**.

Limits: Under **Limits:**, define the normal and the maximum amount of the resource that may be needed. For CARP 1, click on **Normal** and click on 2 to set the normal limit. Click on **Max**, then 3 to set the maximum limit at 3 craftspersons. Click on **Close** to accept the new resource limits and to close the **Resource Dictionary** dialog box.

Draw limits   To show the limits by resource more clearly, click on the **Display...** button of the **Resource Profile/Table** dialog box. When the

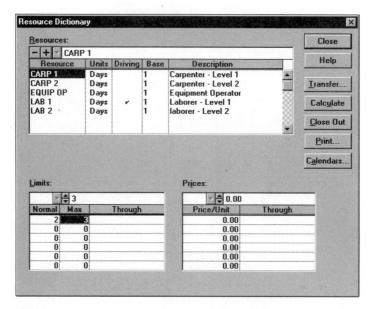

Figure 11-10   Resource Dictionary Dialog Box

Resource Profile Display Options box appears (Figure 11-11), click on the **Draw limits** and the **Emphasize overload with color** check boxes. Then click the **Close** button to accept the change.

Resource Profile/Table Dialog Box   Click on the down arrow under the **Resources** button of the **Resource Profile/Table** dialog box. Then select **CARP 1**. The resource profile in Figure 11-12 shows the

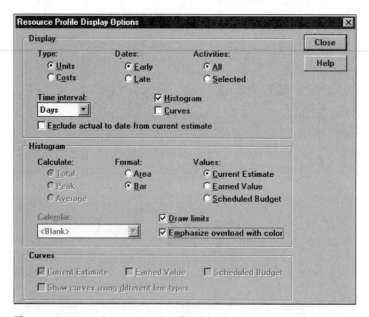

Figure 11-11   Resource Profile Display Options Dialog Box

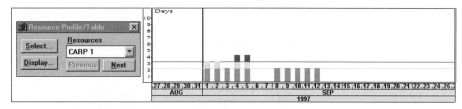

Figure 11-12   Resource Profile—CARP 1

project requirements for CARP 1. Three CARP 1s are required for September 1 and 2 for Activity 100; two CARP 1s are required for September 3 and 8 to 12 for Activities 200, 300, 600, and 700; and four CARP 1s are required for September 4 and 5 for Activities 300 and 500. The resource bar has two colors associated with it. The normal limit set for CARP 1 in the resource dictionary was 2. The difference in color above the 2 shows that the normal limit was exceeded. The resource profile in Figure 11-12 shows horizontal lines indicating the normal and maximum limits as set for CARP 1 in the resource dictionary.

CARP 2

Resource Dictionary Dialog Box   In the **Resource Dictionary** dialog box, select **CARP 2** under **Resources:**.

Limits: Set the normal limit at 2 and the maximum limit at 4 for CARP 2. Click on **Close** to accept the new resource limits and to close the **Resource Dictionary** dialog box.

Resource Profile/Table Dialog Box   Click on the down arrow under the **Resources** button of the **Resource Profile/Table** dialog box. Then select **CARP 2**. The resource profile in Figure 11-13 shows the project requirements for CARP 2. Two CARP 2s are required for September 3 for Activity 200; two CARP 2s are required for September 4 and 5 for Activity 500. The normal limit was not exceeded. The resource profile in Figure 11-13 shows horizontal lines indicating the normal and maximum limits as set for CARP 2 in the resource dictionary.

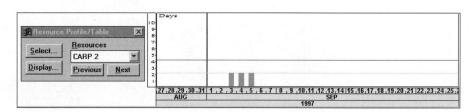

Figure 11-13   Resource Profile—CARP 2

EQUIP OP

Resource Dictionary Dialog Box   In the **Resource Dictionary** dialog box, select **EQUIP OP** under **Resources:**.

Limits: Set the normal limit at 3 and the maximum limit at 5 for EQUIP OP. Click on **Close** to accept the new resource limits and to close the **Resource Dictionary** dialog box.

Resource Profile/Table Dialog Box   Click on the down arrow under the **Resources** button of the **Resource Profile/Table** dialog box. Then select **EQUIP OP**. The resource profile in Figure 11-14 shows the project requirements for EQUIP OP. One EQUIP OP is required for September 4, 5, 8, and 9 for Activity 300; two EQUIP OPs are required for September 10 and 11 for Activity 600; two EQUIP OPs are required for September 12 for Activity 700. The normal limit was not exceeded. The resource profile in Figure 11-14 shows horizontal lines indicating the normal and maximum limits as set for EQUIP OP in the resource dictionary.

LAB 1

Resource Dictionary Dialog Box   In the **Resource Dictionary** dialog box, select **LAB 1** under **Resources:**.

Limits: Set the normal limit at 2 and the maximum limit at 3 for LAB 1. Click on **Close** to accept the new resource limits and to close the **Resource Dictionary** dialog box.

Resource Profile/Table Dialog Box   Click on the down arrow under the **Resources** button of the **Resource Profile/Table** dialog box. Then select **LAB 1**. The resource profile in Figure 11-15 shows the project requirements for LAB 1. Two LAB 1s are required for September 3 for Activity 200; two LAB 1s are required for September 4 for Activities 300 and 400; one LAB 1 is required for September 5, 8, and 9 for Activity 300; and two LAB 1s are required for September 12 for Activity 700. The normal limit was not exceeded. The resource profile in Figure 11-15 shows horizontal lines indicating the normal and maximum limits as set for LAB 1 in the resource dictionary.

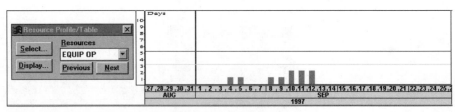

Figure 11-14   Resource Profile—EQUIP OP

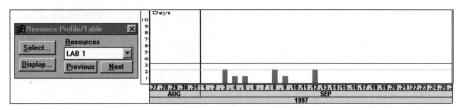

Figure 11-15   Resource Profile—LAB 1

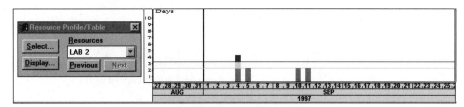

Figure 11-16   Resource Profile—LAB 2

LAB 2

Resource Dictionary Dialog Box   In the **Resource Dictionary** dialog box, select **LAB 2** under **Resources:**.

Limits: Set the normal limit at 2 and the maximum limit at 3 for LAB 2. Click on **Close** to accept the new resource limits and to close the **Resource Dictionary** dialog box.

Resource Profile/Table Dialog Box   Click on the down arrow under the **Resources** button of the **Resource Profile/Table** dialog box. Then select **LAB 2**. The resource profile in Figure 11-16 shows the project requirements for LAB 2. Four LAB 2s are required for September 4 for Activities 400 and 500; two LAB 2s are required for September 5 for Activity 500; two LAB 2s are required for September 10 and 11 for Activity 600. Notice that the normal and maximum limits have been exceeded. The resource profile in Figure 11-16 shows horizontal lines indicating the normal and maximum limits as set for LAB 2 in the resource dictionary.

RESOURCE LEVELING

Resource Leveling Dialog Box

LAB 2 requirements are an ideal place to look for spreading resource requirements to flatten craft and overall labor requirements. Click on the **Tools** main pull-down menu (Figure 11-17), and select **Level....** The **Resource Leveling** dialog box will appear (Figure 11-18).

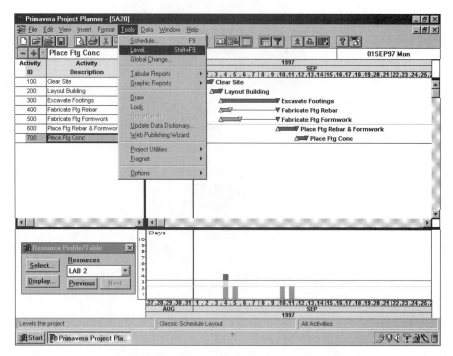

Figure 11-17 Tools Main Pull-Down Menu—Level Option

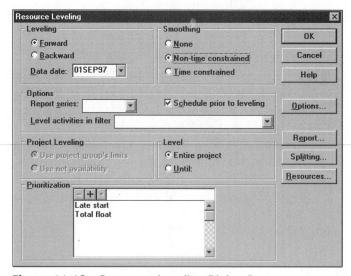

Figure 11-18 Resource Leveling Dialog Box

Resource Selection Dialog Box

To identify the resources to be leveled, click on the **Resources...** button in the **Resource Leveling** dialog box, and the **Resource Selection** dialog box will appear (Figure 11-19).

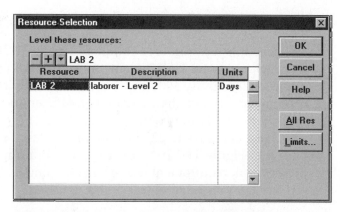

Figure 11-19   Resource Selection Dialog Box

Level these resources: In the **Resource Selection** dialog box, click the **All Res** button to select all resources for leveling or use the **Level these resources:** field to identify selected resources. Using the **Level these resources:** option, click the + button, then the down arrow to expose the options. **LAB 2** was selected for leveling.

Resource Limits Dialog Box

While still in the **Resource Selection** dialog box, click the **Limits...** button to check or change the information input in the resource dictionary. The **Resource Limits** dialog box will appear (Figure 11-20). For LAB 2, the normal limit set is 2; the maximum limit is 3. Now click the **OK** button to accept the resource limits, click **OK** again at the **Resource Selection** dialog box, and click **OK** again on the **Resource Leveling** dialog box for *P3* to level the requirements for LAB 2.

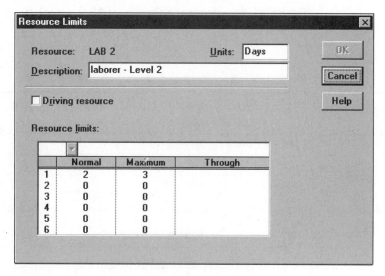

Figure 11-20   Resource Limits Dialog Box

Resource Profiles

The schedule and resource profile in Figure 11-21 is a result of the *P3's* leveling process for LAB 2. Compare Figure 11-21 to Figure 11-17 (p. 304). Notice that Activity 400 now comes after Activity 500. It takes place on September 8 rather than September 4. Examine Table 11-2 (optimized target schedule), and compare it to Table 11-1 on p. 296 (target schedule). Look at Figure 11-22, the total project labor requirements, and compare it to Figure 11-2 (p. 296). The labor buildup is more uniform. The project peak requirements have been flattened.

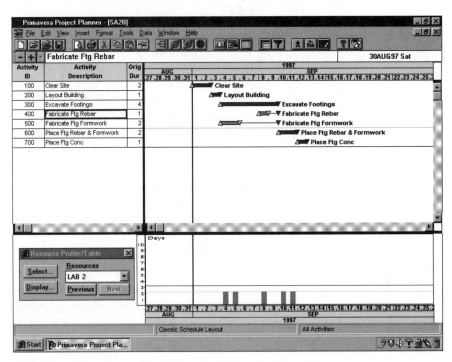

Figure 11-21   Resource Profile—LAB 2, Leveled

	1	2	3	4	5	6	7	8	9	10	11	12
CARP 1	100 – 3	100 – 3	200 – 2	300 – 2 500 – 2	300 – 2 500 – 2			300 – 2	300 – 2	600 – 2	600 – 2	700 – 2
CARP 2			200 – 2	500 – 2	500 – 2							
EQUIP OP				300 – 1	300 – 1			300 – 1	300 – 1	600 – 2	600 – 2	700 – 2
LAB 1			200 – 2	300 – 1	300 – 1			300 – 1 400 – 1	300 – 1			700 – 2
LAB 2				500 – 2	500 – 2			400 – 2		600 – 2	600 – 2	
Total	3	3	6	10	10			7	4	6	6	6

Table 11-2   Craft Requirements by Activity by Day—Leveled Target Schedule

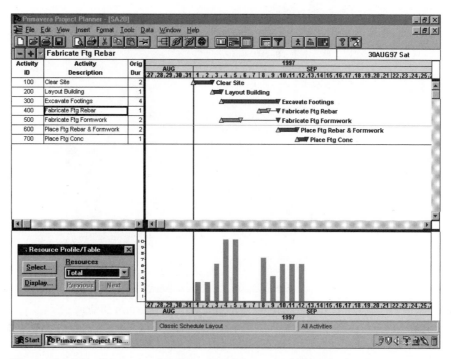

Figure 11-22　Resource Profile—Total Resources Leveled

SCHEDULE COMPRESSION

There may be any number of reasons for compressing or "crashing" a schedule. One reason is to make up time on a project that is behind schedule. Another is orders by the client to accelerate the schedule. The goal is to get the most reduction in time for the extra amount of funds expended.

The typical methods available for compressing the schedule are (a) to use different methods for increased productivity, (b) to work overtime with the present crew sizes, (c) to increase crew sizes, or (d) to modify activities logic to accomplish tasks concurrently.

Overtime

P3 documents schedule compression in several ways. Many contractors use ten-hour workdays and four-day workweeks. This leaves three-day weekends. Fridays are usually makeup days. In other words, they can be used to make up for days lost due to rain or as catch-up days for critical activities that are behind schedule.

To modify the project calendar, click the **Data** main pull-down menu and select **Calendars....** As can be seen from Figure 11-23, the plan for the schedule for the sample project was to work five days per week.

Figure 11-23   Project Calendars Dialog Box

Click on September 6, and then click on the **Work** button; do the same for September 7. Now both weekend days are workdays (Figure 11-24). Compare the new schedule (Figure 11-25) to Figure 11-22 (p. 307). Also compare Table 11-3, which shows the calendar changes, with Table 11-2 (p. 306). The original planned date for completing the project was September 12. The new anticipated completion date is September 10, or a two-day reduction.

Figure 11-24   Project Calendars Dialog Box—Workdays Added

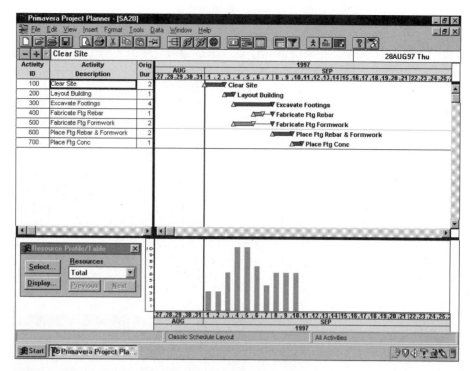

Figure 11-25   Resources Profile—Workdays Added

	1	2	3	4	5	6	7	8	9	10
CARP 1	100 – 3	100 – 3	200 – 2	300 – 2 500 – 2	300 – 2 500 – 2	300 – 2	300 – 2	600 – 2	600 – 2	700 – 2
CARP 2			200 – 2	500 – 2	500 – 2					
EQUIP OP				300 – 1	300 – 1	300 – 1	300 – 1	600 – 2	600 – 2	700 – 2
LAB 1			200 – 2	300 – 1	300 – 1	300 – 1 400 – 1	300 – 1			700 – 2
LAB 2				500 – 2	500 – 2	400 – 2		600 – 2	600 – 2	
Total	3	3	6	10	10	7	4	6	6	6

Table 11-3   Craft Requirements by Activity by Day—Target Schedule with Workdays Added

More Labor

If the contractor hires more craftspersons to increase crew size on a project activity, the crew can get more work done per unit of time and thus reduce the duration of the activity. To reduce the duration of the project, the contractor must choose a critical activity, such as Activity 600, Place Ftg Rebar & Formwork.

Select **Activity Form** from the **View** main pull-down menu to display the duration information (Figure 11-26). Change the **OD** (original duration) and **RD** (remaining duration) fields from 2 days to 1 day, assuming the contractor has added enough people to save 1 day (Figure 11-27). Also change the **EF** and **LF** fields from the end of the day 09SEP97 to 08SEP97, or a 1-day duration. Click on the **Res** (Resources) button of the **Activity** form button bar to expose the **Resources** dialog box (Figure 11-28).

Notice that *P3* automatically kept the **At completion** of 4 total units to complete the activity, but changed the **Units per day** field from 2 to 4 to complete the work. The total units expended is the same in both scenarios. In Figure 11-29, the schedule is calculated to incorporate the duration and profile changes. Compare Figure 11-29 (with Activity 600 duration change) to Figure 11-25 on p. 309 (without Activity 600 dura-

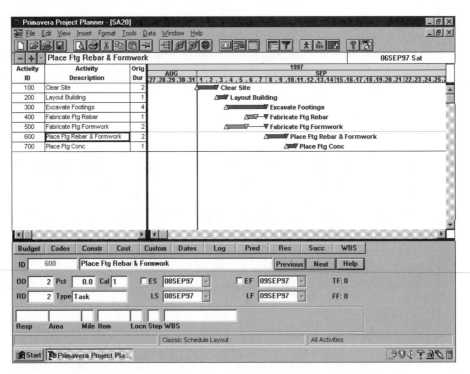

Figure 11-26   Activity Detail—Activity 600

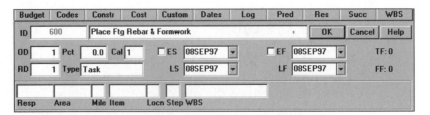

Figure 11-27   Activity Detail—Activity 600 Modified

Resources

EQUIP OP

Resource	EQUIP OP	LAB 2	CARP 1
Cost Acct/Category			
Driving			
Curve			
Units per day	4.00	4.00	4.00
Budgeted quantity	4.00	4.00	4.00
Res Lag/Duration	0	0	0
Percent complete			
Actual this period	0.00	0.00	0.00
Actual to date	0.00	0.00	0.00
To complete	4.00	4.00	4.00
At completion	4.00	4.00	4.00
Variance (units)	0.00	0.00	0.00
Early start	08SEP97	08SEP97	08SEP97
Early finish	08SEP97	08SEP97	08SEP97
Late start	08SEP97	08SEP97	08SEP97
Late finish	08SEP97	08SEP97	08SEP97

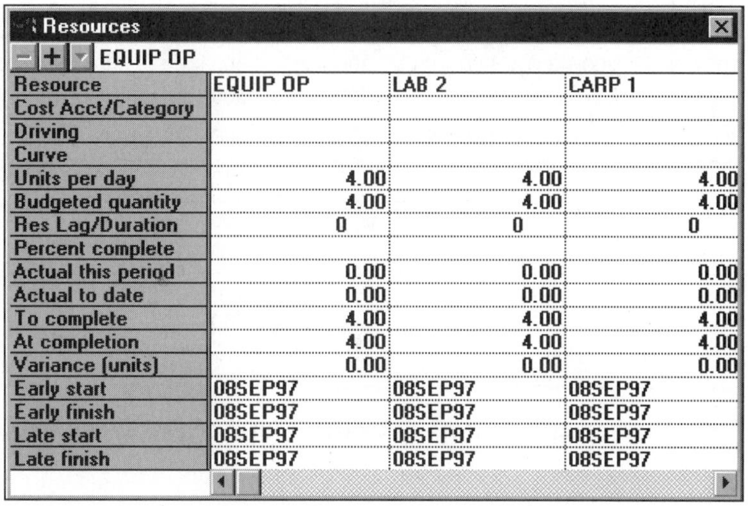

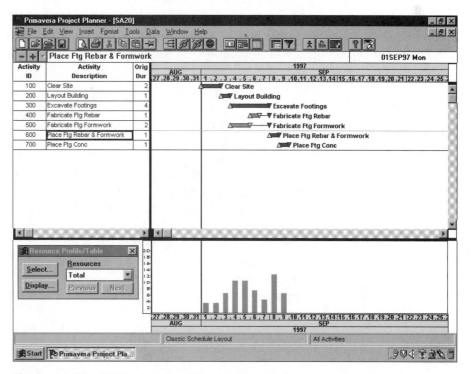

Figure 11-28 Resources Dialog Box—Activity 600

Figure 11-29 Resources Profile—Compressed

tion change). Also compare Table 11-4 with the changed duration and labor requirements for Activity 600 with Table 11-3 (p. 309). The scheduled completion date is now September 9 instead of September 10, for a one-day savings. The number of craftspersons required went from 7 to 12 on September 8. The increased labor was necessary to complete the activity in the shorter duration.

	1	2	3	4	5	6	7	8	9
CARP 1	100 – 3	100 – 3	200 – 2	300 – 2	300 – 2	300 – 2	300 – 2	600 – 4	700 – 2
				500 – 2	500 – 2				
CARP 2			200 – 2	500 – 2	500 – 2				
EQUIP OP				300 – 1	300 – 1	300 – 1	300 – 1	600 – 4	700 – 2
LAB 1			200 – 2	300 – 1	300 – 1	300 – 1	300 – 1		700 – 2
						400 – 1			
LAB 2						400 – 2		600 – 4	
				500 – 2	500 – 2				
Total	3	3	6	10	10	7	4	12	6

Table 11-4 Craft Requirements by Activity by Day—Target Schedule with Duration/Resource Change

The relationship between at completion quantity and units per day used in the preceding paragraph is the *P3* default. This relationship can be turned off by using **Autocost Rules....** From the **Tools** main pull-down menu, select **Options**, then click **Autocost Rules....** By checking **Rule #2**, the resource units per time period will be frozen.

SCHEDULE DECOMPRESSION

The same types of analysis used in schedule compression can be used in schedule decompression. The project is simply being stretched out for whatever reason rather than being compressed. If the project duration must be extended, the most orderly and economical method for doing so should be selected. Since efficient layout, organization, and use of assets are essential whether speeding up or slowing down a project, this chapter's example problem demonstrates schedule optimization.

EXAMPLE PROBLEM: Schedule Optimization

Figures 11-30 and 11-31 are hard copy prints of resource profile reports for a house put together as an example for student use (see the wood frame house drawings in the Appendix). Figures 11-32 and 11-33 are the same reports, but with resources leveled for optimization.

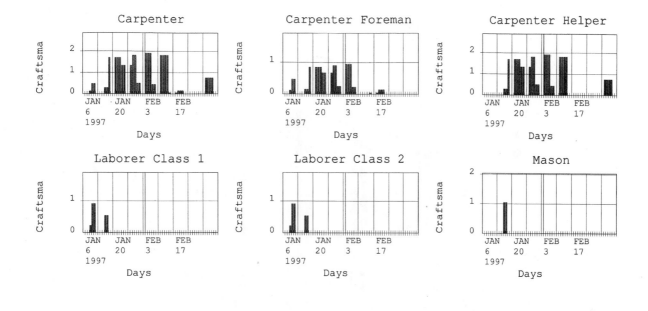

Figure 11-30   Resources Profiles, Not Leveled—Wood Frame House

Figure 11-30 is a printout of the resource profile for all resources using the RC-08 report (resource profiles). Figure 11-31 is a printout of the resource profile for all resources using the RC-02 report (total usage). Figure 11-32 is a printout of total leveled resources using the RC-08 report. Figure 11-33 is a printout of total leveled resources using the RC-02 report. Both the RC-02 and the RC-08 reports are accessed by selecting the **Graphic Reports** and then the **Resource and Cost...** options under the **Tools** main pull-down menu.

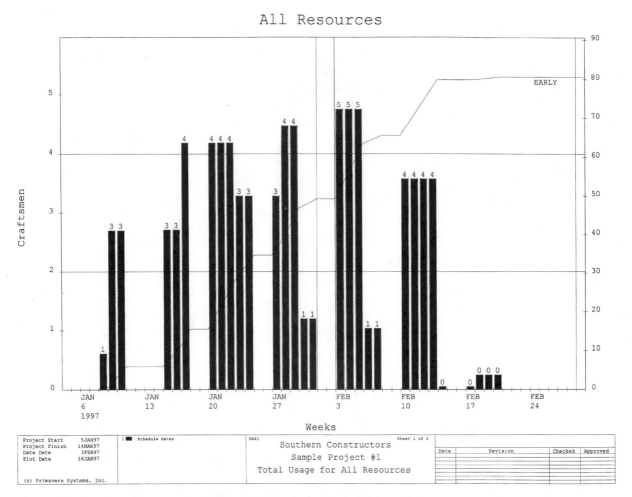

Figure 11-31 Total Resource Profile, Not Leveled—Wood Frame House

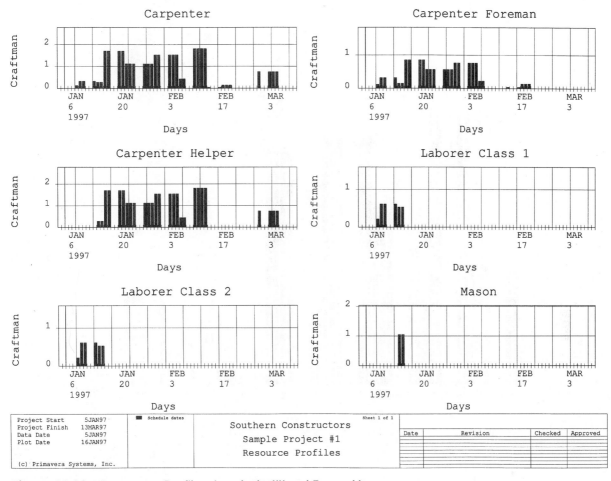

Figure 11-32   Resources Profiles, Leveled—Wood Frame House

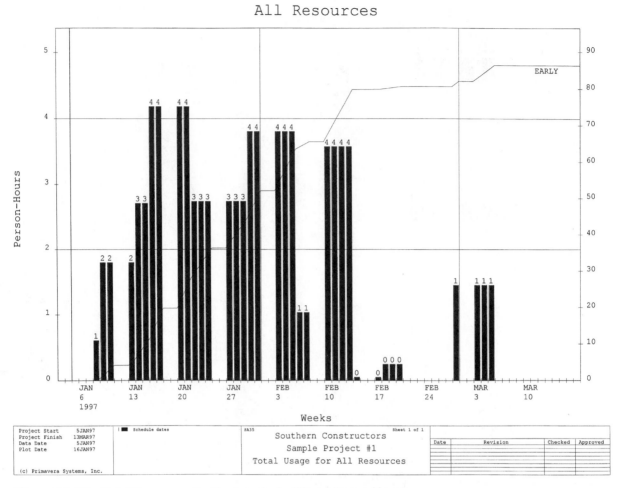

Figure 11-33   Total Resource Profile, Leveled—Wood Frame House

EXERCISES

1. Small Commercial Concrete Block Building—Optimization

Prepare the following reports for the small commercial concrete block building located in the Appendix.

1. Resource profile report
2. Total usage for all resources

2. Large Commercial Building—Optimization

Prepare the following reports for the large commercial building located in the Appendix.

1. Resource profile report
2. Total usage for all resources

Tracking Resources Using *P3*

Objectives

Upon completion of this chapter, the reader should be able to:

- Record resource usage
- Use updated resource tables
- Use updated resource profiles
- Print updated resource reports

TRACKING RESOURCES: ACTUAL VS. PLANNED EXPENDITURES

Once a project is under way, the contractor tracks (monitors) resources to compare actual costs to planned expenditures. The resources can be labor, equipment, materials, or other resources that were assigned to the activity (usually allocated from the original estimate) when the target schedule was constructed. Figure 12-1 is a copy of the target (original) schedule with the resource table in view. Table 12-1 shows the detailed craft requirements by activity by day for the target (original) schedule. The scheduled budget was used for both Figure 12-1 and Table 12-1.

In Chapter 10 the project was updated in terms of current progress. This updated schedule is called the current schedule and is compared to the target schedule to determine progress. Figure 12-2 is a copy of the current schedule. Table 12-1 shows the detailed craft requirements by activity by day for the current schedule as defined by the target schedule. The resource requirements for the target Aand current schedules are the same until actual usage is input or budgets are changed.

Only by tracking and comparing actual usage of resources to the target (original) budget can the contractor determine progress and earned value. Management must gauge the progress gained for the amount of resources expended to detect and solve potential problems of cost and time. On a cost-plus or negotiated project, the owner has access to the contractor's cost information. The resource-loaded and resource-moni-

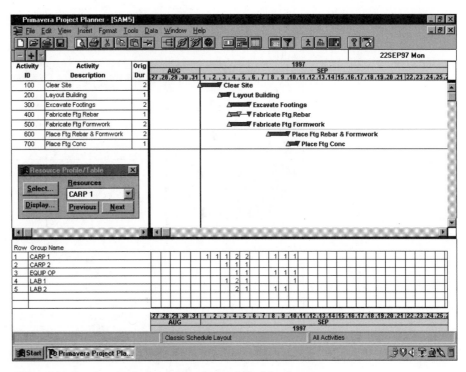

Figure 12-1   Target Schedule—Scheduled Budget

	1	2	3	4	5	6	7	8	9	10
CARP 1	100 - 1	100 - 1	200 - 1	300 - 1 500 - 1	300 - 1 500 - 1			600 - 1	600 - 1	700 - 1
CARP 2			200 - 1	500 - 1	500 - 1					
EQUIP OP				300 - 1	300 - 1			600 - 1	600 - 1	700 - 1
LAB 1			200 - 1	300 - 1 400 - 1	300 - 1					700 - 1
LAB 2				400 - 1 500 - 1	500 - 1			600 - 1	600 - 1	
Total	1	1	3	8	6	0	0	3	3	3

Table 12-1   Craft Requirements by Activity by Day—Target Schedule

tored schedule thus gives the owner valuable information for managing his or her company's time and cash.

This chapter will be used to update resources to match the updated duration, logic, and activity changes made to the schedule in Chapter 10. Notice that Figures 12-1 and 12-2 (the update) and Table 12-1 all show the same resource requirements since all three are based on the scheduled budget from the target schedule. Figure 12-3 is a copy of the current schedule with the current estimate budget selected. Table 12-2 shows the

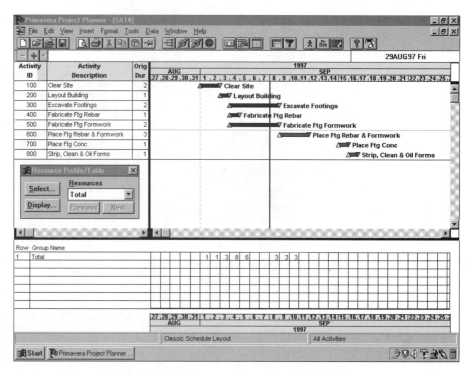

Figure 12-2 Current Schedule—Scheduled Budget

	1	2	3	4	5	6	7	8	9	10	11	12	13	14	15
CARP 1	100 - 1	100 - 1	200 - 1	300 - .67 500 - .67	300 - .67 500 - .67			300 - .67 500 - .67	600 - .67	600 - .67	600 - .67				700 - 1
CARP 2			200 - 1	500 - .67	500 - .67			500 - .67							
EQUIP OP				300 - .67	300 - .67			300 - .67	600 - .67	600 - .67	600 - .67				700 - 1
LAB 1			200 - 1	300 - .67 400 - 1	300 - .67			300 - .67							700 - 1
LAB 2				400 - 1 500 - .67	500 - .67			500 - .67	600 - .67	600 - .67	600 - .67				
Total	1	1	3	6	4	0	0	4	2	2	2				3

Table 12-2 Craft Requirements by Activity by Day—Current Schedule with Updated Durations

detailed craft requirements by activity by day for the current schedule with the current estimate budget selected. Compare Table 12-1 to Table 12-2. Notice that the scheduled budget for the target schedule for Activity 300 for CARP 1 was 1 for September 4 and 1 for September 5. When the duration for Activity 300 was changed to 3 days in the current (updated) schedule, the budget of 2 man-days was maintained. But, now the budget is for 0.67 man-days to be expended on September 4, 5, and 8, for a total of 2 man-days. So, the resource requirements for the current schedule are the same as defined in the target (original) schedule. The next step is to further update the schedule showing the *actual* expenditure of resources and to make forecasts.

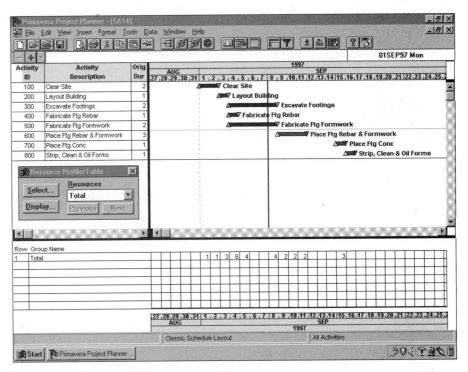

Figure 12-3   Current Schedule—Current Estimate

RESOURCE TABLE

Select the **View** pull-down menu, then **Resource Table**. If when the **Resource Profile/Table** dialog box appears at the bottom of the screen costs and not resources are shown, click on the **Display...** button. The **Resource Table Display Options** dialog box appears (Figure 12-4).

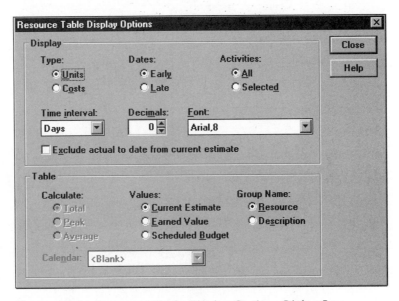

Figure 12-4   Resource Table Display Options Dialog Box

Click on the **Units** box. Make sure that the **Current Estimate** check box under **Values:** is selected. To look at the individual craft requirements, rather than total labor, select CARP 1 or any of the craft selections from the **Resources** pull-down menu in the **Resource Profile/Table** dialog box.

RECORDING EXPENDITURES

The labor resources that were input in Chapter 5 will be tracked in this section. The schedule will be updated to show actual expenditures of resources and to make forecasts.

Activity 100

Activity Form   Select the activity to be updated. In Figure 12-5, Activity 100 has been selected. Look at the **Activity** form window to check the updated information. The original duration is 2, the percent complete is 100%, the actual start is 01SEP97, and the actual finish is 02SEP97.

Resources Dialog Box   Now, return to the activity table window. Click the right mouse button (the mouse pointer must be above the **Activity** form) to expose the **Activity Detail** pull-down menu. Select **Resources**, and the **Resources** dialog box for Activity 100 will appear

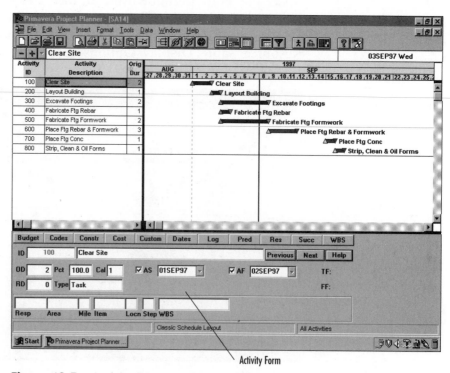

Figure 12-5   Activity Form—Activity 100

(Figure 12-6). Notice that **Budgeted quantity, Actual this period, Actual to date**, and the **At completion** fields are all showing 2 days—one CARP 1 for September 1, and one CARP 1 for September 2.

Actual Labor Expenditures   From project time sheets, actual labor expenditures and projections to completion are gathered (Table 12-3). For Activity 100, two people for two days were actually used, rather

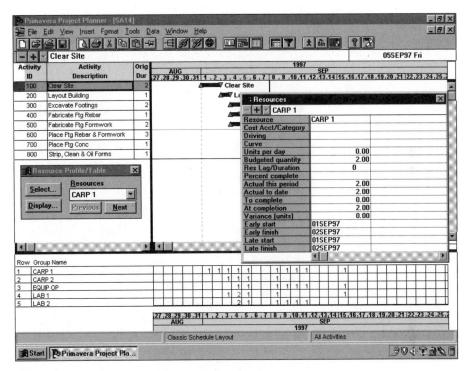

Figure 12-6   Resources Dialog Box—Activity 100

As of Data Date 08SEP97

Activity	Actual Start Date	Percent Complete	Actual Expenditures to Date					Projected Expenditures to Completion				
			CARP 1	CARP 2	EQUIP OP	LAB 1	LAB 2	CARP 1	CARP 2	EQUIP OP	LAB 1	LAB 2
100	SEP 1	100 %	4									
200	SEP 3	100 %	1	1		2						
300	SEP 4	50 %	1		1	1		1		1	3	
400	SEP 4	100 %				2	2					
500	SEP 4	50 %	1	1			1	1	1			1
600		0 %						3		3		3
700		0 %						1		1	1	
800		0 %								1	1	

Table 12-3   Craft Requirements—Actual Expenditures to Date/Projected Expenditures to Completion

than one person for two days; therefore, a total of four days of CARP 1 time was expended.

Percent complete   Click on the **Percent complete** field, and input 100 (Figure 12-7).

Actual to date   Click on the **Actual to date** field, and input 4.

At completion   Click on the **At completion** field, and input 4.

Notice when these changes are made, none of the other fields mentioned above are changed. By clicking on another activity and then clicking back on Activity 100, the change is accepted and the other fields are modified (Figure 12-8).

Budgeted quantity   The **Budgeted quantity** field remains at 2. Since this activity is 100% complete, the **Actual this period, Actual to date**, and **At completion** fields all show 4 days expended.

Variance (units)   The **Variance (units)** field at the bottom of the dialog box has a –2, representing two days overbudget. Since the activity is 100% complete, the **Units per day** field is 0. When we look at an activity with no progress later in the chapter, this field will still have the numbers input in Chapter 5.

Resource Table   Notice also, that the resource table window for Activity 100 has been changed, and 2 CARP 1s are required for September 1 and 2.

Resources		✕
− + ∨ ↕ 4.00		
Resource	CARP 1	
Cost Acct/Category		
Driving		
Curve		
Units per day	0.00	
Budgeted quantity	2.00	
Res Lag/Duration	0	
Percent complete	100.0	
Actual this period	2.00	
Actual to date	4.00	
To complete	0.00	
At completion	4.00	
Variance (units)	0.00	
Early start	01SEP97	
Early finish	02SEP97	
Late start	01SEP97	
Late finish	02SEP97	

Figure 12-7   Resources Dialog Box—Activity 100—Percent Complete

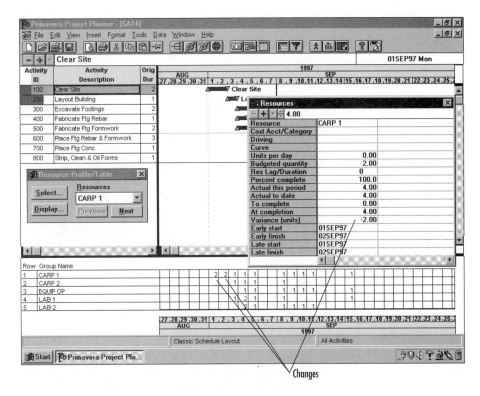

Figure 12-8   Resources Dialog Box—Activity 100 Variance

Activity 200

Activity Form   Click on the next activity to be updated, Activity 200 (Figure 12-9). Look at the **Activity** form window to check the updated information. The original duration is 1, the percent complete is 100%, the actual start is the beginning of the day 03SEP97, and the actual finish is the end of the day 03SEP97.

Resources Dialog Box   Return to the **Resources** dialog box (Figure 12-10).

Actual Labor Expenditures   The budgeted quantity of 1 for the CARP 1 and CARP 2 was correct according to cost information received (Table 12-3, p. 323), but 2 man-days for LAB 1 was expended.

Actual to Date   The **Actual to date** field is changed to 2 for LAB 1.

At completion   The **At completion** field is changed to 2 for LAB 1.

Percent complete   The **Percent complete** fields for all three crafts were changed to 100.

Resource Table   Now look at Figure 12-11 and notice that when the updated resource requirements for this activity are accepted, the new

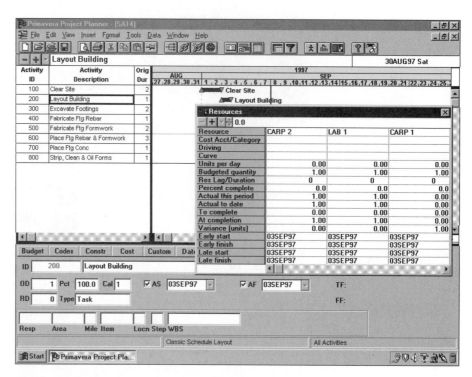

Figure 12-9 Activity Form—Activity 200

Resources			
− + ⌄ ⌵ 100.0			
Resource	CARP 2	LAB 1	CARP 1
Cost Acct/Category			
Driving			
Curve			
Units per day	0.00	0.00	0.00
Budgeted quantity	1.00	1.00	1.00
Res Lag/Duration	0	0	0
Percent complete	100.0	100.0	100.0
Actual this period	1.00	2.00	1.00
Actual to date	1.00	2.00	1.00
To complete	0.00	0.00	0.00
At completion	1.00	2.00	1.00
Variance (units)	0.00	-1.00	0.00
Early start	03SEP97	03SEP97	03SEP97
Early finish	03SEP97	03SEP97	03SEP97
Late start	03SEP97	03SEP97	03SEP97
Late finish	03SEP97	03SEP97	03SEP97

Figure 12-10 Resources Dialog Box—Activity 200 Percent Complete

resource table window shows that requirements for September 3 went from 1 to 2 LAB 1s.

Activity 300

Activity Form Click on the next activity to be updated, Activity 300 (Figure 12-12). Look at the **Activity** form window to check the updated

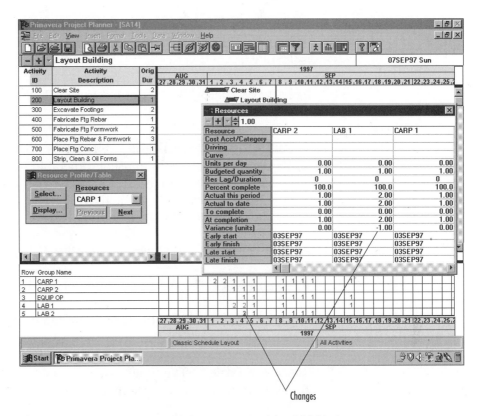

Figure 12-11   Resources Dialog Box—Activity 200 Variance

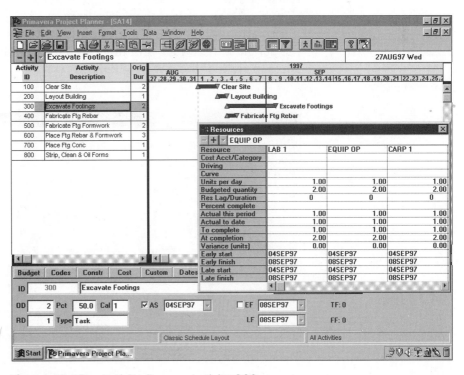

Figure 12-12   Activity Form—Activity 300

information. The original duration is 2, the remaining duration is 1, the percent complete is 50%, the actual start is 04SEP97, and the early finish is 08SEP97.

Resources Dialog Box   Return to the **Resources** dialog box (Figure 12-13).

Actual Labor Expenditures   Since this activity is not complete, the **Units per day** field is still in use. From Table 12-3 (p. 323), the budgeted quantity of 1 for LAB 1 has already been expended. The projection is for 3 more units to complete the activity. For EQUIP OP and CARP 1, the original usage projections are holding up.

Percent complete   The **Percent complete** fields are changed to 50.

At completion   The **At completion** field is changed to 4 for LAB 1.

Variance   Notice that when this is updated the **Variance (units)** field for LAB 1 changes to –2.

Resource Table   The new resource table window (Figure 12-14) shows requirements for September 8 went from 1 to 3 LAB 1s.

Activity 400

Activity Form   Click on the next activity to be updated, Activity 400 (Figure 12-15). Look at the **Activity** form window to check the updated information. The original duration is 1, the percent complete is 100%, the actual start is the beginning of the day 04SEP97, and the actual finish is the end of the day 04SEP97.

Resources			
− + ▾ ⬍ 3.00			
Resource	LAB 1	EQUIP OP	CARP 1
Cost Acct/Category			
Driving			
Curve			
Units per day	3.00	1.00	1.00
Budgeted quantity	2.00	2.00	2.00
Res Lag/Duration	0	0	0
Percent complete	50.0	50.0	50.0
Actual this period	1.00	1.00	1.00
Actual to date	1.00	1.00	1.00
To complete	3.00	1.00	1.00
At completion	4.00	2.00	2.00
Variance (units)	-2.00	0.00	0.00
Early start	04SEP97	04SEP97	04SEP97
Early finish	08SEP97	08SEP97	08SEP97
Late start	04SEP97	04SEP97	04SEP97
Late finish	08SEP97	08SEP97	08SEP97

Figure 12-13   Resources Dialog Box—Activity 300 Percent Complete

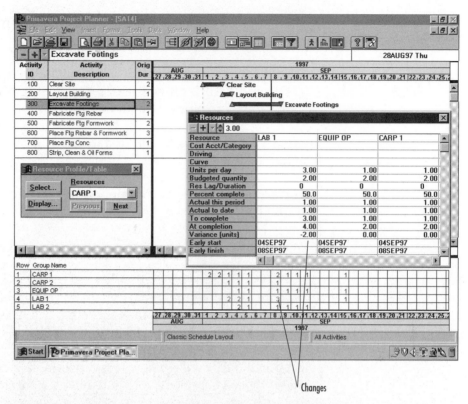

Figure 12-14   Resources Dialog Box—Activity 300 Variance

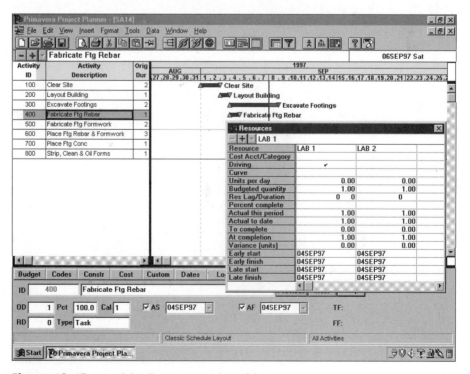

Figure 12-15   Activity Form—Activity 400

Resources Dialog Box Return to the **Resources** dialog box (Figure 12-16).

Actual Labor Expenditures The budgeted quantity of 1 unit for the LAB 1 and LAB 2 proved not to be enough. The actual expenditures were 2 units each for LAB 1 and LAB 2.

Actual to date The **Actual to date** fields were changed to 2 for LAB 1 and LAB 2.

At completion The **At completion** fields were changed to 2 for LAB 1 and LAB 2.

Percent complete The **Percent complete** fields for both crafts were changed to 100.

Variance (units) Notice that when this is updated, resource requirements for this activity are accepted. There is a –1 in the **Variance (units)** fields for both LAB 1 and LAB 2.

Resource Table The new resource table window (Figure 12-17) shows that requirements for September 4 went from 2 to 3 LAB 1s and 2 to 3 LAB 2s.

Activity 500

Activity Form Click on the next activity to be updated, Activity 500 (Figure 12-18). Look at the **Activity** form window to check the updated information. The original duration is 2, the remaining duration is 1, the

Resources			
– + ▾ LAB 1			
Resource	LAB 1	LAB 2	
Cost Acct/Category			
Driving	✔		
Curve			
Units per day	0.00	0.00	
Budgeted quantity	1.00	1.00	
Res Lag/Duration	0 0	0	
Percent complete	100.0	100.0	
Actual this period	2.00	2.00	
Actual to date	2.00	2.00	
To complete	0.00	0.00	
At completion	2.00	2.00	
Variance (units)	-1.00	-1.00	
Early start	04SEP97	04SEP97	
Early finish	04SEP97	04SEP97	
Late start	04SEP97	04SEP97	
Late finish	04SEP97	04SEP97	

Figure 12-16 Resources Dialog Box—Activity 400 Percent Complete

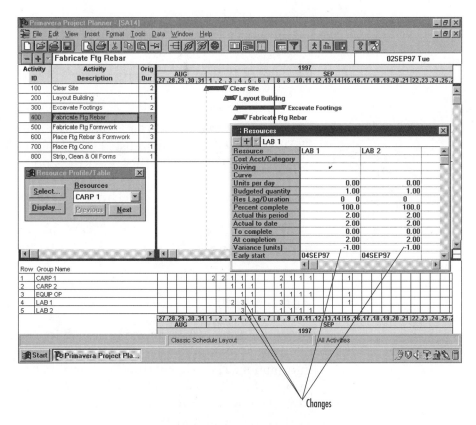

Figure 12-17   Resources Dialog Box—Activity 400 Variance

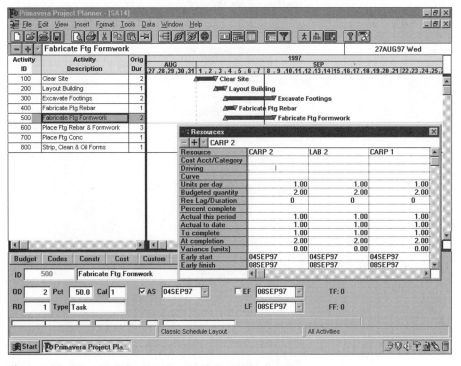

Figure 12-18   Activity Form—Activity 500

percent complete is 50%, the actual start is 04SEP97, and the early finish is 08SEP97.

Resources Dialog Box   Return to the **Resources** dialog box (Figure 12-19).

Actual Labor Expenditures   Since this activity is not complete, the **Units per day** field is still in use. From Table 12-3 (p. 323), the budgeted quantity fields of 2 for CARP 1, CARP 2, and LAB 2 are good.

Percent complete   The only change is to modify the **Percent complete** fields to 50.

Resource Table   The new resource table window shows no changes.

Activity 600

Activity Form   Click on the next activity to be updated, Activity 600 (Figure 12-20). Look at the **Activity** form window to check the activity information. The original duration input in Chapter 5 was 2. This activity was changed to a 3-day duration in Chapter 10 in the updating exercise. The remaining duration is 3, the percent complete is 0%, the early start is 09SEP97, and the early finish is 11SEP97. The **Units per day**

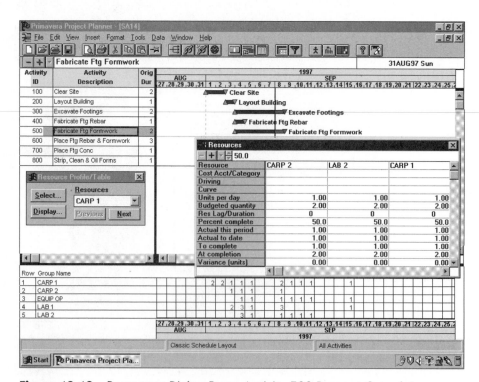

Figure 12-19   Resources Dialog Box—Activity 500 Percent Complete

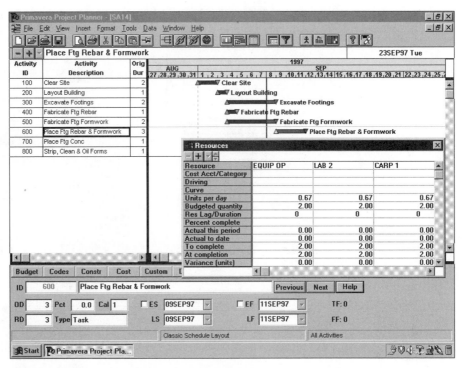

Figure 12-20   Activity Form—Activity 600

fields are set at 0.67, and the **At completion** fields are at 2 for EQUIP OP, LAB 2, and CARP 1.

Resources Dialog Box   Return to the **Resources** dialog box (Figure 12-21).

Actual Labor Expenditure   Since no progress is claimed for this activity the **Percent complete** field is left at 0.

Units per day   The **Units per day** field was changed to 1, which caused the **Budgeted quantity** and the **At completion** fields to change to 3.

Resource Table   The resource table and window shows 1 EQUIP OP, 1 LAB 2, and 1 CARP 1 for September 9, 10, 11. This would equal 3 units each and not 2. *P3* has rounded the 0.67 units to 1 for each day. Notice that with this change, the resource table window did not change.

Activity 700

Click on the next activity, Activity 700 (Figure 12-22). Since no progress is claimed for this activity the **Percent complete** field is left at 0. There are no changes.

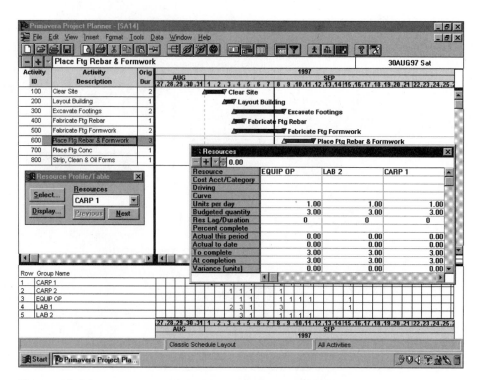

Figure 12-21 Resources Dialog Box—Activity 600 Units per Day

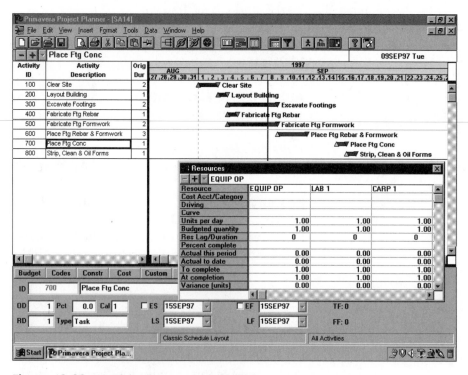

Figure 12-22 Activity Form—Activity 700

Activity 800

Activity Form   Click on the last activity to be updated, Activity 800 (Figure 12-23). Look at the **Activity** form window to check the activity information. This activity was added in the update to the schedule in Chapter 10. The original duration shown is 1. The percent complete is 0%, the early start is the beginning of the day 16SEP97, and the early finish is the end of the day 16SEP97.

Resources Dialog Box   Return to the **Resources** dialog box (Figure 12-24).

Input budget   In Chapter 10, no resources were added when activity 800 was added to the schedule.

Units per day   The **Units per day** field was set at 1 for EQUIP OP and LAB 1.

Budgeted quantity   The **Budgeted quantity** field was set at 1 for EQUIP OP and LAB 1.

To complete   The **To complete** field was set at 1 for EQUIP OP and LAB 1.

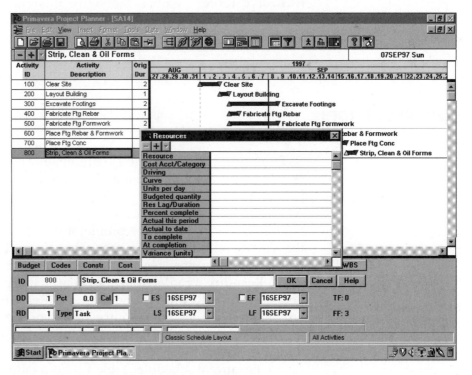

Figure 12-23   Activity Form—Activity 800

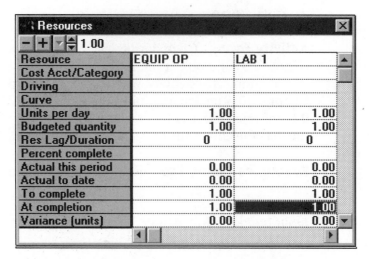

Figure 12-24   Resources Dialog Box—Activity 800, Units per Day

At completion   The **At completion** field was set at 1 for EQUIP OP and LAB 1.

Percent complete   Since no progress is claimed for this activity, the **Percent complete** field is left at 0.

Resource Table   Notice that with this change, the resource table window now shows 1 EQUIP OP and 1 LAB 1 for September 16 (Figure 12-25).

RESOURCE PROFILE

So far in this chapter, resource requirements have been presented in the on-screen representation. Sometimes a graphical representation is more beneficial.

Figure 12-26 is the resource profile of the current schedule. It includes the activity and duration update changes made in Chapter 10 and the resource update changes made in this chapter. To view this screen, select **Resource Profile** from the **View** main pull-down menu. Select **Total** from the **Resources** menu of the **Resource Profile/Table** dialog box. Figure 12-27 is the resource profile of the target schedule as put together in Chapter 5. Compare Figure 12-26 to Figure 12-27 to determine the change in labor requirements from the target (original) schedule to the current schedule. Table 12-4 is a tabular representation of the resource requirements of the current schedule.

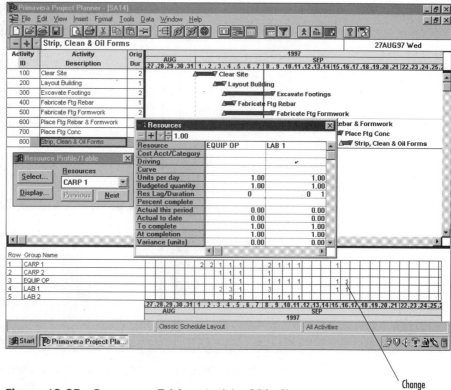

Figure 12-25   Resources Table—Activity 800, Changes

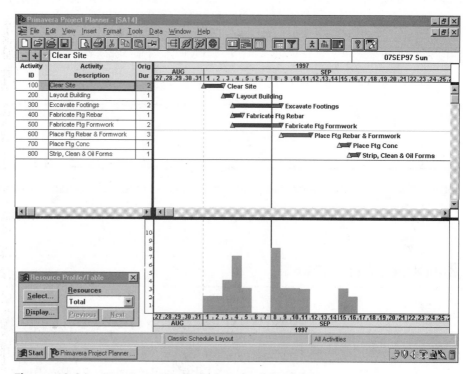

Figure 12-26   Resource Profile—Current Schedule

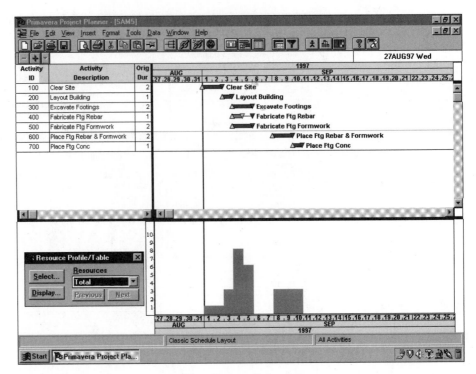

Figure 12-27 Resource Profile—Target Schedule

	1	2	3	4	5	6	7	8	9	10	11	12	13	14	15	16
CARP 1	100 - 2	100 - 2	200 - 1	300 - 1	500 - 1			300 - 1 500 - 1	600 - 1	600 - 1	600 - 1				700 - 1	
CARP 2			200 - 1		500 - 1			500 - 1								
EQUIP OP				300 - 1				300 - 1	600 - 1	600 - 1	600 - 1				700 - 1	800 - 1
LAB 1			200 - 2	300 - 1 400 - 2				300 - 3							700 - 1	800 - 1
LAB 2				400 - 2	500 - 1			500 - 1	600 - 1	600 - 1	600 - 1					
Total	2	2	4	7	3	0	0	8	3	3	3	0	0	0	3	2

Table 12-4 Craft Requirements by Activity by Day—Current Schedule with Updated Resources

RESOURCE REPORTS

Evaluating Resource Usage by Activity

The best reports for evaluating resource usage are the resource control reports. Select **Tabular Reports** from the **Tools** main pull-down menu. Then select **Resource** and then the **Control...** options. The **Resource**

Control Reports title selection box will appear (Figure 12-28). Select **RC-01, Resource Control - Detail By Activity**, and click on the <u>R</u>un button. Figure 12-29 is a printout of the resource control activity report. The information given for each activity, using Activity 100 as an example, is as follows:

Activity ID:	100
Activity Description:	Clear Site
Remaining Duration:	0
Actual Start:	1SEP97
Actual Finish:	2SEP97
Resource:	CARP 1
Unit of Measure:	Days
Budget:	2
Percent Complete:	100
Actual to Date:	4
Actual this Period:	4
Estimate to Complete:	0
Forecast:	4
Variance:	–2

Evaluating Resource Usage by Resource

Another very useful report is the resource control report sorted by resource. Select **RC-04, Resource Control - Detail By Resource**, and click on the <u>R</u>un button (Figure 12-30). Figure 12-31 is a printout of the resource control report. This report gives most of the information in the report above, except it is sorted by resource. There is no remaining duration, actual start, or actual finish provided in this report (which were provided in Figure 12-29).

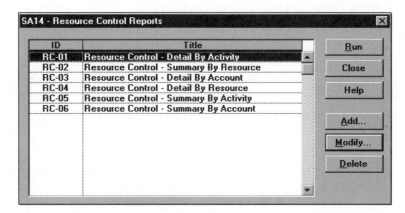

Figure 12-28   Resource Control Reports Title Selection Box—Resource Control - Detail by Activity

```
------------------------------------------------------------------------------------------------------------
Southern Constructors               PRIMAVERA PROJECT PLANNER          Sample Schedule

REPORT DATE 23JAN97  RUN NO.  16     RESOURCE CONTROL ACTIVITY REPORT   START DATE 1SEP97  FIN DATE 11SEP97
            10:17
Resource Control - Detail By Activity                                  DATA DATE  8SEP97    PAGE NO.   1
------------------------------------------------------------------------------------------------------------
                  COST      ACCOUNT  UNIT          PCT    ACTUAL    ACTUAL    ESTIMATE TO
ACTIVITY ID RESOURCE ACCOUNT CATEGORY MEAS  BUDGET  CMP   TO DATE  THIS PERIOD COMPLETE   FORECAST   VARIANCE
------------------------------------------------------------------------------------------------------------
       100  Clear Site
            RD   0 AS  1SEP97 AF  2SEP97

            CARP 1               Days   2.00 100.0    4.00      4.00       .00       4.00     -2.00
                                      ----------- ----- --------- --------- ---------- ---------- --------
            TOTAL :                     2.00 200.0    4.00      4.00       .00       4.00     -2.00

       200  Layout Building
            RD   0 AS  3SEP97 AF  3SEP97

            CARP 2               Days   1.00 100.0    1.00      1.00       .00       1.00       .00
            LAB 1                Days   1.00 100.0    2.00      2.00       .00       2.00     -1.00
            CARP 1               Days   1.00 100.0    1.00      1.00       .00       1.00       .00
                                      ----------- ----- --------- --------- ---------- ---------- --------
            TOTAL :                     3.00 133.3    4.00      4.00       .00       4.00     -1.00

       300  Excavate Footings
            RD   1 AS  4SEP97 EF  8SEP97     LF  8SEP97 TF   0

            LAB 1                Days   2.00  50.0    1.00      1.00      3.00       4.00     -2.00
            EQUIP OP             Days   2.00  50.0    1.00      1.00      1.00       2.00       .00
            CARP 1               Days   2.00  50.0    1.00      1.00      1.00       2.00       .00
                                      ----------- ----- --------- --------- ---------- ---------- --------
            TOTAL :                     6.00  50.0    3.00      3.00      5.00       8.00     -2.00

       400  Fabricate Ftg Rebar
            RD   0 AS  4SEP97 AF  4SEP97

            LAB 1                Days   1.00 100.0    2.00      2.00       .00       2.00     -1.00
            LAB 2                Days   1.00 100.0    2.00      2.00       .00       2.00     -1.00
                                      ----------- ----- --------- --------- ---------- ---------- --------
            TOTAL :                     2.00 200.0    4.00      4.00       .00       4.00     -2.00

       500  Fabricate Ftg Formwork
            RD   1 AS  4SEP97 EF  8SEP97     LF  8SEP97 TF   0

            CARP 2               Days   2.00  50.0    1.00      1.00      1.00       2.00       .00
            LAB 2                Days   2.00  50.0    1.00      1.00      1.00       2.00       .00
            CARP 1               Days   2.00  50.0    1.00      1.00      1.00       2.00       .00
                                      ----------- ----- --------- --------- ---------- ---------- --------
            TOTAL :                     6.00  50.0    3.00      3.00      3.00       6.00       .00

       600  Place Ftg Rebar & Formwork
            RD   3 ES  9SEP97 EF 11SEP97 LS  9SEP97 LF 11SEP97 TF   0

            EQUIP OP             Days   3.00   .0     .00       .00      3.00       3.00       .00
            LAB 2                Days   3.00   .0     .00       .00      3.00       3.00       .00
            CARP 1               Days   3.00   .0     .00       .00      3.00       3.00       .00
                                      ----------- ----- --------- --------- ---------- ---------- --------
            TOTAL :                     9.00   .0     .00       .00      9.00       9.00       .00

       700  Place Ftg Conc
            RD   1 ES 15SEP97 EF 15SEP97 LS 15SEP97 LF 15SEP97 TF   0

            EQUIP OP             Days   1.00   .0     .00       .00      1.00       1.00       .00
            LAB 1                Days   1.00   .0     .00       .00      1.00       1.00       .00
            CARP 1               Days   1.00   .0     .00       .00      1.00       1.00       .00
                                      ----------- ----- --------- --------- ---------- ---------- --------
            TOTAL :                     3.00   .0     .00       .00      3.00       3.00       .00

       800  Strip, Clean & Oil Forms
            RD   1 ES 16SEP97 EF 16SEP97 LS 16SEP97 LF 16SEP97 TF   0

            EQUIP OP             Days   1.00   .0     .00       .00      1.00       1.00       .00
            LAB 1                Days   1.00   .0     .00       .00      1.00       1.00       .00
                                      ----------- ----- --------- --------- ---------- ---------- --------
            TOTAL :                     2.00   .0     .00       .00      2.00       2.00       .00

                                      ----------- ----- --------- --------- ---------- ---------- --------
            REPORT TOTALS              33.00  54.5   18.00     18.00     22.00      40.00     -7.00
------------------------------------------------------------------------------------------------------------
```

Figure 12-29 Resource Control Activity Report Print—Detail by Activity

Once a project is under way, tracking (monitoring) resources and comparing actual to planned expenditures are tools in controlling the use of resources. Updating the resource requirements and comparing actual usage back to the original schedule is a valuable tool in determining physical progress and its relationship to the original budget. Observe

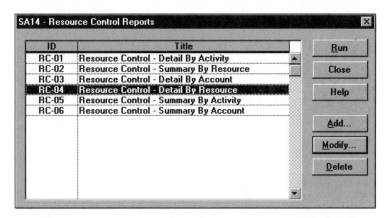

Figure 12-30   Resource Control Reports Title Selection Box—Resource Control - Detail by Resource

the report totals line at the bottom of the resource control report. The budget (target) is 33 man-days (unit meas). The actual to date (expenditures) is 18 man-days, or 54.5 percent complete in terms of the target budget. The estimate to complete is 22 man-days. The forecast at completion is 40 man-days. The variance (actual compared to the target budget) is a –7, or 40 (forecast) minus 33 (target budget). The project is forecast to overrun the target budget by 7 man-days.

Accurate and timely cost/schedule information is essential in controlling project time and resources.

EXAMPLE PROBLEM: TRACKING RESOURCES

Figures 12-32a, b, c, and d are printouts of the current schedule with updated durations (Chapter 10) of the resource control report showing detail by activity (RC-01) for the wood frame house located in the Appendix.

Figures 12-33a, b, and c are printouts of the current schedule with updated resources of the resource control report showing detail by activity (RC-01) for the wood frame house located in the Appendix.

```
--------------------------------------------------------------------------------
Southern Constructors                    PRIMAVERA PROJECT PLANNER          Sample Schedule

REPORT DATE  23JAN97  RUN NO.   19          RESOURCE CONTROL REPORT         START DATE  1SEP97  FIN DATE 11SEP97
             10:21
Resource Control - Detail By Resource                                       DATA DATE   8SEP97  PAGE NO.    1
--------------------------------------------------------------------------------
                          COST      ACCOUNT  UNIT            PCT    ACTUAL    ACTUAL     ESTIMATE TO
ACTIVITY ID RESOURCE      ACCOUNT   CATEGORY MEAS   BUDGET   CMP    TO DATE   THIS PERIOD COMPLETE  FORECAST   VARIANCE
----------- --------      -------   -------- ----   ------   ----   -------   ----------- --------  --------   --------

            CARP 1    - Carpenter - Level 1

        100 CARP 1                          Days    2.00  100.0     4.00      4.00        .00       4.00      -2.00
        200 CARP 1                          Days    1.00  100.0     1.00      1.00        .00       1.00        .00
        300 CARP 1                          Days    2.00   50.0     1.00      1.00       1.00       2.00        .00
        500 CARP 1                          Days    2.00   50.0     1.00      1.00       1.00       2.00        .00
        600 CARP 1                          Days    3.00     .0      .00       .00       3.00       3.00        .00
        700 CARP 1                          Days    1.00     .0      .00       .00       1.00       1.00        .00
                                                  -----  ----- ------------ ------------ -------- -------- ------------
      TOTAL CARP 1                          Days   11.00   63.6     7.00      7.00       6.00      13.00      -2.00
                                                  ========================================================================

            CARP 2    - Carpenter - Level 2

        200 CARP 2                          Days    1.00  100.0     1.00      1.00        .00       1.00        .00
        500 CARP 2                          Days    2.00   50.0     1.00      1.00       1.00       2.00        .00
                                                  -----  ----- ------------ ------------ -------- -------- ------------
      TOTAL CARP 2                          Days    3.00   66.7     2.00      2.00       1.00       3.00        .00
                                                  ========================================================================

            EQUIP OP - Equipment Operator

        300 EQUIP OP                        Days    2.00   50.0     1.00      1.00       1.00       2.00        .00
        600 EQUIP OP                        Days    3.00     .0      .00       .00       3.00       3.00        .00
        700 EQUIP OP                        Days    1.00     .0      .00       .00       1.00       1.00        .00
        800 EQUIP OP                        Days    1.00     .0      .00       .00       1.00       1.00        .00
                                                  -----  ----- ------------ ------------ -------- -------- ------------
      TOTAL EQUIP OP                        Days    7.00   14.3     1.00      1.00       6.00       7.00        .00
                                                  ========================================================================

            LAB 1     - Laborer - Level 1

        200 LAB 1                           Days    1.00  100.0     2.00      2.00        .00       2.00      -1.00
        300 LAB 1                           Days    2.00   50.0     1.00      1.00       3.00       4.00      -2.00
        400 LAB 1                           Days    1.00  100.0     2.00      2.00        .00       2.00      -1.00
        700 LAB 1                           Days    1.00     .0      .00       .00       1.00       1.00        .00
        800 LAB 1                           Days    1.00     .0      .00       .00       1.00       1.00        .00
                                                  -----  ----- ------------ ------------ -------- -------- ------------
      TOTAL LAB 1                           Days    6.00   83.3     5.00      5.00       5.00      10.00      -4.00
                                                  ========================================================================

            LAB 2     - laborer - Level 2

        400 LAB 2                           Days    1.00  100.0     2.00      2.00        .00       2.00      -1.00
        500 LAB 2                           Days    2.00   50.0     1.00      1.00       1.00       2.00        .00
        600 LAB 2                           Days    3.00     .0      .00       .00       3.00       3.00        .00
                                                  -----  ----- ------------ ------------ -------- -------- ------------
      TOTAL LAB 2                           Days    6.00   50.0     3.00      3.00       4.00       7.00      -1.00
                                                  ========================================================================

                  REPORT TOTALS                   33.00   54.5    18.00     18.00      22.00      40.00      -7.00
                                                  ========================================================================
```

Figure 12-31 Resource Control Activity Report Print—Detail By Resource

```
--------------------------------------------------------------------------------------------------
Student Constructors                    PRIMAVERA PROJECT PLANNER        Sample Project #1

REPORT DATE 26JAN97 RUN NO.  27         RESOURCE CONTROL ACTIVITY REPORT   START DATE  5JAN97  FIN DATE 13MAR97
            8:30
Resource Control - Detail By Activity                                     DATA DATE  5JAN97   PAGE NO.   1
--------------------------------------------------------------------------------------------------
```

ACTIVITY ID	RESOURCE	COST ACCOUNT	ACCOUNT CATEGORY	UNIT MEAS	BUDGET	PCT CMP	ACTUAL TO DATE	ACTUAL THIS PERIOD	ESTIMATE TO COMPLETE	FORECAST	VARIANCE
100	Clear Site										
	RD 2 ES 6JAN97 EF 7JAN97 LS 6JAN97 LF 7JAN97 TF 0										
	SUB		S SUBCNT'R		.00	.0	.00	.00	.00	.00	.00
	TOTAL :				.00	.0	.00	.00	.00	.00	.00
200	Building Layout										
	RD 1 ES 8JAN97 EF 8JAN97 LS 8JAN97 LF 8JAN97 TF 0										
	CARPENTR		L LABOR	day	.10	.0	.00	.00	.10	.10	.00
	CRPN FOR		L LABOR	day	.10	.0	.00	.00	.10	.10	.00
	LAB CL 1		L LABOR	day	.20	.0	.00	.00	.20	.20	.00
	LAB CL 2		L LABOR	day	.20	.0	.00	.00	.20	.20	.00
	MATERIAL		M MATERIAL		.00	.0	.00	.00	.00	.00	.00
	SUB		S SUBCNT'R		.00	.0	.00	.00	.00	.00	.00
	TOTAL :				.60	.0	.00	.00	.60	.60	.00
300	Form/Pour Footings										
	RD 3 ES 9JAN97 EF 13JAN97 LS 9JAN97 LF 13JAN97 TF 0										
	CARPENTR		L LABOR	day	.90	.0	.00	.00	.90	.90	.00
	CRPN FOR		L LABOR	day	.90	.0	.00	.00	.90	.90	.00
	LAB CL 1		L LABOR	day	1.79	.0	.00	.00	1.79	1.79	.00
	LAB CL 2		L LABOR	day	1.79	.0	.00	.00	1.79	1.79	.00
	MATERIAL		M MATERIAL		.00	.0	.00	.00	.00	.00	.00
	SUB		S SUBCNT'R		.00	.0	.00	.00	.00	.00	.00
	TOTAL :				5.38	.0	.00	.00	5.38	5.38	.00
400	Pier Masonry										
	RD 2 ES 14JAN97 EF 15JAN97 LS 14JAN97 LF 15JAN97 TF 0										
	CARPENTR		L LABOR	day	.51	.0	.00	.00	.51	.51	.00
	CRPN FOR		L LABOR	day	.26	.0	.00	.00	.26	.26	.00
	CRPN HLP		L LABOR	day	.51	.0	.00	.00	.51	.51	.00
	LAB CL 1		L LABOR	day	1.03	.0	.00	.00	1.03	1.03	.00
	LAB CL 2		L LABOR	day	1.03	.0	.00	.00	1.03	1.03	.00
	MASON		L LABOR	day	2.06	.0	.00	.00	2.06	2.06	.00
	MATERIAL		M MATERIAL		.00	.0	.00	.00	.00	.00	.00
	TOTAL :				5.40	.0	.00	.00	5.40	5.40	.00
500	Wood Floor System										
	RD 4 ES 16JAN97 EF 21JAN97 LS 16JAN97 LF 21JAN97 TF 0										
	CARPENTR		L LABOR	day	6.68	.0	.00	.00	6.68	6.68	.00
	CRPN FOR		L LABOR	day	3.34	.0	.00	.00	3.34	3.34	.00
	CRPN HLP		L LABOR	day	6.68	.0	.00	.00	6.68	6.68	.00
	MATERIAL		M MATERIAL		.00	.0	.00	.00	.00	.00	.00
	TOTAL :				16.70	.0	.00	.00	16.70	16.70	.00
600	Rough Framing Walls										
	RD 6 ES 22JAN97 EF 29JAN97 LS 22JAN97 LF 29JAN97 TF 0										
	CARPENTR		L LABOR	day	6.55	.0	.00	.00	6.55	6.55	.00
	CRPN FOR		L LABOR	day	3.28	.0	.00	.00	3.28	3.28	.00
	CRPN HLP		L LABOR	day	6.55	.0	.00	.00	6.55	6.55	.00
	MATERIAL		M MATERIAL		.00	.0	.00	.00	.00	.00	.00
	TOTAL :				16.38	.0	.00	.00	16.38	16.38	.00
700	Rough Framing Roof										
	RD 4 ES 30JAN97 EF 4FEB97 LS 30JAN97 LF 4FEB97 TF 0										
	CARPENTR		L LABOR	day	7.51	.0	.00	.00	7.51	7.51	.00
	CRPN FOR		L LABOR	day	3.72	.0	.00	.00	3.72	3.72	.00

Figure 12-32a   Resource Control Activity Report—Wood Frame House, page 1

```
-------------------------------------------------------------------------------------------------------------
Student Constructors                    PRIMAVERA PROJECT PLANNER           Sample Project #1

REPORT DATE  26JAN97  RUN NO.   27      RESOURCE CONTROL ACTIVITY REPORT    START DATE  5JAN97  FIN DATE 13MAR97
             8:30
Resource Control - Detail By Activity                                      DATA DATE   5JAN97    PAGE NO.    2
-------------------------------------------------------------------------------------------------------------
```

ACTIVITY ID	RESOURCE	COST ACCOUNT	ACCOUNT CATEGORY	UNIT MEAS	BUDGET	PCT CMP	ACTUAL TO DATE	ACTUAL THIS PERIOD	ESTIMATE TO COMPLETE	FORECAST	VARIANCE
	CRPN HLP		L LABOR	day	7.59	.0	.00	.00	7.59	7.59	.00
	D, 50 HP		L LABOR	day	.16	.0	.00	.00	.16	.16	.00
	MATERIAL		M MATERIAL		.00	.0	.00	.00	.00	.00	.00
	TOTAL :				18.98	.0	.00	.00	18.98	18.98	.00
800	Doors & Windows										
	RD 4 ES 7FEB97 EF 12FEB97 LS 14FEB97 LF 19FEB97 TF 5										
	CARPENTR		L LABOR	day	7.13	.0	.00	.00	7.13	7.13	.00
	CRPN HLP		L LABOR	day	7.13	.0	.00	.00	7.13	7.13	.00
	MATERIAL		M MATERIAL		.00	.0	.00	.00	.00	.00	.00
	SUB		S SUBCNT'R		.00	.0	.00	.00	.00	.00	.00
	TOTAL :				14.26	.0	.00	.00	14.26	14.26	.00
900	Ext Wall Board										
	RD 2 ES 5FEB97 EF 6FEB97 LS 6FEB97 LF 7FEB97 TF 1										
	CARPENTR		L LABOR	day	.82	.0	.00	.00	.82	.82	.00
	CRPN FOR		L LABOR	day	.41	.0	.00	.00	.41	.41	.00
	CRPN HLP		L LABOR	day	.82	.0	.00	.00	.82	.82	.00
	MATERIAL		M MATERIAL		.00	.0	.00	.00	.00	.00	.00
	TOTAL :				2.05	.0	.00	.00	2.05	2.05	.00
1000	Ext Wall Insulation										
	RD 1 ES 13FEB97 EF 13FEB97 LS 13FEB97 LF 13FEB97 TF 0										
	SUB		S SUBCNT'R		.00	.0	.00	.00	.00	.00	.00
	TOTAL :				.00	.0	.00	.00	.00	.00	.00
1100	Rough Plumbing										
	RD 4 ES 7FEB97 EF 12FEB97 LS 10FEB97 LF 13FEB97 TF 1										
	SUB		S SUBCNT'R		.00	.0	.00	.00	.00	.00	.00
	TOTAL :				.00	.0	.00	.00	.00	.00	.00
1200	Rough HVAC										
	RD 3 ES 5FEB97 EF 7FEB97 LS 5FEB97 LF 7FEB97 TF 0										
	SUB		S SUBCNT'R		.00	.0	.00	.00	.00	.00	.00
	TOTAL :				.00	.0	.00	.00	.00	.00	.00
1300	Rough Elect										
	RD 3 ES 10FEB97 EF 12FEB97 LS 10FEB97 LF 12FEB97 TF 0										
	SUB		S SUBCNT'R		.00	.0	.00	.00	.00	.00	.00
	TOTAL :				.00	.0	.00	.00	.00	.00	.00
1400	Shingles										
	RD 3 ES 13FEB97 EF 17FEB97 LS 17FEB97 LF 19FEB97 TF 2										
	MATERIAL		M MATERIAL		.00	.0	.00	.00	.00	.00	.00
	SUB		S SUBCNT'R		.00	.0	.00	.00	.00	.00	.00
	TOTAL :				.00	.0	.00	.00	.00	.00	.00
1500	Ext Siding										
	RD 3 ES 17FEB97 EF 19FEB97 LS 24FEB97 LF 26FEB97 TF 5										
	CARPENTR		L LABOR	day	.36	.0	.00	.00	.36	.36	.00
	CRPN FOR		L LABOR	day	.36	.0	.00	.00	.36	.36	.00
	MATERIAL		M MATERIAL		.00	.0	.00	.00	.00	.00	.00
	SUB		S SUBCNT'R		.00	.0	.00	.00	.00	.00	.00
	TOTAL :				.72	.0	.00	.00	.72	.72	.00

Figure 12-32b Resource Control Activity Report—Wood Frame House, page 2

```
---------------------------------------------------------------------------------------------
Student Constructors                   PRIMAVERA PROJECT PLANNER        Sample Project #1

REPORT DATE  26JAN97  RUN NO.   27     RESOURCE CONTROL ACTIVITY REPORT   START DATE  5JAN97  FIN DATE 13MAR97
             8:30
Resource Control - Detail By Activity                                   DATA DATE   5JAN97   PAGE NO.    3
---------------------------------------------------------------------------------------------
                       COST    ACCOUNT  UNIT            PCT    ACTUAL     ACTUAL     ESTIMATE TO
ACTIVITY ID RESOURCE  ACCOUNT  CATEGORY MEAS   BUDGET   CMP   TO DATE  THIS PERIOD   COMPLETE   FORECAST    VARIANCE
----------- --------  -------- -------- ----  -------- ----- -------- ------------ ------------ ---------- ------------

     1600  Ext Finish Carpentry
           RD   2 ES 13FEB97  EF 14FEB97  LS 20FEB97  LF 21FEB97  TF   5

           CARPENTR    L LABOR    day       .04   .0      .00         .00          .04        .04         .00
           CRPN FOR    L LABOR    day       .04   .0      .00         .00          .04        .04         .00
           MATERIAL    M MATERIAL           .00   .0      .00         .00          .00        .00         .00
           SUB         S SUBCNT'R           .00   .0      .00         .00          .00        .00         .00
                                        ------------ ----- -------- ------------ ------------ ---------- ------------
           TOTAL :                          .08   .0      .00         .00          .08        .08         .00

     1700  Hang Drywall
           RD   4 ES 14FEB97  EF 19FEB97  LS 14FEB97  LF 19FEB97  TF   0

           LABOR       L LABOR              .00   .0      .00         .00          .00        .00         .00
           MATERIAL    M MATERIAL           .00   .0      .00         .00          .00        .00         .00
           SUB         S SUBCNT'R           .00   .0      .00         .00          .00        .00         .00
                                        ------------ ----- -------- ------------ ------------ ---------- ------------
           TOTAL :                          .00   .0      .00         .00          .00        .00         .00

     1800  Finish Drywall
           RD   4 ES 20FEB97  EF 25FEB97  LS 20FEB97  LF 25FEB97  TF   0

           LABOR       L LABOR              .00   .0      .00         .00          .00        .00         .00
           MATERIAL    M MATERIAL           .00   .0      .00         .00          .00        .00         .00
           SUB         S SUBCNT'R           .00   .0      .00         .00          .00        .00         .00
                                        ------------ ----- -------- ------------ ------------ ---------- ------------
           TOTAL :                          .00   .0      .00         .00          .00        .00         .00

     1900  Cabinets
           RD   2 ES 26FEB97  EF 27FEB97  LS 26FEB97  LF 27FEB97  TF   0

           SUB         S SUBCNT'R           .00   .0      .00         .00          .00        .00         .00
                                        ------------ ----- -------- ------------ ------------ ---------- ------------
           TOTAL :                          .00   .0      .00         .00          .00        .00         .00

     2000  Ext Paint
           RD   3 ES 20FEB97  EF 24FEB97  LS 27FEB97  LF  3MAR97  TF   5

           SUB         S SUBCNT'R           .00   .0      .00         .00          .00        .00         .00
                                        ------------ ----- -------- ------------ ------------ ---------- ------------
           TOTAL :                          .00   .0      .00         .00          .00        .00         .00

     2100  Int Finish Carpentry
           RD   4 ES 28FEB97  EF  5MAR97  LS  3MAR97  LF  6MAR97  TF   1

           CARPENTR    L LABOR    day      2.89   .0      .00         .00         2.89       2.89         .00
           CRPN HLP    L LABOR    day      2.89   .0      .00         .00         2.89       2.89         .00
           MATERIAL    M MATERIAL           .00   .0      .00         .00          .00        .00         .00
                                        ------------ ----- -------- ------------ ------------ ---------- ------------
           TOTAL :                         5.78   .0      .00         .00         5.78       5.78         .00

     2200  Int Paint
           RD   3 ES  4MAR97  EF  6MAR97  LS  4MAR97  LF  6MAR97  TF   0

           SUB         S SUBCNT'R           .00   .0      .00         .00          .00        .00         .00
                                        ------------ ----- -------- ------------ ------------ ---------- ------------
           TOTAL :                          .00   .0      .00         .00          .00        .00         .00

     2300  Finish Plumbing
           RD   2 ES 28FEB97  EF  3MAR97  LS 28FEB97  LF  3MAR97  TF   0

           SUB         S SUBCNT'R           .00   .0      .00         .00          .00        .00         .00
                                        ------------ ----- -------- ------------ ------------ ---------- ------------
           TOTAL :                          .00   .0      .00         .00          .00        .00         .00
```

Figure 12-32c Resource Control Activity Report—Wood Frame House, page 3

```
----------------------------------------------------------------------------------------------------------------
Student Constructors                      PRIMAVERA PROJECT PLANNER         Sample Project #1

REPORT DATE  26JAN97  RUN NO.   27        RESOURCE CONTROL ACTIVITY REPORT   START DATE  5JAN97  FIN DATE 13MAR97
             8:30
Resource Control - Detail By Activity                                       DATA DATE   5JAN97  PAGE NO.    4

----------------------------------------------------------------------------------------------------------------
                        COST    ACCOUNT UNIT              PCT      ACTUAL       ACTUAL   ESTIMATE TO
ACTIVITY ID RESOURCE   ACCOUNT  CATEGORY MEAS   BUDGET    CMP     TO DATE   THIS PERIOD   COMPLETE   FORECAST   VARIANCE
----------- -------- --------   -------- ----  --------- -----  --------- ------------  ----------- ---------- ----------

      2400  Finish HVAC
            RD    3 ES 26FEB97  EF 28FEB97 LS 27FEB97 LF  3MAR97  TF    1

            SUB               S SUBCNT'R          .00    .0        .00         .00          .00        .00        .00
                                            ------------ ----- ------------ ------------ ------------ ------------ ------------
            TOTAL :                               .00    .0        .00         .00          .00        .00        .00

      2500  Finish Elect
            RD    2 ES 26FEB97  EF 27FEB97 LS 28FEB97 LF  3MAR97  TF    2

            SUB               S SUBCNT'R          .00    .0        .00         .00          .00        .00        .00
                                            ------------ ----- ------------ ------------ ------------ ------------ ------------
            TOTAL :                               .00    .0        .00         .00          .00        .00        .00

      2600  Flooring
            RD    3 ES  7MAR97  EF 11MAR97 LS  7MAR97 LF 11MAR97  TF    0

            SUB               S SUBCNT'R          .00    .0        .00         .00          .00        .00        .00
                                            ------------ ----- ------------ ------------ ------------ ------------ ------------
            TOTAL :                               .00    .0        .00         .00          .00        .00        .00

      2700  Grading & Landscaping
            RD    4 ES  4MAR97  EF  7MAR97 LS 10MAR97 LF 13MAR97  TF    4

            SUB               S SUBCNT'R          .00    .0        .00         .00          .00        .00        .00
                                            ------------ ----- ------------ ------------ ------------ ------------ ------------
            TOTAL :                               .00    .0        .00         .00          .00        .00        .00

      2800  Punch List
            RD    2 ES 12MAR97  EF 13MAR97 LS 12MAR97 LF 13MAR97  TF    0

            TOTAL :                               .00    .0        .00         .00          .00        .00        .00

                                            ------------ ----- ------------ ------------ ------------ ------------ ------------
                         REPORT TOTALS           86.33   .0        .00         .00         86.33      86.33        .00
================================================================================================================
```

Figure 12-32d   Resource Control Activity Report—Wood Frame House, page 4

```
------------------------------------------------------------------------------------------------------------
Student Constructors                    PRIMAVERA PROJECT PLANNER         Sample Project #1

REPORT DATE 26JAN97  RUN NO.   35       RESOURCE CONTROL ACTIVITY REPORT   START DATE  5JAN97  FIN DATE 14MAR97
             8:23
Resource Control - Detail By Activity                                     DATA DATE  3FEB97   PAGE NO.    1
------------------------------------------------------------------------------------------------------------
```

ACTIVITY ID RESOURCE	COST ACCOUNT	ACCOUNT CATEGORY	UNIT MEAS	BUDGET	PCT CMP	ACTUAL TO DATE	ACTUAL THIS PERIOD	ESTIMATE TO COMPLETE	FORECAST	VARIANCE
100 Clear Site										
RD 0 AS 6JAN97 AF 8JAN97										
TOTAL :				.00	.0	.00	.00	.00	.00	.00
200 Building Layout										
RD 0 AS 8JAN97 AF 8JAN97										
CARPENTR	L	LABOR	day	.10	100.0	.20	.20	.00	.20	-.10
CRPN FOR	L	LABOR	day	.10	100.0	.10	.10	.00	.10	.00
LAB CL 1	L	LABOR	day	.20	100.0	.30	.30	.00	.30	-.10
LAB CL 2	L	LABOR	day	.20	100.0	.30	.30	.00	.30	-.10
TOTAL :				.60	150.0	.90	.90	.00	.90	-.30
300 Form/Pour Footings										
RD 0 AS 9JAN97 AF 10JAN97										
CARPENTR	L	LABOR	day	.90	100.0	1.00	1.00	.00	1.00	-.10
CRPN FOR	L	LABOR	day	.90	100.0	1.00	1.00	.00	1.00	-.10
LAB CL 1	L	LABOR	day	1.79	100.0	2.00	2.00	.00	2.00	-.21
LAB CL 2	L	LABOR	day	1.79	100.0	2.00	2.00	.00	2.00	-.21
TOTAL :				5.38	111.5	6.00	6.00	.00	6.00	-.62
400 Pier Masonry										
RD 0 AS 15JAN97 AF 16JAN97										
CARPENTR	L	LABOR	day	.51	100.0	.60	.60	.00	.60	-.09
CRPN FOR	L	LABOR	day	.26	100.0	.20	.20	.00	.20	.06
CRPN HLP	L	LABOR	day	.51	100.0	.60	.60	.00	.60	-.09
LAB CL 1	L	LABOR	day	1.03	100.0	1.00	1.00	.00	1.00	.03
LAB CL 2	L	LABOR	day	1.03	100.0	1.50	1.50	.00	1.50	-.47
MASON	L	LABOR	day	2.06	100.0	2.06	2.06	.00	2.06	.00
TOTAL :				5.40	110.4	5.96	5.96	.00	5.96	-.56
500 Wood Floor System										
RD 0 AS 17JAN97 AF 22JAN97										
CARPENTR	L	LABOR	day	6.68	100.0	5.00	5.00	.00	5.00	1.68
CRPN FOR	L	LABOR	day	3.34	100.0	3.00	3.00	.00	3.00	.34
CRPN HLP	L	LABOR	day	6.68	100.0	5.00	5.00	.00	5.00	1.68
TOTAL :				16.70	77.8	13.00	13.00	.00	13.00	3.70
600 Rough Framing Walls										
RD 0 AS 23JAN97 AF 29JAN97										
CARPENTR	L	LABOR	day	6.55	100.0	5.00	5.00	.00	5.00	1.55
CRPN FOR	L	LABOR	day	3.28	100.0	4.00	4.00	.00	4.00	-.72
CRPN HLP	L	LABOR	day	6.55	100.0	5.00	5.00	.00	5.00	1.55
TOTAL :				16.38	85.5	14.00	14.00	.00	14.00	2.38
700 Rough Framing Roof										
RD 3 AS 28JAN97 EF 5FEB97 LF 5FEB97 TF 0										
CARPENTR	L	LABOR	day	7.51	25.0	1.88	1.88	5.63	7.51	.00
CRPN FOR	L	LABOR	day	3.72	25.0	.93	.93	2.79	3.72	.00
CRPN HLP	L	LABOR	day	7.59	25.0	1.90	1.90	5.69	7.59	.00
TOTAL :				18.82	25.0	4.71	4.71	14.11	18.82	.00
800 Doors & Windows										
RD 4 ES 10FEB97 EF 13FEB97 LS 17FEB97 LF 20FEB97 TF 5										
CARPENTR	L	LABOR	day	7.13	.0	.00	.00	7.13	7.13	.00
CRPN HLP	L	LABOR	day	7.13	.0	.00	.00	7.13	7.13	.00
TOTAL :				14.26	.0	.00	.00	14.26	14.26	.00

Figure 12-33a Updated Resource Control Activity Report—Wood Frame House, page 1

```
---------------------------------------------------------------------------------------------------------------
Student Constructors                    PRIMAVERA PROJECT PLANNER            Sample Project #1

REPORT DATE  26JAN97  RUN NO.   35      RESOURCE CONTROL ACTIVITY REPORT     START DATE  5JAN97  FIN DATE 14MAR97
             8:23
Resource Control - Detail By Activity                                       DATA DATE  3FEB97   PAGE NO.   2
---------------------------------------------------------------------------------------------------------------

                           COST     ACCOUNT  UNIT              PCT      ACTUAL       ACTUAL    ESTIMATE TO
ACTIVITY ID RESOURCE      ACCOUNT   CATEGORY MEAS   BUDGET     CMP     TO DATE    THIS PERIOD   COMPLETE    FORECAST    VARIANCE
----------- --------      -------   -------- ----   ------     ---     -------    -----------  ----------   --------    --------

       900  Ext Wall Board
            RD    2 ES  6FEB97  EF  7FEB97  LS  7FEB97  LF 10FEB97  TF   1

            CARPENTR             L LABOR    day      .82      .0        .00          .00         .82         .82         .00
            CRPN FOR             L LABOR    day      .41      .0        .00          .00         .41         .41         .00
            CRPN HLP             L LABOR    day      .82      .0        .00          .00         .82         .82         .00
                                                 ----------- -----  ------------  ------------  ----------  ---------   ------------
            TOTAL :                               2.05      .0        .00          .00         2.05        2.05         .00

      1000  Ext Wall Insulation
            RD    1 ES 14FEB97  EF 14FEB97  LS 14FEB97  LF 14FEB97  TF   0

            TOTAL :                                .00      .0        .00          .00          .00         .00         .00

      1100  Rough Plumbing
            RD    4 ES 10FEB97  EF 13FEB97  LS 11FEB97  LF 14FEB97  TF   1

            TOTAL :                                .00      .0        .00          .00          .00         .00         .00

      1200  Rough HVAC
            RD    3 ES  6FEB97  EF 10FEB97  LS  6FEB97  LF 10FEB97  TF   0

            TOTAL :                                .00      .0        .00          .00          .00         .00         .00

      1300  Rough Elect
            RD    3 ES 11FEB97  EF 13FEB97  LS 11FEB97  LF 13FEB97  TF   0

            TOTAL :                                .00      .0        .00          .00          .00         .00         .00

      1400  Shingles
            RD    3 ES 14FEB97  EF 18FEB97  LS 18FEB97  LF 20FEB97  TF   2

            TOTAL :                                .00      .0        .00          .00          .00         .00         .00

      1500  Ext Siding
            RD    3 ES 18FEB97  EF 20FEB97  LS 25FEB97  LF 27FEB97  TF   5

            CARPENTR             L LABOR    day      .36      .0        .00          .00         .36         .36         .00
            CRPN FOR             L LABOR    day      .36      .0        .00          .00         .36         .36         .00
                                                 ----------- -----  ------------  ------------  ----------  ---------   ------------
            TOTAL :                                .72      .0        .00          .00         .72         .72         .00

      1600  Ext Finish Carpentry
            RD    2 ES 14FEB97  EF 17FEB97  LS 21FEB97  LF 24FEB97  TF   5

            CARPENTR             L LABOR    day      .04      .0        .00          .00         .04         .04         .00
            CRPN FOR             L LABOR    day      .04      .0        .00          .00         .04         .04         .00
                                                 ----------- -----  ------------  ------------  ----------  ---------   ------------
            TOTAL :                                .08      .0        .00          .00         .08         .08         .00

      1700  Hang Drywall
            RD    4 ES 17FEB97  EF 20FEB97  LS 17FEB97  LF 20FEB97  TF   0

            TOTAL :                                .00      .0        .00          .00          .00         .00         .00

      1800  Finish Drywall
            RD    4 ES 21FEB97  EF 26FEB97  LS 21FEB97  LF 26FEB97  TF   0

            TOTAL :                                .00      .0        .00          .00          .00         .00         .00

      1900  Cabinets
            RD    2 ES 27FEB97  EF 28FEB97  LS 27FEB97  LF 28FEB97  TF   0

            TOTAL :                                .00      .0        .00          .00          .00         .00         .00
```

Figure 12-33b Updated Resource Control Activity Report—Wood Frame House, page 2

```
------------------------------------------------------------------------------------------------------
Student Constructors                    PRIMAVERA PROJECT PLANNER        Sample Project #1

REPORT DATE  26JAN97  RUN NO.   35      RESOURCE CONTROL ACTIVITY REPORT  START DATE  5JAN97  FIN DATE 14MAR97
             8:23
Resource Control - Detail By Activity                                    DATA DATE  3FEB97   PAGE NO.   3
------------------------------------------------------------------------------------------------------

                      COST     ACCOUNT UNIT          PCT    ACTUAL    ACTUAL   ESTIMATE TO
ACTIVITY ID RESOURCE  ACCOUNT  CATEGORY MEAS  BUDGET CMP   TO DATE  THIS PERIOD  COMPLETE   FORECAST   VARIANCE
----------- --------  -------  -------- ----  ------ ---   -------  -----------  --------   --------   --------

      2000  Ext Paint
            RD    3 ES 21FEB97  EF 25FEB97  LS 28FEB97  LF  4MAR97  TF   5

            TOTAL :                              .00   .0      .00        .00       .00        .00        .00

      2100  Int Finish Carpentry
            RD    4 ES  3MAR97  EF  6MAR97  LS  4MAR97  LF  7MAR97  TF   1

            CARPENTR      L LABOR   day     2.89   .0      .00        .00      2.89       2.89        .00
            CRPN HLP      L LABOR   day     2.89   .0      .00        .00      2.89       2.89        .00
                                        ----------- ----- ----------- ----------- ---------- ---------- ----------
            TOTAL :                     5.78   .0      .00        .00      5.78       5.78        .00

      2200  Int Paint
            RD    3 ES  5MAR97  EF  7MAR97  LS  5MAR97  LF  7MAR97  TF   0

            TOTAL :                              .00   .0      .00        .00       .00        .00        .00

      2300  Finish Plumbing
            RD    2 ES  3MAR97  EF  4MAR97  LS  3MAR97  LF  4MAR97  TF   0

            TOTAL :                              .00   .0      .00        .00       .00        .00        .00

      2400  Finish HVAC
            RD    3 ES 27FEB97  EF  3MAR97  LS 28FEB97  LF  4MAR97  TF   1

            TOTAL :                              .00   .0      .00        .00       .00        .00        .00

      2500  Finish Elect
            RD    2 ES 27FEB97  EF 28FEB97  LS  3MAR97  LF  4MAR97  TF   2

            TOTAL :                              .00   .0      .00        .00       .00        .00        .00

      2600  Flooring
            RD    3 ES 10MAR97  EF 12MAR97  LS 10MAR97  LF 12MAR97  TF   0

            TOTAL :                              .00   .0      .00        .00       .00        .00        .00

      2700  Grading & Landscaping
            RD    4 ES  5MAR97  EF 10MAR97  LS 11MAR97  LF 14MAR97  TF   4

            TOTAL :                              .00   .0      .00        .00       .00        .00        .00

      2800  Punch List
            RD    2 ES 13MAR97  EF 14MAR97  LS 13MAR97  LF 14MAR97  TF   0

            TOTAL :                              .00   .0      .00        .00       .00        .00        .00

                                        ----------- ----- ----------- ----------- ---------- ---------- ----------
            REPORT TOTALS               86.17 51.7    44.57      44.57     37.00      81.57       4.60
======================================================================================================
```

Figure 12-33c Updated Resource Control Activity Report—Wood Frame House, page 3

EXERCISES

1. Small Commercial Concrete Block Building— Resource Tracking

Prepare an updated resource control activity report for the small commercial concrete block building located in the Appendix.

2. Large Commercial Building—Resource Tracking

Prepare an updated resource control activity report for the large commercial building located in the Appendix.

13 | Tracking Costs Using *P3*

Objectives

Upon completion of this chapter, the reader should be able to:

- Use updated cost tables
- Input current costs
- Use earned value costs
- Use scheduled budget costs
- Record expenditures
- Use updated cost profile
- Print updated cost reports

TRACKING COSTS: COMPARING ACTUAL TO PLANNED EXPENDITURES

Once a project is under way, tracking (monitoring) of costs and comparing actual to planned expenditures are tools in controlling costs. The costs can either be total costs or the accumulation of the resource costs of manpower, equipment, materials and/or other resources. These costs are usually assigned to the activity (allocated from the original estimate) when the target schedule is constructed. By tracking actual costs and comparing them to the original budget, cost progress and earned value can be determined.

Once the construction managers know whether the project is making or losing money and once they figure out where the problem lies, they can pass on the information to the right people in time to do something about it.

On a cost-plus or negotiated project, the owner, too, can use the cost-loaded and then cost-monitored schedule as valuable information for managing cash assets. This chapter shows how to use *P3* to update costs to match the updated duration, logic, and activity changes made to the schedule in Chapter 10.

COST TABLE

Updating the Target Schedule

Figure 6-24 (p. 136) shows the target (original) schedule with the original or scheduled budget. Figure 13-1 shows the current schedule as updated in Chapter 10, with progress, modified logic, and new activities input. The next step is to further update the schedule showing the actual expenditure of costs and to make forecasts.

The easiest way to evaluate costs on screen is through the use of the resource table. Select the **View** main pull-down menu, then **Resource Table**. The resource table appears at the bottom of the screen (Figure 13-2). If resources and not costs appear, select the **Display...** button. The **Resource Table Display Options** dialog box will appear (Figure 13-3). Click on the **Costs** option. Since the costs in Chapter 6 were input at the total cost level, select **Total** from the **Resources** pull-down menu in the **Resource Profile/Table** dialog box. In this chapter, the total costs will be input.

Actual Costs of Work Performed (Current Estimate)   Under   **Values:** in the **Resource Table Display Options** dialog box, there are three

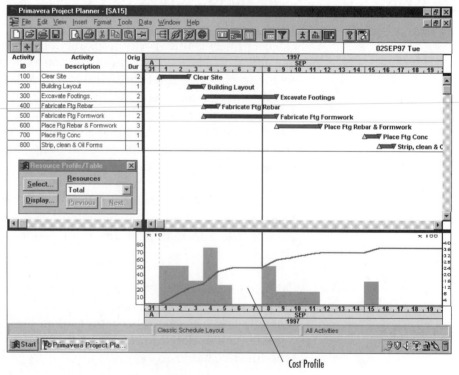

Figure 13-1   Current Schedule—Cost Not Updated

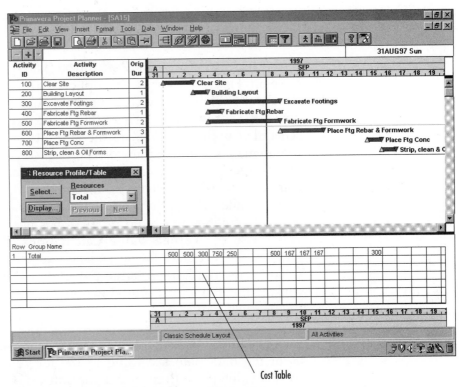

Figure 13-2   Cost Table—Current Estimate

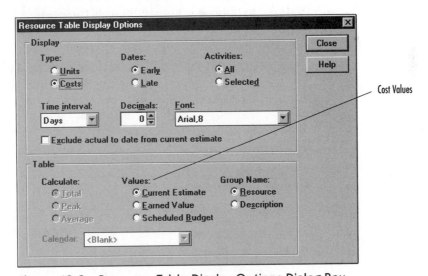

Figure 13-3   Resource Table Display Options Dialog Box

options. They are the **Current Estimate**, **Earned Value**, and **Scheduled Budget**. The resource table in Figure 13-2 has the option **Current Estimate** selected. This option updates the activity information with actual costs of work performed (ACWP) as they are incurred. The comparison of the planned (scheduled budget) to the ACWP makes cost forecasting possible. Cost forecasting is projecting the future cost based on costs to date and management's knowledge of the project.

Budgeted Cost for Work Performed (Earned Value) The resource table in Figure 13-4 has the option **Earned Value** selected in the **Resource Table Display Options** dialog box. Earned value is the budgeted cost for work performed (BCWP). BCWP is a comparison of the scheduled budget (BCWS) to the current estimate. The earned value represents the value of the work performed rather than the actual cost of work performed. *P3* calculates earned value as a product of percent complete times budgeted cost. It represents the portion of the budget that was allocated to the work that was actually accomplished. Notice that the activities where no progress has been made show no earned value. Use the **Earned Value Calculations** dialog box to determine whether the current or the target schedule will be used for earned value calculations. The **Earned Value Calculations** dialog box is obtained by clicking the **Tools** main pull-down menu, then selecting **Options**, and then **Earned Value....**

The scheduled budget reflects the work that should have been finished as of the data date according to the target (original) schedule. The earned value (BCWP) measures the value of the work actually accomplished as of the data date.

From the **Tools** main pull-down menu, select **Options** and then **Earned Value...** to choose whether *P3* uses the budget from the current estimate or the Target 1 schedule to calculate the earned value.

Budgeted Cost of Work Scheduled (Scheduled Budget) The resource table in Figure 13-5 has the **Scheduled Budget** option selected in

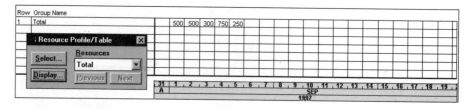

Figure 13-4 Cost Table—Earned Value

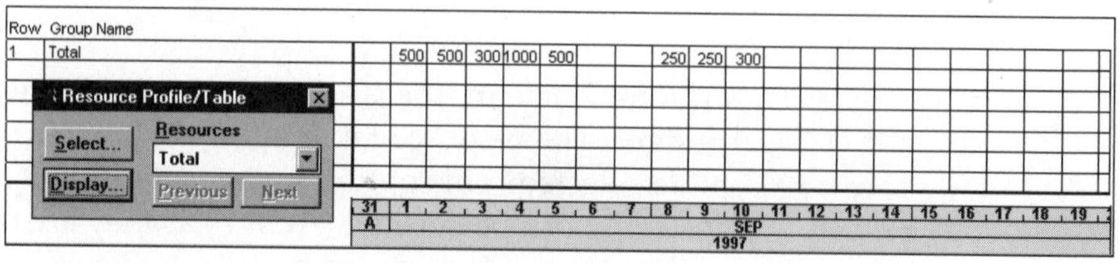

Figure 13-5 Cost Table—Scheduled Budget

the **Resource Table Display Options** dialog box. The scheduled budget (target schedule budget) is the budgeted cost of work scheduled (BCWS). Notice that cost projections for each day match that of Figure 6-23 (p. 135) from the original "un-updated" target schedule. These costs show none of the duration changes, the activities added, or the actual progress input. The scheduled budget shows how costs were to be spent according to the target (original) schedule.

Tabular Representation

Table 13-1 is a tabular representation of the cost modifications made to Activities 300, 400, and 500. The scheduled budget (original target cost budget developed in Chapter 6) is compared to the current estimate (current schedule updated with actual duration and logic changes in Chapter 10).

RECORDING EXPENDITURES

Gathering Costs by Activity

Costs must be gathered by activity. This is not the way most contractors' cost accounting systems work. The usual construction cost system gathers costs by cost account code. Labor time sheets, work measurement reports, purchase orders, and all other cost accounting documents are coded by cost account code. The estimate is organized the same way. To

Activity	Scheduled Budget (Target Schedule)		Current Estimate (Current Schedule)		
	September 4	September 5	September 4	September 5	September 8
300	$200	$200	$100	$100	$200
400	$500		$500		
500	$300	$300	$150	$150	$300
Total	$1,000	$500	$750	$250	$500
Combined Total		$1,500			$1,500

Table 13-1 Cost Comparison of Scheduled Budget to Current Estimate by Activity

be able to gather costs by activity means that another step must be added to the level of information gathered in the field. The documents must be coded with the cost accounting code and the activity ID for cost gathering purposes. Another element that makes the process of gathering costs by activity difficult is the fact that indirect costs and profit are not typically placed in any specific activity, but are prorated (or spread) over all activities. The following section records the actual cost expenditures for the eight-activity example schedule for activities where physical progress has been claimed during the updating process in Chapter 10.

Current Estimate

As discussed earlier in this chapter, there are three versions of the cost information kept in *P3*. The first version, the current estimate, is the current forecast of project costs. It includes the actual to date input of costs (ACWP) for activities with physical progress (monies have been spent on these activities).

Activity 100

Cost Dialog Box   Select the **Current Estimate** option from the **Resource Table Display Options** dialog box. Click on the first activity to be updated with cost information, Activity 100, Clear Site. Click the right mouse button and select **Activity Detail** and then **Cost**. The **Cost** dialog box for Activity 100 will appear (Figure 13-6). Keep in mind that originally costs were loaded in the target schedule.

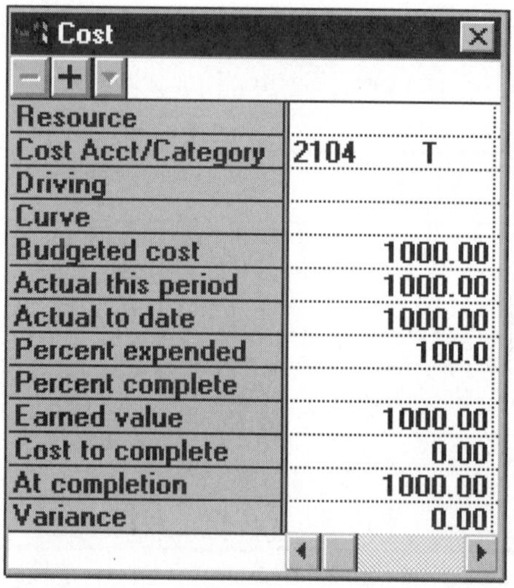

Cost	
Resource	
Cost Acct/Category	2104 T
Driving	
Curve	
Budgeted cost	1000.00
Actual this period	1000.00
Actual to date	1000.00
Percent expended	100.0
Percent complete	
Earned value	1000.00
Cost to complete	0.00
At completion	1000.00
Variance	0.00

Figure 13-6   Cost Dialog Box— Activity 100 Budgeted Cost

Budgeted cost   The amount of 1000 was input in the **Budgeted cost** field, and it was automatically entered by *P3* in the **At completion** field. The current schedule was updated with the physical progress in Chapter 10: notice the impact on the other fields. The **Actual this period**, **Actual to date**, and the **Earned value** fields have all been automatically updated by *P3* to show the $1000 value as input in the target schedule. The **Percent expended** field also shows 100.

Actual to date   The $1000 represents the original budget for this activity, or budgeted cost for work scheduled. The $1400 is the total sum of actual costs as gathered by the project cost accounting system for this activity. In Figure 13-7 the **Actual to date** field is changed to 1400 to reflect the actual job cost of this activity, or actual cost for work planned. *P3* automatically accepts the updated information when the mouse is clicked on another activity and then clicked back on Activity 100.

Percent expended   In Figure 13-8, the **Actual to date** and the **At completion** fields now reflect 1400, and the **Percent expended** field reflects 140. This means the at completion cost is 40% over the current budgeted cost. There is a –400 variance, or an overbudget projection of $400. Since the budgeted cost has already been exceeded, the budget for the current estimate needs to be changed to $1400 to reflect current projections. Unless the **Budgeted cost** field is changed to reflect the overage, the variance number will remain.

Resource Table   Notice that in the resource table the costs for September 1 and 2 are $700 each for a total of $1400. Before this cost update, the values were $500 for each day for a total of $1000.

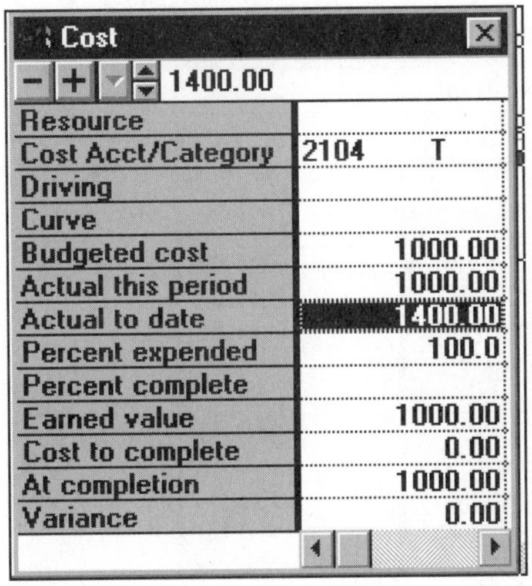

Figure 13-7   Cost Dialog Box—Activity 100 Actual to Date

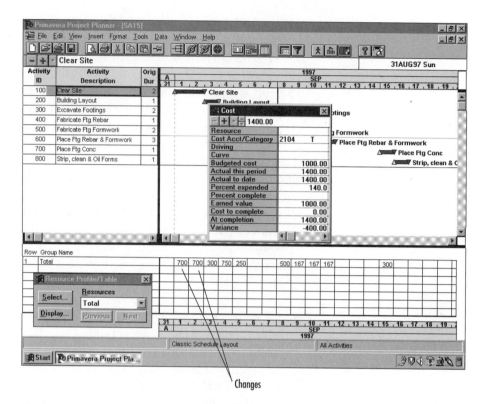

Figure 13-8   Cost Table—Activity 100 Changes to Current Estimate

Activity 200

Cost Dialog Box   Click on the next activity to be updated with cost information, Activity 200, Building Layout. The **Cost** dialog box for Activity 200 will appear (Figure 13-9). Compare Figure 13-9 to Figure 6-16 (p. 130), which shows the costs loaded in the target schedule. The cost of

Cost	300.00
Resource	
Cost Acct/Category	1306 T
Driving	
Curve	
Budgeted cost	300.00
Actual this period	300.00
Actual to date	300.00
Percent expended	100.0
Percent complete	
Earned value	300.00
Cost to complete	0.00
At completion	300.00
Variance	0.00

Figure 13-9   Cost Dialog Box—Activity 200 Budgeted Cost

300 was input in the **Budgeted cost** field. Now, the 300 appears in the **At completion**, **Actual this period**, **Actual to date**, and the **Earned value** fields. The **Percent expended** field shows 100.

Budgeted cost   In Figure 13-10 the **Budgeted cost** field is changed to 250 to reflect the actual job cost of this activity. The **Budgeted cost** field was changed rather than the **Actual to date** field—since the activity is 100% complete, the current estimate needs to be changed to reflect the actual final cost for this activity.

Variance   Notice what *P3* automatically does when the Activity 200 updated cost information is accepted. In Figure 13-11, the **Actual to date** and the **Actual this period** fields reflect 250, and the **Percent**

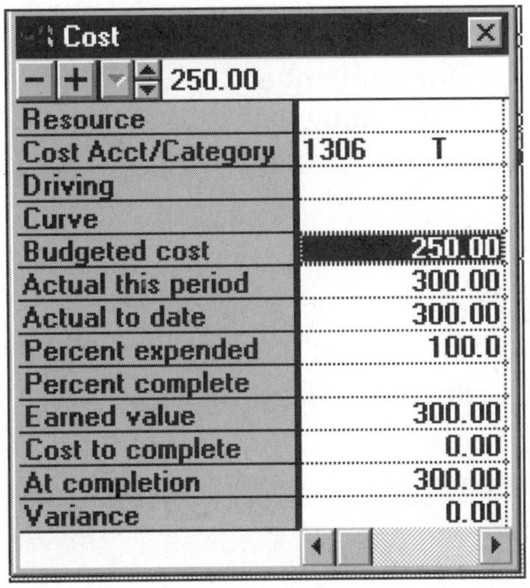

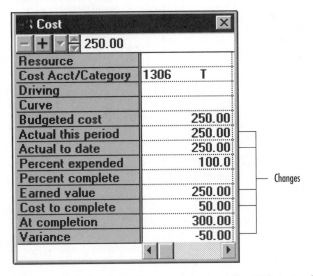

Figure 13-10   Cost Dialog Box—Activity 200 Changes to Budgeted Cost

Figure 13-11   Cost Dialog Box—Activity 200 Actual to Date

expended field reflects 100. This means that at completion 100% of cost has been expended in the current budget. There is a –50 variance, or an underbudget projection of $50.

At completion In Figure 13-12, the **At completion** is changed to $250. Now the activity can be said to be closed. The current estimate of the budget has been updated with actual costs. The activity has 100% of costs expended and is 100% complete.

With Activity 200, the **Budgeted cost** field was modified rather than the **Actual to date** field. The reason for this is that the percent complete is calculated by dividing the **Actual to date** field by the **Budgeted cost** field. Therefore, unless the **Budgeted cost** field is modified, it is impossible to attain a value of 100% expended.

Resource Table Notice the values for Activity 200 in the resource table. The cost for September 3 is $250. Before the update with actual costs was made, the value was $300.

Activity 300

Cost Dialog Box Click on the next activity to be updated with cost information, Activity 300, Excavate Footings. The **Cost** dialog box for Activity 300 will appear (Figure 13-13).

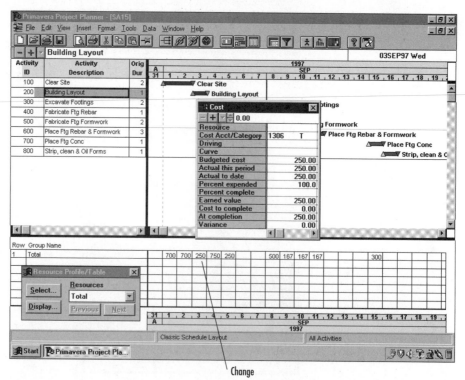

Change

Figure 13-12 Cost Table—Activity 200 Changes to Current Estimate

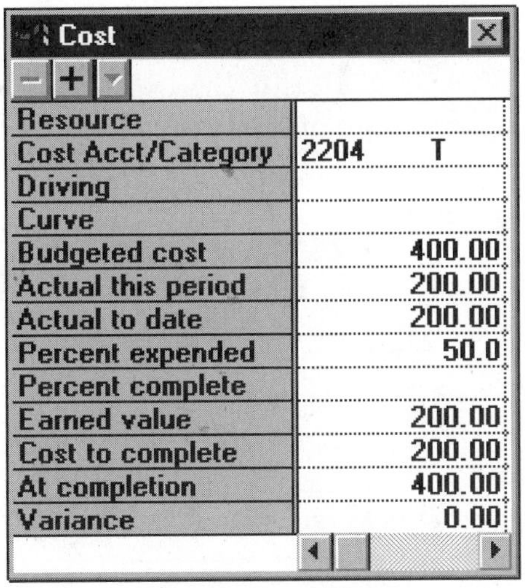

Figure 13-13   Cost Dialog Box—Activity 300 Budgeted Cost

Budgeted cost   A cost of 400 was input in the **Budgeted cost** field. Since the activity was designated as 50% complete when updated in Chapter 10, the **Percent expended** field shows 50 and 200 appears in the **Actual this period**, **Actual to date**, and the **Earned value** fields. The other $200 left in the budget appears in the **Cost to complete** field.

Actual to date   In Figure 13-14, the **Actual to date** field is changed to 300 to reflect the actual job cost to attain 50% completion for this activity. Notice what *P3* automatically does when the Activity 300 updated cost information is accepted in Figure 13-15. Here the **Actual to date** field was modified rather than the **Budgeted cost** field. The budget of $400 was not changed. Putting in an actual to

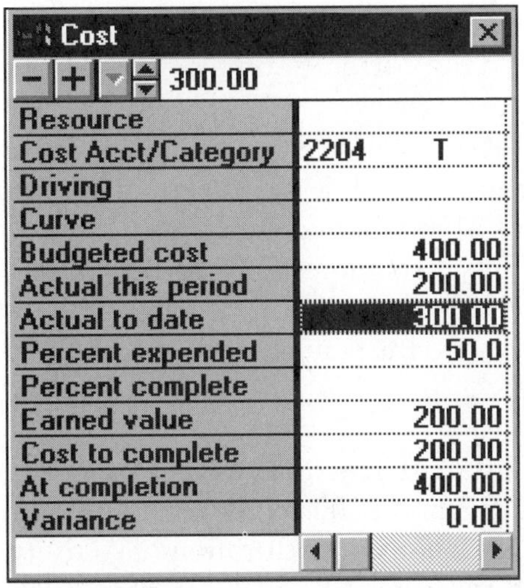

Figure 13-14   Cost Dialog Box—Activity 300 Actual to Date

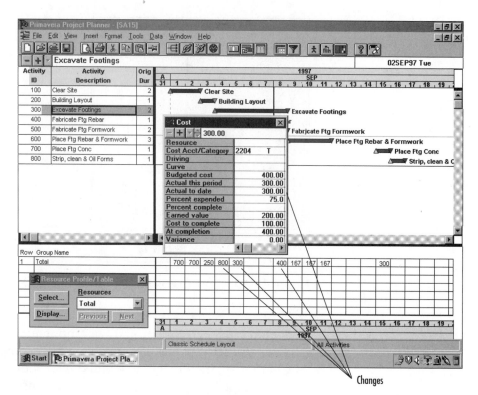

Figure 13-15 Cost Table—Activity 300 Changes to Current Estimate

date of $300 means that three-quarters of the money was spent to attain 50% completion.

At completion   Notice that when this field is modified, the **At completion** field does not change. The **Actual to date** field value of 300 is divided by the **Budgeted cost** field of 400 to produce a percent expended of 75%. Notice there is no variance since the percent expended is less than the at completion costs. Here, the projection is to complete the activity for the remaining $100, so the budget is not changed in the current estimate.

Resource Table   Notice the total costs in the resource table. The cost is $800 for September 4, $300 for September 5, and $400 for September 8, for a total of $1500. September 6 and 7 are weekend (nonwork) days. Previously, the costs were $750 for September 4, $250 for September 5, and $500 for September 8, for the same total of $1500. The total costs for the three days are the same before and after the cost update with actual costs.

The original (target schedule) plan was to spend $200 for Activity 300 on September 3 and $200 on September 4. During the progress update in Chapter 10, the original duration of Activity 300 was changed from 2 days to 2 days expended and 1 day remaining to complete the activity. Physical progress was put at 50% complete. The **Cost** dialog box for

Activity 300 showed an **Actual this period** of $200, or $100 on September 4 $100 on September 5, for the $200 expended (Figure 13-13, p. 361). The 50% left ($200) to be expended was shown on September 8.

In Figure 13-14 (p. 361), an actual cost of $300 was input, or 75% is the percent expended with $100 left to be expended. In Figure 13-15 (p. 363) the $300 cost expended for Activity 300 is spread over two days: $150 on September 4, $150 on September 5. The 25% left ($100) to be expended is shown on September 8. The cash flow plan was increased by $50 on September 4 (from $750 to $800), increased by $50 on September 5 (from $250 to $300), and decreased by $100 on September 8 (from $500 to $400). The original budget of $400 did not change; it has just been spread on different days than the original (target) schedule.

Activity 400

Cost Dialog Box   Click on the next activity to be updated with cost information, Activity 400, Fabricate Ftg Rebar. The **Cost** dialog box for Activity 400 will appear (Figure 13-16).

Budgeted cost   A cost of 500 was input in the **Budgeted cost** field and appears in the **Actual this period**, **Actual to date**, **Earned value**, and **At completion** fields. The **Percent expended** field shows 100. The **Budgeted cost** field is left at 500 to reflect the actual job cost of this activity. This means that at completion 100% of cost has been expended in the current budget.

Variance   There is a 0 variance.

Resource Table   Since no change was made for this activity, the resource table will not change.

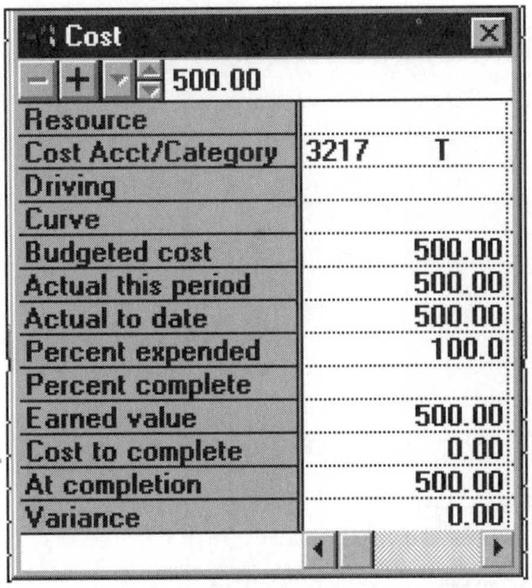

Cost	☒
− + ▾ ⇕ 500.00	
Resource	
Cost Acct/Category	3217 T
Driving	
Curve	
Budgeted cost	500.00
Actual this period	500.00
Actual to date	500.00
Percent expended	100.0
Percent complete	
Earned value	500.00
Cost to complete	0.00
At completion	500.00
Variance	0.00

Figure 13-16   Cost Dialog Box—Activity 400 Budgeted Cost

Activity 500

Cost Dialog Box   Click on the next activity to be updated with cost information, Activity 500, Fabricate Ftg Formwork. The **Cost** dialog box for Activity 500 will appear (Figure 13-17).

Budgeted cost   A cost of 600 was input in the **Budgeted cost** field. Since the activity was designated as 50 complete when updated in Chapter 10, the **Percent expended** field shows 50 and $300 appears in the **Actual this period**, **Actual to date**, and the **Earned value** fields. The other $300 left in the budget appears in the **Cost to complete** field.

At completion   In Figure 13-18, the **Budgeted cost** and the **At completion** fields are changed to 800 to reflect the new budget for this activity. Actual job cost shows that the actual to date is $500, and it is placed in the proper field. Notice what *P3* automatically does when the Activity 500 updated cost information is accepted (Figure 13-19). The **Actual this period** now reflects the $500 actually spent. The **Percent expended** shows 62.5, or the actual cost of $500 divided by the budgeted cost of $800.

Earned value   The earned value is $400, or 50% (physical progress input for this activity in Chapter 10) of the budgeted cost of $800. Notice there is no variance since the at completion costs and the budgeted cost are the same.

Resource Table   In Figure 13-19, the total costs are $900 on September 4, $400 on September 5, and $400 on September 8, for a total of $1700. September 6 and 7 are weekend (nonwork) days. In Figure 13-15

Cost	
− + ▽ ⬆ 300.00	
Resource	
Cost Acct/Category	3158 T
Driving	
Curve	
Budgeted cost	600.00
Actual this period	300.00
Actual to date	300.00
Percent expended	50.0
Percent complete	
Earned value	300.00
Cost to complete	300.00
At completion	600.00
Variance	0.00

Figure 13-17   Cost Dialog Box—Activity 500 Budgeted Cost

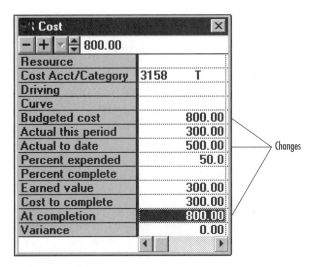

Figure 13-18   Cost Dialog Box—Activity 500 Actual to Date

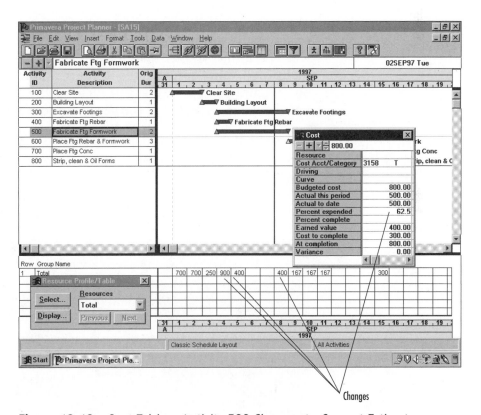

Figure 13-19   Cost Table—Activity 500 Changes to Current Estimate

(p. 362), the costs are $800 for September 4, $300 for September 5, and $400 for September 8, for a total of $1500. This $200 increase was the increase in the budgeted cost of Activity 500 from $600 to $800. The original (target schedule) plan was to spend $300 for Activity 500 on September 3 and $300 on September 4.

The **Actual this period** field has now changed. Previously, it was $300, or $150 on September 4 and $150 on September 5. The 50% left ($300) to be expended was shown on September 8. Now, an actual cost of $500 was input, or 62.5% is the percent expended of the $800 at completion cost. The costs are spread for Activity 500 as $250 on September 4 and $250 on September 5. The 37.5% left ($300) to be expended is shown on September 8. The cash flow plan was increased by $100 (from $150 to $250) on September 4 and 5 and stayed the same at $300 on September 8.

All changes made relating to actual costs in this chapter have been made to the current estimate version of the cost information. Compare the current estimate before costs were updated (Figure 13-2, p. 353) to the updated cost information input (Figure 13-19, p. 365). Table 13-2 compares these two figures.

Earned Value

The second of the three versions of the cost information kept in *P3* is the earned value cost information. The earned value represents the value of work performed rather than the actual cost of work performed. Select the **Earned Value** option in the **Resource Table Display Options** dialog box (Figure 13-20). Figure 13-21 shows the resource table when the **Earned Value** option has been selected. Notice that the activities where no progress has been made show no earned value. Compare Figure 13-4 (p. 354), which shows earned value before costs were updated, to Figure 13-21, which has all the updated cost information. Table 13-3 compares the two figures.

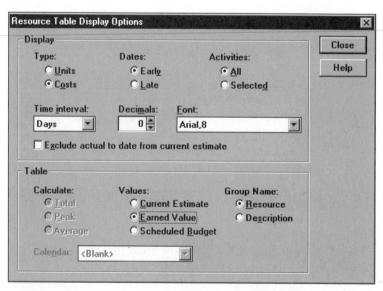

Figure 13-20 Resource Table Display Options Dialog Box—Earned Value Option

September	Current Estimate (Current Schedule) Before Cost Update		Current Estimate (Current Schedule) After Cost Update	
1	100 - $500	$500	100 - $700	$700
2	100 - $500	$500	100 - $700	$700
3	200 - $300	$300	200 - $250	$250
4	300 - $100, 400 - $500, 500 - $150	$750	300 - $150, 400 - $500, 500 - $250	$900
5	300 - $100, 500 - $150	$250	300 - $150, 500 - $250	$400
6				
7				
8	300 - $200, 500 - $300	$500	300 - $100, 500 - $300	$400
9	600 - $167	$167	600 - $167	$167
10	600 - $167	$167	600 - $167	$167
11	600 - $167	$167	600 - $167	$167
12				
13				
14				
15	700 - $300	$300	700 - $300	$300
Total		$3,600		$4,150

$4,150 – $3,600 = $550

Activity 100   + $400
Activity 200   – $ 50
Activity 500   + $200

Table 13-2   Cost Comparison of Current Estimate—Before and After Update

Scheduled Budget

The third of the three versions of cost information kept in *P3* is the scheduled budget. The scheduled budget shows costs that were input in the target schedule or changes made to the project budget. From the **Resource Table Display Options** dialog box, select the option **Scheduled Budget**. Figure 13-22 shows the cost table when the **Scheduled Budget** option has been selected. The budgeted cost for Activity 200

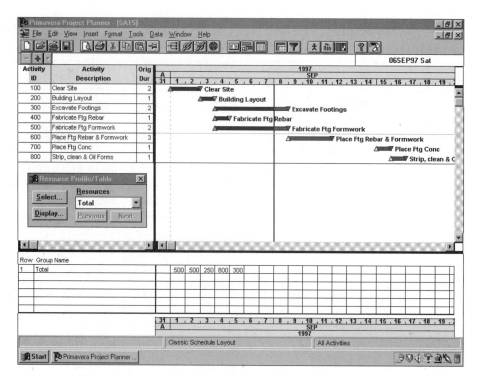

Figure 13-21 Cost Table—Earned Value

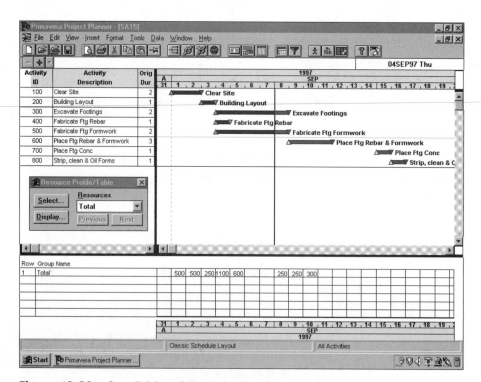

Figure 13-22 Cost Table—Scheduled Budget

September	Current Estimate (Current Schedule) Before Cost Update		Earned Value Before Cost Update		Current Estimate (Current Schedule) After Cost Update		Earned Value After Cost Update	
1	100 - $500	$500	100 - $500	$500	100 - $700	$700	100 - $500	$500
2	100 - $500	$500	100 - $500	$500	100 - $700	$700	100 - $500	$500
3	200 - $300	$300	200 - $300	$300	200 - $250	$250	200 - $250	$250
4	300 - $100, 400 - $500, 500 - $150	$750	300 - $100, 400 - $500, 500 - $150	$750	300 - $150, 400 - $500, 500 - $250	$900	300 - $100, 400 - $500, 500 - $200	$800
5	300 - $100, 500 - $150	$250	300 - $100, 500 - $150	$250	300 - $150, 500 - $250	$400	300 - $100, 500 - $200	$300
6								
7								
8	300 - $200, 500 - $300	$500			300 - $100, 500 - $300	$400		
9	600 - $167	$167			600 - $167	$167		
10	600 - $167	$167			600 - $167	$167		
11	600 - $167	$167			600 - $167	$167		
12								
13								
14								
15	700 - $300	$300			700 - $300	$300		
Total		$3,600		$2,300		$4,150		$2,350

Table 13-3   Cost Comparison of Current Estimate to Earned Value—Before and After Update

was changed from $300 to $250. This change is reflected on September 3 in Figure 13-22 (p. 368). The budgeted cost for Activity 500 was changed from $600 to $800. This change is reflected on September 4 and September 5, with the $200 increase spread as $100 for each of the original two-days of duration.

COST PROFILE

So far in this chapter, cost information has been presented in a tabular format in the on-screen representation. Sometimes a graphical representation is more beneficial. Click on the **View** main pull-down menu. Select **Resource Profile**, and the graphical representation of the cost information will appear (Figure 13-23). If a cost profile doesn't appear on the screen, click on the **Display** button of the **Resource Profile/Table** dialog box. The **Resource Profile Display Options** dialog box will appear (Figure 13-24). Select the **Costs** option and the **Histogram** check box for the cost profile as shown in Figure 13-23. To get the cumulative curve as shown in Figure 13-25, select the **Curves** check box. Also, under the **Curves** portion, select the **Current Estimate, Scheduled Budget, and Show curves using different line types** check boxes.

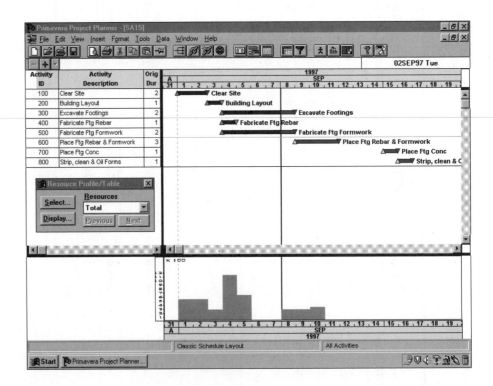

Figure 13-23 Cost Profile—Histogram of Scheduled Budget

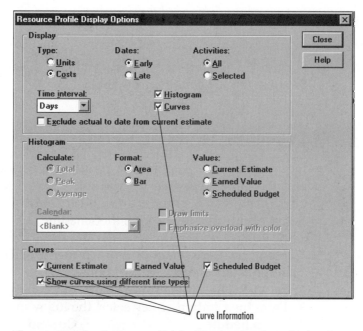

Curve Information

Figure 13-24   Resource Table Display Options Dialog Box—Curves

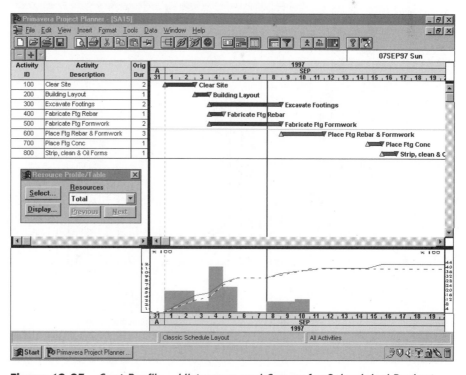

Figure 13-25   Cost Profile—Histogram and Curves for Scheduled Budget

P3 Cost Reports

A number of built-in *P3* cost reports are very useful in evaluating progress and updating information.

Current vs. Target Area Profile

Figure 13-26 is a graphic area profile comparing the current schedule and the target schedule. To run this report follow the steps outlined below. The ??????? entry is a "wild card" entry that tells *P3* to select all cost accounts that are used in the schedule.

1. From the **Tools** main pull-down menu, select the **Graphic Reports** and **Resource and Cost...** options.
2. Select **RC-04, Area Profile - Current vs. Target.**
3. From the **Content** screen select the following options...

     Schedule   **C**urrent

                                          Target **1**

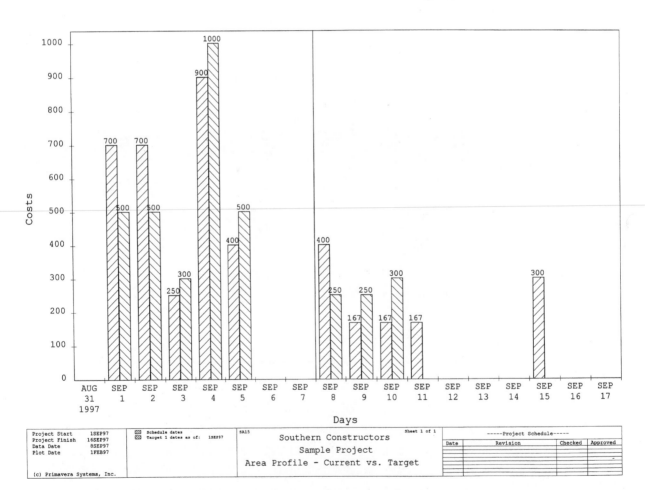

Figure 13-26   Area Profile—Current vs. Target Report

Curve	Histogram
Dates	Early
Show Data	Cost
Format	Bar
Use/Cost	Total

4. At the **Resource Selection** screen, select 1 for the **Group #** field, EQ for the **Profile if** field, ???????????? for the **Low Value Cost Account** field, Fill for the **Pattern** field, and 1 for the **Pen** field.

Cumulative Cash Flow Report

Figure 13-27 graphically compares the current schedule and the target schedule. To run this report, select the following:

1. From the **Tools** main pull-down menu, select the **Graphic Reports** and **Resource and Cost...** options.
2. Select **RC-05, Cumulative Cash Flow.**

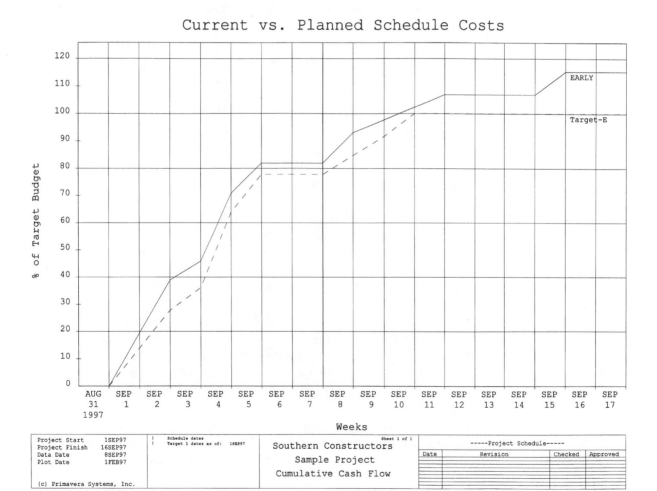

Figure 13-27 Cumulative Cash Flow Report

3. From the **Content** screen select the following options...

Schedule	**Current**
	Target <u>1</u>
Curve	**Cumulative**
Dates	**Earl<u>y</u>**
Show Data	**<u>C</u>ost**
Cumulative	**Display % on <u>Y</u> axis**
Base % on:	**Target 1 early**

4. At the **Resource Selection** screen, select 1 for the **Group #** field, EQ for the **Profile if** field, ???????????? for the **Low Value Cost Account** field, Fill for the **Pattern** field, and 2 for the **Pen** field.

Cost Control Activity Report

Figure 13-28 is a tabular cost control activity report comparing the current schedule and the target schedule. It includes the forecast and variance of the activity's current budget compared to the target budget. Compare Figure 13-28 to Figure 9-31 (p. 234), which is the same report run before the updated cost changes. To run this report, follow the steps outlined below:

1. From the **<u>T</u>ools** main pull-down menu, select **<u>T</u>abular Reports**, then **<u>C</u>ost** and **<u>C</u>ontrol...**.
2. Select **CC-07, Cost Control - Detailed by Activity.**
3. At the **Format** screen select **Activity ID** in the **Report <u>o</u>rganized by:** field.

Earned Value Report

Figure 13-29 is a tabular report comparing the current schedule and the target schedule. It includes the forecast and variance of the activity's current budget compared to the target budget. Compare Figure 13-29 to Figure 9-36 (p. 237), which is the same report run before the updated cost changes. To run this report, follow the steps outlined below:

1. From the **<u>T</u>ools** main pull-down menu, select **<u>T</u>abular Reports**, then **<u>C</u>ost** and **<u>E</u>arned Value...**.
2. Select **CE-01, Earned Value Report - Cost**
3. At the **Resource Selection** screen, select EQ for the **Profile if** field and ???????????? for the **Low Value Cost Account** field.

Once a project is under way, tracking (monitoring) of costs and comparing actual to planned expenditures is a tool in controlling costs. These costs are usually assigned to the activity (allocated from the original

```
--------------------------------------------------------------------------------------------
Southern Constructors              PRIMAVERA PROJECT PLANNER        Schedule Update

REPORT DATE  1FEB97  RUN NO.  70   COST CONTROL ACTIVITY REPORT     START DATE  1SEP97  FIN DATE 16SEP97
             11:02
Cost Control - Detailed by Activity                                DATA DATE  8SEP97   PAGE NO.    1
--------------------------------------------------------------------------------------------
                    COST     ACCOUNT UNIT          PCT    ACTUAL    ACTUAL   ESTIMATE TO
ACTIVITY ID RESOURCE ACCOUNT CATEGORY MEAS BUDGET  CMP   TO DATE  THIS PERIOD COMPLETE  FORECAST  VARIANCE
---------- -------- ------- -------- ---- ------- ----- -------- ------------ --------- -------- --------

   100  Clear Site
        RD   0 AS  1SEP97

                2104        T Tot Cost 1000.00 100.0  1400.00   1400.00      .00   1400.00  -400.00
                                       ------- -----  --------  --------   ------  -------- --------
        TOTAL :                        1000.00 140.0  1400.00   1400.00      .00   1400.00  -400.00

   200  Building Layout
        RD   0 AS  3SEP97  AF  3SEP97

                1306        T Tot Cost  250.00 100.0   250.00    250.00      .00    250.00      .00
                                       ------- -----  --------  --------   ------  -------- --------
        TOTAL :                         250.00 100.0   250.00    250.00      .00    250.00      .00

   300  Excavate Footings
        RD   1 AS  4SEP97  EF  8SEP97       LF  8SEP97 TF   0

                2204        T Tot Cost  400.00  50.0   300.00    300.00   100.00    400.00      .00
                                       ------- -----  --------  --------   ------  -------- --------
        TOTAL :                         400.00  75.0   300.00    300.00   100.00    400.00      .00

   400  Fabricate Ftg Rebar
        RD   0 AS  4SEP97  AF  4SEP97

                3217        T Tot Cost  500.00 100.0   500.00    500.00      .00    500.00      .00
                                       ------- -----  --------  --------   ------  -------- --------
        TOTAL :                         500.00 100.0   500.00    500.00      .00    500.00      .00

   500  Fabricate Ftg Formwork
        RD   1 AS  4SEP97  EF  8SEP97       LF  8SEP97 TF   0

                3158        T Tot Cost  800.00  50.0   500.00    500.00   300.00    800.00      .00
                                       ------- -----  --------  --------   ------  -------- --------
        TOTAL :                         800.00  62.5   500.00    500.00   300.00    800.00      .00

   600  Place Ftg Rebar & Formwork
        RD   3 ES  9SEP97  EF 11SEP97 LS  9SEP97  LF 11SEP97 TF   0

                3158         200.00  .0     .00       .00   200.00    200.00      .00
                3217         300.00  .0     .00       .00   300.00    300.00      .00
                            ------- -----  --------  --------   ------  -------- --------
        TOTAL :              500.00  .0     .00       .00   500.00    500.00      .00

   700  Place Ftg Conc
        RD   1 ES 15SEP97  EF 15SEP97 LS 15SEP97  LF 15SEP97 TF   0

                3130         300.00  .0     .00       .00   300.00    300.00      .00
                            ------- -----  --------  --------   ------  -------- --------
        TOTAL :              300.00  .0     .00       .00   300.00    300.00      .00

   800  Strip, clean & Oil Forms
        RD   1 ES 16SEP97  EF 16SEP97 LS 16SEP97  LF 16SEP97 TF   0

        TOTAL :                 .00  .0     .00       .00      .00       .00      .00

                            ------- -----  --------  --------   ------  -------- --------
        REPORT TOTALS       3750.00 78.7  2950.00   2950.00  1200.00   4150.00  -400.00
============================================================================================
```

Figure 13-28   Cost Control Activity Report

estimate) when the target schedule is constructed. This chapter was used to update costs to match the updated duration, logic, and activity changes made to the schedule in Chapter 10.

```
-----------------------------------------------------------------------------------------------
Southern Constructors                    PRIMAVERA PROJECT PLANNER          Schedule Update

REPORT DATE   1FEB97  RUN NO.   73        EARNED VALUE REPORT - COST        START DATE  1SEP97  FIN DATE 16SEP97
              11:11
Earned Value Report - Cost                                                  DATA DATE   8SEP97   PAGE NO.    1
-----------------------------------------------------------------------------------------------
  COST                       PCT  .........CUMULATIVE TO DATE.........  ........VARIANCE.........  ......AT COMPLETION......
 ACCOUNT  RESOURCE ACTIVITY ID CMP    ACWP         BCWP         BCWS       COST     SCHEDULE    BUDGET       ESTIMATE
-----------------------------------------------------------------------------------------------

                  ????????????
  1306      T         200  100.0   250.00       250.00       250.00       .00         .00      250.00        250.00
  2104      T         100  100.0  1400.00      1000.00      1000.00    -400.00        .00     1000.00       1400.00
  2204      T         300   50.0   300.00       200.00       400.00    -100.00     -200.00     400.00        400.00
  3130                700    .0     .00          .00          .00         .00         .00      300.00        300.00
  3158                600    .0     .00          .00          .00         .00         .00      200.00        200.00
  3158      T         500   50.0   500.00       400.00       800.00    -100.00     -400.00     800.00        800.00
  3217                600    .0     .00          .00          .00         .00         .00      300.00        300.00
  3217      T         400  100.0   500.00       500.00       500.00       .00         .00      500.00        500.00
                           -----  ----------  ----------  ----------  ----------  ----------  ----------  ----------
 ????????????       TOTAL   62.7  2950.00      2350.00      2950.00    -600.00     -600.00    3750.00       4150.00
=================================================================================================

          REPORT    TOTALS   62.7  2950.00      2350.00      2950.00    -600.00     -600.00    3750.00       4150.00
=================================================================================================
```

Figure 13-29   Earned Value Report—Cost

EXAMPLE PROBLEM: Tracking Costs

Figures 13-30a, b, and c are printouts of the current schedule with updated durations from Chapter 10 for the wood frame house drawings located in the Appendix. It was run using the CC-01 report (Cost Control - Summary by Activity), which is accessed by selecting the **Tabular Reports** and then the **Resource** and **Control...** options under the **Tools** main pull-down menu.

Figures 13-31a and b are printouts of the current schedule with updated durations from Chapter 10 for the wood frame house drawings located in the Appendix. It was run using the CC-02 report (Cost Control - Detailed by Resource), which is accessed by selecting the **Tabular Reports** and then the **Resource** and **Control...** options under the **Tools** main pull-down menu.

Figures 13-32a, b, and c are printouts of the current schedule with updated actual costs for the wood frame house drawings located in the Appendix. It was run using the CC-01 report (Cost Control - Summary by Activity), which is accessed by selecting the **Tabular Reports** and then the **Resource** and **Control...** options under the **Tools** main pull-down menu.

Figures 13-33a and b are printouts of the current schedule with updated actual costs for the wood frame house drawings located in the Appendix. It was run using the CC-02 report (Cost Control - Detailed by Resource), which is accessed by selecting the **Tabular Reports** and then the **Resource** and **Control...** options under the **Tools** main pull-down menu.

```
-------------------------------------------------------------------------------------------------------------
Student Constructors                      PRIMAVERA PROJECT PLANNER          Sample Project #1

REPORT DATE   2FEB97  RUN NO.   38        COST CONTROL ACTIVITY REPORT       START DATE  5JAN97  FIN DATE 14MAR97
              9:01
Cost Control - Summary by Activity                                          DATA DATE  3FEB97    PAGE NO.    1
-------------------------------------------------------------------------------------------------------------
```

ACTIVITY ID	BUDGET	PCT CMP	ACTUAL TO DATE	ACTUAL THIS PERIOD	ESTIMATE TO COMPLETE	FORECAST	VARIANCE
100 Clear Site RD 0 AS 6JAN97 AF 8JAN97							
TOTAL :	1280.00	100.0	1280.00	1280.00	.00	1280.00	.00
200 Building Layout RD 0 AS 8JAN97 AF 8JAN97							
TOTAL :	130.49	117.0	152.63	152.63	.04	152.67	-22.18
300 Form/Pour Footings RD 0 AS 9JAN97 AF 10JAN97							
TOTAL :	1174.32	104.0	1220.80	1220.80	.00	1220.80	-46.48
400 Pier Masonry RD 0 AS 15JAN97 AF 16JAN97							
TOTAL :	967.29	103.5	1001.35	1001.35	.00	1001.35	-34.06
500 Wood Floor System RD 0 AS 17JAN97 AF 22JAN97							
TOTAL :	4181.04	92.2	3856.25	3856.25	.00	3856.25	324.79
600 Rough Framing Walls RD 0 AS 23JAN97 AF 29JAN97							
TOTAL :	3323.99	94.3	3135.61	3135.61	.00	3135.61	188.38
700 Rough Framing Roof RD 3 AS 28JAN97 EF 5FEB97 LF 5FEB97 TF 0							
TOTAL :	3468.40	25.0	867.10	867.10	2601.53	3468.63	-.23
800 Doors & Windows RD 4 ES 10FEB97 EF 13FEB97 LS 17FEB97 LF 20FEB97 TF 5							
TOTAL :	3995.40	.0	.00	.00	3995.40	3995.40	.00
900 Ext Wall Board RD 2 ES 6FEB97 EF 7FEB97 LS 7FEB97 LF 10FEB97 TF 1							
TOTAL :	409.97	.0	.00	.00	409.97	409.97	.00
1000 Ext Wall Insulation RD 1 ES 14FEB97 EF 14FEB97 LS 14FEB97 LF 14FEB97 TF 0							
TOTAL :	385.32	.0	.00	.00	385.32	385.32	.00
1100 Rough Plumbing RD 4 ES 10FEB97 EF 13FEB97 LS 11FEB97 LF 14FEB97 TF 1							
TOTAL :	750.00	.0	.00	.00	750.00	750.00	.00
1200 Rough HVAC RD 3 ES 6FEB97 EF 10FEB97 LS 6FEB97 LF 10FEB97 TF 0							
TOTAL :	1168.75	.0	.00	.00	1168.75	1168.75	.00
1300 Rough Elect RD 3 ES 11FEB97 EF 13FEB97 LS 11FEB97 LF 13FEB97 TF 0							
TOTAL :	940.00	.0	.00	.00	940.00	940.00	.00

Figure 13-30a Cost Control Activity Report Before Cost Update—Wood Frame House, page 1

```
--------------------------------------------------------------------------------------------------------
Student Constructors                    PRIMAVERA PROJECT PLANNER        Sample Project #1

REPORT DATE   2FEB97  RUN NO.   38      COST CONTROL ACTIVITY REPORT     START DATE  5JAN97  FIN DATE 14MAR97
              9:01
Cost Control - Summary by Activity                                      DATA DATE  3FEB97    PAGE NO.    2
--------------------------------------------------------------------------------------------------------
                                             PCT     ACTUAL     ACTUAL      ESTIMATE TO
ACTIVITY ID                       BUDGET     CMP     TO DATE  THIS PERIOD    COMPLETE    FORECAST    VARIANCE
----------- ------- ----------- -------- ---- ----------- ----- ------------ ----------- ------------ ------------

    1400  Shingles
          RD   3 ES 14FEB97  EF 18FEB97  LS 18FEB97  LF 20FEB97  TF   2
          TOTAL :                1091.29   .0        .00       .00      1091.29     1091.29        .00

    1500  Ext Siding
          RD   3 ES 18FEB97  EF 20FEB97  LS 25FEB97  LF 27FEB97  TF   5
          TOTAL :                1710.54   .0        .00       .00      1710.54     1710.54        .00

    1600  Ext Finish Carpentry
          RD   2 ES 14FEB97  EF 17FEB97  LS 21FEB97  LF 24FEB97  TF   5
          TOTAL :                 190.06   .0        .00       .00       190.06      190.06        .00

    1700  Hang Drywall
          RD   4 ES 17FEB97  EF 20FEB97  LS 17FEB97  LF 20FEB97  TF   0
          TOTAL :                1844.60   .0        .00       .00      1844.60     1844.60        .00

    1800  Finish Drywall
          RD   4 ES 21FEB97  EF 26FEB97  LS 21FEB97  LF 26FEB97  TF   0
          TOTAL :                 790.54   .0        .00       .00       790.54      790.54        .00

    1900  Cabinets
          RD   2 ES 27FEB97  EF 28FEB97  LS 27FEB97  LF 28FEB97  TF   0
          TOTAL :                1618.00   .0        .00       .00      1618.00     1618.00        .00

    2000  Ext Paint
          RD   3 ES 21FEB97  EF 25FEB97  LS 28FEB97  LF  4MAR97  TF   5
          TOTAL :                 525.00   .0        .00       .00       525.00      525.00        .00

    2100  Int Finish Carpentry
          RD   4 ES  3MAR97  EF  6MAR97  LS  4MAR97  LF  7MAR97  TF   1
          TOTAL :                1203.15   .0        .00       .00      1203.15     1203.15        .00

    2200  Int Paint
          RD   3 ES  5MAR97  EF  7MAR97  LS  5MAR97  LF  7MAR97  TF   0
          TOTAL :                4725.00   .0        .00       .00      4725.00     4725.00        .00

    2300  Finish Plumbing
          RD   2 ES  3MAR97  EF  4MAR97  LS  3MAR97  LF  4MAR97  TF   0
          TOTAL :                3000.00   .0        .00       .00      3000.00     3000.00        .00

    2400  Finish HVAC
          RD   3 ES 27FEB97  EF  3MAR97  LS 28FEB97  LF  4MAR97  TF   1
          TOTAL :                3506.25   .0        .00       .00      3506.25     3506.25        .00

    2500  Finish Elect
          RD   2 ES 27FEB97  EF 28FEB97  LS  3MAR97  LF  4MAR97  TF   2
          TOTAL :                1410.00   .0        .00       .00      1410.00     1410.00        .00

    2600  Flooring
          RD   3 ES 10MAR97  EF 12MAR97  LS 10MAR97  LF 12MAR97  TF   0
          TOTAL :                1583.34   .0        .00       .00      1583.34     1583.34        .00
```

Figure 13-30b   Cost Control Activity Report Before Cost Update—Wood Frame House, page 2

```
------------------------------------------------------------------------------------------------------------------
Student Constructors                       PRIMAVERA PROJECT PLANNER              Sample Project #1

REPORT DATE   2FEB97  RUN NO.   38         COST CONTROL ACTIVITY REPORT          START DATE  5JAN97  FIN DATE 14MAR97
              9:01
Cost Control - Summary by Activity                                              DATA DATE   3FEB97   PAGE NO.    3

------------------------------------------------------------------------------------------------------------------
                                                    PCT     ACTUAL      ACTUAL    ESTIMATE TO
ACTIVITY ID                                BUDGET   CMP    TO DATE    THIS PERIOD  COMPLETE    FORECAST    VARIANCE
---------- -------- ------------ -------- ----  ------------ -----  ------------ ------------ ------------ ------------ ------------

     2700  Grading & Landscaping
           RD    4 ES   5MAR97 EF 10MAR97  LS 11MAR97  LF 14MAR97  TF    4

           TOTAL :                         600.00   .0        .00         .00      600.00      600.00          .00

     2800  Punch List
           RD    2 ES 13MAR97 EF 14MAR97  LS 13MAR97  LF 14MAR97  TF    0

           TOTAL :                            .00   .0        .00         .00         .00         .00          .00


                                         ------------ ----- ------------ ------------ ------------ ------------ ------------
                    REPORT TOTALS         45972.74  25.0   11513.74    11513.74    34048.78    45562.52      410.22
==================================================================================================================
```

Figure 13-30c   Cost Control Activity Report Before Cost Update—Wood Frame House, page 3

```
---------------------------------------------------------------------------------------------
Student Constructors                    PRIMAVERA PROJECT PLANNER        Sample Project #1
REPORT DATE   2FEB97  RUN NO.   41         COST CONTROL REPORT           START DATE  5JAN97  FIN DATE 14MAR97
              9:06
Cost Control - Detailed by Resource                                     DATA DATE  3FEB97    PAGE NO.   1
---------------------------------------------------------------------------------------------

                    COST     ACCOUNT  UNIT          PCT    ACTUAL     ACTUAL     ESTIMATE TO
ACTIVITY ID RESOURCE ACCOUNT CATEGORY MEAS  BUDGET  CMP    TO DATE    THIS PERIOD COMPLETE   FORECAST    VARIANCE
----------- -------- -------- -------- ----  ------- ----   --------   ---------- ---------- --------    --------
```

	CARPENTR - Carpenter								
200 CARPENTR		L LABOR	day	9.96	100.0	20.00	20.00	.00	20.00	-10.04
300 CARPENTR		L LABOR	day	89.62	100.0	100.00	100.00	.00	100.00	-10.38
400 CARPENTR		L LABOR	day	51.25	100.0	60.00	60.00	.00	60.00	-8.75
500 CARPENTR		L LABOR	day	667.61	100.0	500.00	500.00	.00	500.00	167.61
600 CARPENTR		L LABOR	day	655.03	100.0	500.00	500.00	.00	500.00	155.03
700 CARPENTR		L LABOR	day	751.15	25.0	187.79	187.79	563.21	751.00	.15
800 CARPENTR		L LABOR	day	712.50	.0	.00	.00	712.50	712.50	.00
900 CARPENTR		L LABOR	day	82.08	.0	.00	.00	82.08	82.08	.00
1500 CARPENTR		L LABOR	day	36.00	.0	.00	.00	36.00	36.00	.00
1600 CARPENTR		L LABOR	day	4.00	.0	.00	.00	4.00	4.00	.00
2100 CARPENTR		L LABOR	day	289.28	.0	.00	.00	289.28	289.28	.00
TOTAL CARPENTR			day	3348.48	40.8	1367.79	1367.79	1687.07	3054.86	293.62

	CRPN FOR - Carpenter Foreman									
200 CRPN FOR		L LABOR	day	10.76	100.0	10.76	10.76	.04	10.80	-.04
300 CRPN FOR		L LABOR	day	96.80	100.0	108.00	108.00	.00	108.00	-11.20
400 CRPN FOR		L LABOR	day	27.68	100.0	21.60	21.60	.00	21.60	6.08
500 CRPN FOR		L LABOR	day	360.51	100.0	324.00	324.00	.00	324.00	36.51
600 CRPN FOR		L LABOR	day	353.72	100.0	432.00	432.00	.00	432.00	-78.28
700 CRPN FOR		L LABOR	day	401.29	25.0	100.32	100.32	301.44	401.76	-.47
900 CRPN FOR		L LABOR	day	44.33	.0	.00	.00	44.33	44.33	.00
1500 CRPN FOR		L LABOR	day	38.88	.0	.00	.00	38.88	38.88	.00
1600 CRPN FOR		L LABOR	day	4.32	.0	.00	.00	4.32	4.32	.00
TOTAL CRPN FOR			day	1338.29	74.5	996.68	996.68	389.01	1385.69	-47.40

	CRPN HLP - Carpenter Helper									
400 CRPN HLP		L LABOR	day	36.90	100.0	43.20	43.20	.00	43.20	-6.30
500 CRPN HLP		L LABOR	day	480.67	100.0	360.00	360.00	.00	360.00	120.67
600 CRPN HLP		L LABOR	day	471.63	100.0	360.00	360.00	.00	360.00	111.63
700 CRPN HLP		L LABOR	day	546.57	25.0	136.64	136.64	409.84	546.48	.09
800 CRPN HLP		L LABOR	day	513.00	.0	.00	.00	513.00	513.00	.00
900 CRPN HLP		L LABOR	day	59.10	.0	.00	.00	59.10	59.10	.00
2100 CRPN HLP		L LABOR	day	208.29	.0	.00	.00	208.29	208.29	.00
TOTAL CRPN HLP			day	2316.16	38.9	899.84	899.84	1190.23	2090.07	226.09

	D, 50 HP - Dozier, 50 HP									
700 D, 50 HP		L LABOR	day	76.80	25.0	19.20	19.20	57.60	76.80	.00
TOTAL D, 50 HP			day	76.80	25.0	19.20	19.20	57.60	76.80	.00

	LAB CL 1 - Laborer Class 1									
200 LAB CL 1		L LABOR	day	12.75	100.0	19.20	19.20	.00	19.20	-6.45
300 LAB CL 1		L LABOR	day	114.72	100.0	128.00	128.00	.00	128.00	-13.28
400 LAB CL 1		L LABOR	day	65.77	100.0	64.00	64.00	.00	64.00	1.77
TOTAL LAB CL 1			day	193.24	109.3	211.20	211.20	.00	211.20	-17.96

	LAB CL 2 - Laborer Class 2									
200 LAB CL 2		L LABOR	day	11.15	100.0	16.80	16.80	.00	16.80	-5.65
300 LAB CL 2		L LABOR	day	100.38	100.0	112.00	112.00	.00	112.00	-11.62
400 LAB CL 2		L LABOR	day	57.55	100.0	84.00	84.00	.00	84.00	-26.45
TOTAL LAB CL 2			day	169.08	125.9	212.80	212.80	.00	212.80	-43.72

Figure 13-31a Cost Control Report Before Cost Update—Wood Frame House, page 1

```
Student Constructors                  PRIMAVERA PROJECT PLANNER        Sample Project #1
REPORT DATE   2FEB97  RUN NO.   41        COST CONTROL REPORT          START DATE  5JAN97  FIN DATE 14MAR97
              9:06
Cost Control - Detailed by Resource                                    DATA DATE  3FEB97   PAGE NO.    2
------------------------------------------------------------------------------------------------------------
                       COST    ACCOUNT  UNIT            PCT    ACTUAL     ACTUAL    ESTIMATE TO
ACTIVITY ID RESOURCE  ACCOUNT  CATEGORY MEAS  BUDGET    CMP   TO DATE  THIS PERIOD  COMPLETE   FORECAST   VARIANCE
------------------------------------------------------------------------------------------------------------
            LABOR    -
   1700 LABOR            L LABOR         101.92   .0      .00       .00    101.92    101.92      .00
   1800 LABOR            L LABOR          43.68   .0      .00       .00     43.68     43.68      .00
   TOTAL LABOR                           145.60   .0      .00       .00    145.60    145.60      .00
============================================================================================================
            MASON    - Mason
    400 MASON          L LABOR    day   180.87 100.0   181.28    181.28       .00    181.28     -.41
   TOTAL MASON                   day   180.87 100.2   181.28    181.28       .00    181.28     -.41
============================================================================================================
            MATERIAL -
    200 MATERIAL       M MATERIAL        70.87 100.0    70.87     70.87      .00      70.87      .00
    300 MATERIAL       M MATERIAL       637.80 100.0   637.80    637.80      .00     637.80      .00
    400 MATERIAL       M MATERIAL       547.27 100.0   547.27    547.27      .00     547.27      .00
    500 MATERIAL       M MATERIAL      2672.25 100.0  2672.25   2672.25      .00    2672.25      .00
    600 MATERIAL       M MATERIAL      1843.61 100.0  1843.61   1843.61      .00    1843.61      .00
    700 MATERIAL       M MATERIAL      1692.59  25.0   423.15    423.15  1269.44    1692.59      .00
    800 MATERIAL       M MATERIAL      2522.90   .0      .00       .00  2522.90    2522.90      .00
    900 MATERIAL       M MATERIAL       224.46   .0      .00       .00   224.46     224.46      .00
   1400 MATERIAL       M MATERIAL       834.52   .0      .00       .00   834.52     834.52      .00
   1500 MATERIAL       M MATERIAL        36.86   .0      .00       .00    36.86      36.86      .00
   1600 MATERIAL       M MATERIAL         4.10   .0      .00       .00     4.10       4.10      .00
   1700 MATERIAL       M MATERIAL       477.25   .0      .00       .00   477.25     477.25      .00
   1800 MATERIAL       M MATERIAL       204.53   .0      .00       .00   204.53     204.53      .00
   2100 MATERIAL       M MATERIAL       705.58   .0      .00       .00   705.58     705.58      .00
   TOTAL MATERIAL                     12474.59  49.7  6194.95   6194.95  6279.64   12474.59      .00
============================================================================================================
            SUB      -
    100 SUB            S SUBCNT'R     1280.00 100.0  1280.00   1280.00      .00    1280.00      .00
    200 SUB            S SUBCNT'R       15.00 100.0    15.00     15.00      .00      15.00      .00
    300 SUB            S SUBCNT'R      135.00 100.0   135.00    135.00      .00     135.00      .00
    800 SUB            S SUBCNT'R      247.00   .0      .00       .00   247.00     247.00      .00
   1000 SUB            S SUBCNT'R      385.32   .0      .00       .00   385.32     385.32      .00
   1100 SUB            S SUBCNT'R      750.00   .0      .00       .00   750.00     750.00      .00
   1200 SUB            S SUBCNT'R     1168.75   .0      .00       .00  1168.75    1168.75      .00
   1300 SUB            S SUBCNT'R      940.00   .0      .00       .00   940.00     940.00      .00
   1400 SUB            S SUBCNT'R      256.77   .0      .00       .00   256.77     256.77      .00
   1500 SUB            S SUBCNT'R     1598.80   .0      .00       .00  1598.80    1598.80      .00
   1600 SUB            S SUBCNT'R      177.64   .0      .00       .00   177.64     177.64      .00
   1700 SUB            S SUBCNT'R     1265.43   .0      .00       .00  1265.43    1265.43      .00
   1800 SUB            S SUBCNT'R      542.33   .0      .00       .00   542.33     542.33      .00
   1900 SUB            S SUBCNT'R     1618.00   .0      .00       .00  1618.00    1618.00      .00
   2000 SUB            S SUBCNT'R      525.00   .0      .00       .00   525.00     525.00      .00
   2200 SUB            S SUBCNT'R     4725.00   .0      .00       .00  4725.00    4725.00      .00
   2300 SUB            S SUBCNT'R     3000.00   .0      .00       .00  3000.00    3000.00      .00
   2400 SUB            S SUBCNT'R     3506.25   .0      .00       .00  3506.25    3506.25      .00
   2500 SUB            S SUBCNT'R     1410.00   .0      .00       .00  1410.00    1410.00      .00
   2600 SUB            S SUBCNT'R     1583.34   .0      .00       .00  1583.34    1583.34      .00
   2700 SUB            S SUBCNT'R      600.00   .0      .00       .00   600.00     600.00      .00
   TOTAL SUB                        25729.63   5.6  1430.00   1430.00 24299.63   25729.63      .00
============================================================================================================
            REPORT TOTALS          45972.74  25.0 11513.74  11513.74 34048.78   45562.52   410.22
============================================================================================================
```

Figure 13-31b Cost Control Report Before Cost Update—Wood Frame House, page 2

```
----------------------------------------------------------------------------------------------------
Student Constructors                    PRIMAVERA PROJECT PLANNER         Sample Project #1

REPORT DATE   2FEB97  RUN NO.   43       COST CONTROL ACTIVITY REPORT      START DATE  5JAN97  FIN DATE 14MAR97
              10:05
Cost Control - Summary by Activity                                        DATA DATE   3FEB97   PAGE NO.    1
----------------------------------------------------------------------------------------------------
```

ACTIVITY ID	BUDGET	PCT CMP	ACTUAL TO DATE	ACTUAL THIS PERIOD	ESTIMATE TO COMPLETE	FORECAST	VARIANCE
100 Clear Site RD 0 AS 6JAN97 AF 8JAN97							
TOTAL :	1280.00	100.0	1280.00	1280.00	70.00	1350.00	-70.00
200 Building Layout RD 0 AS 8JAN97 AF 8JAN97							
TOTAL :	130.49	117.0	152.63	152.63	94.17	246.80	-116.31
300 Form/Pour Footings RD 0 AS 9JAN97 AF 10JAN97							
TOTAL :	1174.32	104.0	1220.80	1220.80	10.00	1230.80	-56.48
400 Pier Masonry RD 0 AS 15JAN97 AF 16JAN97							
TOTAL :	967.29	103.5	1001.35	1001.35	28.55	1029.90	-62.61
500 Wood Floor System RD 0 AS 17JAN97 AF 22JAN97							
TOTAL :	4181.04	92.2	3856.25	3856.25	.00	3856.25	324.79
600 Rough Framing Walls RD 0 AS 23JAN97 AF 29JAN97							
TOTAL :	3323.99	94.3	3135.61	3135.61	.00	3135.61	188.38
700 Rough Framing Roof RD 3 AS 28JAN97 EF 5FEB97 LF 5FEB97 TF 0							
TOTAL :	3468.40	25.0	867.10	867.10	2619.19	3486.29	-17.89
800 Doors & Windows RD 4 ES 10FEB97 EF 13FEB97 LS 17FEB97 LF 20FEB97 TF 5							
TOTAL :	3995.40	.0	.00	.00	3995.40	3995.40	.00
900 Ext Wall Board RD 2 ES 6FEB97 EF 7FEB97 LS 7FEB97 LF 10FEB97 TF 1							
TOTAL :	409.97	.0	.00	.00	409.97	409.97	.00
1000 Ext Wall Insulation RD 1 ES 14FEB97 EF 14FEB97 LS 14FEB97 LF 14FEB97 TF 0							
TOTAL :	385.32	.0	.00	.00	385.32	385.32	.00
1100 Rough Plumbing RD 4 ES 10FEB97 EF 13FEB97 LS 11FEB97 LF 14FEB97 TF 1							
TOTAL :	750.00	.0	.00	.00	750.00	750.00	.00
1200 Rough HVAC RD 3 ES 6FEB97 EF 10FEB97 LS 6FEB97 LF 10FEB97 TF 0							
TOTAL :	1168.75	.0	.00	.00	1168.75	1168.75	.00
1300 Rough Elect RD 3 ES 11FEB97 EF 13FEB97 LS 11FEB97 LF 13FEB97 TF 0							
TOTAL :	940.00	.0	.00	.00	940.00	940.00	.00

Figure 13-32a Cost Control Activity Report After Cost Update—Wood Frame House, page 1

```
----------------------------------------------------------------------------------------------------
Student Constructors                  PRIMAVERA PROJECT PLANNER        Sample Project #1

REPORT DATE  2FEB97  RUN NO.  43        COST CONTROL ACTIVITY REPORT    START DATE  5JAN97  FIN DATE 14MAR97
             10:05
Cost Control - Summary by Activity                                     DATA DATE  3FEB97   PAGE NO.   2
----------------------------------------------------------------------------------------------------
                                               PCT    ACTUAL    ACTUAL    ESTIMATE TO
ACTIVITY ID                          BUDGET    CMP    TO DATE  THIS PERIOD  COMPLETE   FORECAST   VARIANCE
----------- -------- ----------- -------- ---- ------------ ----- ------------ ------------ ------------ ------------ ------------
   1400  Shingles
         RD   3 ES 14FEB97  EF 18FEB97  LS 18FEB97  LF 20FEB97  TF   2
         TOTAL :                       1091.29   .0       .00       .00     1091.29    1091.29         .00

   1500  Ext Siding
         RD   3 ES 18FEB97  EF 20FEB97  LS 25FEB97  LF 27FEB97  TF   5
         TOTAL :                       1710.54   .0       .00       .00     1710.54    1710.54         .00

   1600  Ext Finish Carpentry
         RD   2 ES 14FEB97  EF 17FEB97  LS 21FEB97  LF 24FEB97  TF   5
         TOTAL :                        190.06   .0       .00       .00      190.06     190.06         .00

   1700  Hang Drywall
         RD   4 ES 17FEB97  EF 20FEB97  LS 17FEB97  LF 20FEB97  TF   0
         TOTAL :                       1844.60   .0       .00       .00     1844.60    1844.60         .00

   1800  Finish Drywall
         RD   4 ES 21FEB97  EF 26FEB97  LS 21FEB97  LF 26FEB97  TF   0
         TOTAL :                        790.54   .0       .00       .00      790.54     790.54         .00

   1900  Cabinets
         RD   2 ES 27FEB97  EF 28FEB97  LS 27FEB97  LF 28FEB97  TF   0
         TOTAL :                       1618.00   .0       .00       .00     1618.00    1618.00         .00

   2000  Ext Paint
         RD   3 ES 21FEB97  EF 25FEB97  LS 28FEB97  LF  4MAR97  TF   5
         TOTAL :                        525.00   .0       .00       .00      525.00     525.00         .00

   2100  Int Finish Carpentry
         RD   4 ES  3MAR97  EF  6MAR97  LS  4MAR97  LF  7MAR97  TF   1
         TOTAL :                       1203.15   .0       .00       .00     1203.15    1203.15         .00

   2200  Int Paint
         RD   3 ES  5MAR97  EF  7MAR97  LS  5MAR97  LF  7MAR97  TF   0
         TOTAL :                       4725.00   .0       .00       .00     4725.00    4725.00         .00

   2300  Finish Plumbing
         RD   2 ES  3MAR97  EF  4MAR97  LS  3MAR97  LF  4MAR97  TF   0
         TOTAL :                       3000.00   .0       .00       .00     3000.00    3000.00         .00

   2400  Finish HVAC
         RD   3 ES 27FEB97  EF  3MAR97  LS 28FEB97  LF  4MAR97  TF   1
         TOTAL :                       3506.25   .0       .00       .00     3506.25    3506.25         .00

   2500  Finish Elect
         RD   2 ES 27FEB97  EF 28FEB97  LS  3MAR97  LF  4MAR97  TF   2
         TOTAL :                       1410.00   .0       .00       .00     1410.00    1410.00         .00

   2600  Flooring
         RD   3 ES 10MAR97  EF 12MAR97  LS 10MAR97  LF 12MAR97  TF   0
         TOTAL :                       1583.34   .0       .00       .00     1583.34    1583.34         .00
```

Figure 13-32b Cost Control Activity Report After Cost Update—Wood Frame House, page 2

```
-------------------------------------------------------------------------------------------------------
Student Constructors                    PRIMAVERA PROJECT PLANNER            Sample Project #1
REPORT DATE   2FEB97  RUN NO.   43      COST CONTROL ACTIVITY REPORT         START DATE  5JAN97  FIN DATE 14MAR97
              10:05
Cost Control - Summary by Activity                                          DATA DATE  3FEB97   PAGE NO.   3
-------------------------------------------------------------------------------------------------------

                                                     PCT     ACTUAL      ACTUAL    ESTIMATE TO
ACTIVITY ID                               BUDGET     CMP    TO DATE   THIS PERIOD   COMPLETE   FORECAST    VARIANCE
----------- -------------------- ----    --------  -----  ---------- ----------- ----------- ----------- -----------
     2700  Grading & Landscaping
           RD   4 ES  5MAR97 EF 10MAR97 LS 11MAR97  LF 14MAR97 TF   4

           TOTAL :                        600.00     .0        .00        .00       600.00      600.00        .00

     2800  Punch List
           RD   2 ES 13MAR97 EF 14MAR97 LS 13MAR97  LF 14MAR97 TF   0

           TOTAL :                           .00     .0        .00        .00          .00         .00        .00

                                        ---------- ----- ---------- ----------- ----------- ----------- -----------
           REPORT TOTALS                45972.74  25.0   11513.74    11513.74    34269.12    45782.86     189.88
                                        ========== ===== ========== =========== =========== =========== ===========
```

Figure 13-32c   Cost Control Activity Report After Cost Update—Wood Frame House, page 3

```
--------------------------------------------------------------------------------------------
Student Constructors                    PRIMAVERA PROJECT PLANNER        Sample Project #1

REPORT DATE   2FEB97  RUN NO.   46         COST CONTROL REPORT        START DATE  5JAN97   FIN DATE 14MAR97
              10:09
Cost Control - Detailed by Resource                                  DATA DATE  3FEB97    PAGE NO.    1
--------------------------------------------------------------------------------------------

                         COST      ACCOUNT UNIT            PCT    ACTUAL      ACTUAL     ESTIMATE TO
ACTIVITY ID RESOURCE    ACCOUNT   CATEGORY MEAS   BUDGET   CMP    TO DATE   THIS PERIOD  COMPLETE    FORECAST   VARIANCE

             CARPENTR - Carpenter

      200 CARPENTR        L LABOR   day      9.96 100.0    20.00       20.00       .00      20.00    -10.04
      300 CARPENTR        L LABOR   day     89.62 100.0   100.00      100.00       .00     100.00    -10.38
      400 CARPENTR        L LABOR   day     51.25 100.0    60.00       60.00       .00      60.00     -8.75
      500 CARPENTR        L LABOR   day    667.61 100.0   500.00      500.00       .00     500.00    167.61
      600 CARPENTR        L LABOR   day    655.03 100.0   500.00      500.00       .00     500.00    155.03
      700 CARPENTR        L LABOR   day    751.15  25.0   187.79      187.79    563.21     751.00       .15
      800 CARPENTR        L LABOR   day    712.50   .0       .00         .00    712.50     712.50       .00
      900 CARPENTR        L LABOR   day     82.08   .0       .00         .00     82.08      82.08       .00
     1500 CARPENTR        L LABOR   day     36.00   .0       .00         .00     36.00      36.00       .00
     1600 CARPENTR        L LABOR   day      4.00   .0       .00         .00      4.00       4.00       .00
     2100 CARPENTR        L LABOR   day    289.28   .0       .00         .00    289.28     289.28       .00
                                         ------- -----  --------    --------  --------   --------  --------
      TOTAL CARPENTR                day   3348.48  40.8  1367.79     1367.79   1687.07    3054.86    293.62
                                         ======= =====  ========    ========  ========   ========  ========

             CRPN FOR - Carpenter Foreman

      200 CRPN FOR        L LABOR   day     10.76 100.0    10.76       10.76       .04      10.80      -.04
      300 CRPN FOR        L LABOR   day     96.80 100.0   108.00      108.00       .00     108.00    -11.20
      400 CRPN FOR        L LABOR   day     27.68 100.0    21.60       21.60       .00      21.60      6.08
      500 CRPN FOR        L LABOR   day    360.51 100.0   324.00      324.00       .00     324.00     36.51
      600 CRPN FOR        L LABOR   day    353.72 100.0   432.00      432.00       .00     432.00    -78.28
      700 CRPN FOR        L LABOR   day    401.29  25.0   100.32      100.32    301.44     401.76      -.47
      900 CRPN FOR        L LABOR   day     44.33   .0       .00         .00     44.33      44.33       .00
     1500 CRPN FOR        L LABOR   day     38.88   .0       .00         .00     38.88      38.88       .00
     1600 CRPN FOR        L LABOR   day      4.32   .0       .00         .00      4.32       4.32       .00
                                         ------- -----  --------    --------  --------   --------  --------
      TOTAL CRPN FOR                day   1338.29  74.5   996.68      996.68    389.01    1385.69    -47.40
                                         ======= =====  ========    ========  ========   ========  ========

             CRPN HLP - Carpenter Helper

      400 CRPN HLP        L LABOR   day     36.90 100.0    43.20       43.20       .00      43.20     -6.30
      500 CRPN HLP        L LABOR   day    480.67 100.0   360.00      360.00       .00     360.00    120.67
      600 CRPN HLP        L LABOR   day    471.63 100.0   360.00      360.00       .00     360.00    111.63
      700 CRPN HLP        L LABOR   day    546.57  25.0   136.64      136.64    409.84     546.48       .09
      800 CRPN HLP        L LABOR   day    513.00   .0       .00         .00    513.00     513.00       .00
      900 CRPN HLP        L LABOR   day     59.10   .0       .00         .00     59.10      59.10       .00
     2100 CRPN HLP        L LABOR   day    208.29   .0       .00         .00    208.29     208.29       .00
                                         ------- -----  --------    --------  --------   --------  --------
      TOTAL CRPN HLP                day   2316.16  38.9   899.84      899.84   1190.23    2090.07    226.09
                                         ======= =====  ========    ========  ========   ========  ========

             D, 50 HP - Dozier, 50 HP

      700 D, 50 HP        L LABOR   day     76.80  25.0    19.20       19.20     57.60      76.80       .00
                                         ------- -----  --------    --------  --------   --------  --------
      TOTAL D, 50 HP                day     76.80  25.0    19.20       19.20     57.60      76.80       .00
                                         ======= =====  ========    ========  ========   ========  ========

             LAB CL 1 - Laborer Class 1

      200 LAB CL 1        L LABOR   day     12.75 100.0    19.20       19.20       .00      19.20     -6.45
      300 LAB CL 1        L LABOR   day    114.72 100.0   128.00      128.00       .00     128.00    -13.28
      400 LAB CL 1        L LABOR   day     65.77 100.0    64.00       64.00       .00      64.00      1.77
                                         ------- -----  --------    --------  --------   --------  --------
      TOTAL LAB CL 1                day    193.24 109.3   211.20      211.20       .00     211.20    -17.96
                                         ======= =====  ========    ========  ========   ========  ========

             LAB CL 2 - Laborer Class 2

      200 LAB CL 2        L LABOR   day     11.15 100.0    16.80       16.80       .00      16.80     -5.65
      300 LAB CL 2        L LABOR   day    100.38 100.0   112.00      112.00       .00     112.00    -11.62
      400 LAB CL 2        L LABOR   day     57.55 100.0    84.00       84.00       .00      84.00    -26.45
                                         ------- -----  --------    --------  --------   --------  --------
      TOTAL LAB CL 2                day    169.08 125.9   212.80      212.80       .00     212.80    -43.72
                                         ======= =====  ========    ========  ========   ========  ========
```

Figure 13-33a   Cost Control Report After Cost Update—Wood Frame House, page 1

```
------------------------------------------------------------------------------------------------------------------
Student Constructors                    PRIMAVERA PROJECT PLANNER              Sample Project #1

REPORT DATE   2FEB97  RUN NO.   46          COST CONTROL REPORT               START DATE  5JAN97  FIN DATE 14MAR97
              10:09
Cost Control - Detailed by Resource                                          DATA DATE   3FEB97  PAGE NO.    2
------------------------------------------------------------------------------------------------------------------

                         COST     ACCOUNT  UNIT            PCT    ACTUAL     ACTUAL    ESTIMATE TO
ACTIVITY ID RESOURCE   ACCOUNT   CATEGORY  MEAS   BUDGET   CMP   TO DATE  THIS PERIOD   COMPLETE    FORECAST    VARIANCE
----------- --------   -------   --------  ----   ------   ---   -------  -----------  -----------  --------    --------

            LABOR    -

    1700 LABOR                   L LABOR          101.92   .0       .00        .00       101.92      101.92        .00
    1800 LABOR                   L LABOR           43.68   .0       .00        .00        43.68       43.68        .00
                                                --------  ----   -------  -----------  -----------  --------    --------
         TOTAL LABOR                              145.60   .0       .00        .00       145.60      145.60        .00
                                                ==================================================================

            MASON    - Mason

     400 MASON                   L LABOR    day   180.87 100.0   181.28     181.28          .00      181.28       -.41
                                                --------  ----   -------  -----------  -----------  --------    --------
         TOTAL MASON                        day   180.87 100.2   181.28     181.28          .00      181.28       -.41
                                                ==================================================================

            MATERIAL -

     200 MATERIAL                M MATERIAL        70.87 100.0    70.87      70.87        84.13      155.00     -84.13
     300 MATERIAL                M MATERIAL       637.80 100.0   637.80     637.80          .00      637.80        .00
     400 MATERIAL                M MATERIAL       547.27 100.0   547.27     547.27        28.55      575.82     -28.55
     500 MATERIAL                M MATERIAL      2672.25 100.0  2672.25    2672.25          .00     2672.25        .00
     600 MATERIAL                M MATERIAL      1843.61 100.0  1843.61    1843.61          .00     1843.61        .00
     700 MATERIAL                M MATERIAL      1692.59  25.0   423.15     423.15      1287.10     1710.25     -17.66
     800 MATERIAL                M MATERIAL      2522.90   .0       .00        .00      2522.90     2522.90        .00
     900 MATERIAL                M MATERIAL       224.46   .0       .00        .00       224.46      224.46        .00
    1400 MATERIAL                M MATERIAL       834.52   .0       .00        .00       834.52      834.52        .00
    1500 MATERIAL                M MATERIAL        36.86   .0       .00        .00        36.86       36.86        .00
    1600 MATERIAL                M MATERIAL         4.10   .0       .00        .00         4.10        4.10        .00
    1700 MATERIAL                M MATERIAL       477.25   .0       .00        .00       477.25      477.25        .00
    1800 MATERIAL                M MATERIAL       204.53   .0       .00        .00       204.53      204.53        .00
    2100 MATERIAL                M MATERIAL       705.58   .0       .00        .00       705.58      705.58        .00
                                                --------  ----   -------  -----------  -----------  --------    --------
         TOTAL MATERIAL                         12474.59  49.7  6194.95    6194.95      6409.98    12604.93    -130.34
                                                ==================================================================

            SUB      -

     100 SUB                     S SUBCNT'R      1280.00 100.0  1280.00    1280.00        70.00     1350.00     -70.00
     200 SUB                     S SUBCNT'R        15.00 100.0    15.00      15.00        10.00       25.00     -10.00
     300 SUB                     S SUBCNT'R       135.00 100.0   135.00     135.00        10.00      145.00     -10.00
     800 SUB                     S SUBCNT'R       247.00   .0       .00        .00       247.00      247.00        .00
    1000 SUB                     S SUBCNT'R       385.32   .0       .00        .00       385.32      385.32        .00
    1100 SUB                     S SUBCNT'R       750.00   .0       .00        .00       750.00      750.00        .00
    1200 SUB                     S SUBCNT'R      1168.75   .0       .00        .00      1168.75     1168.75        .00
    1300 SUB                     S SUBCNT'R       940.00   .0       .00        .00       940.00      940.00        .00
    1400 SUB                     S SUBCNT'R       256.77   .0       .00        .00       256.77      256.77        .00
    1500 SUB                     S SUBCNT'R      1598.80   .0       .00        .00      1598.80     1598.80        .00
    1600 SUB                     S SUBCNT'R       177.64   .0       .00        .00       177.64      177.64        .00
    1700 SUB                     S SUBCNT'R      1265.43   .0       .00        .00      1265.43     1265.43        .00
    1800 SUB                     S SUBCNT'R       542.33   .0       .00        .00       542.33      542.33        .00
    1900 SUB                     S SUBCNT'R      1618.00   .0       .00        .00      1618.00     1618.00        .00
    2000 SUB                     S SUBCNT'R       525.00   .0       .00        .00       525.00      525.00        .00
    2200 SUB                     S SUBCNT'R      4725.00   .0       .00        .00      4725.00     4725.00        .00
    2300 SUB                     S SUBCNT'R      3000.00   .0       .00        .00      3000.00     3000.00        .00
    2400 SUB                     S SUBCNT'R      3506.25   .0       .00        .00      3506.25     3506.25        .00
    2500 SUB                     S SUBCNT'R      1410.00   .0       .00        .00      1410.00     1410.00        .00
    2600 SUB                     S SUBCNT'R      1583.34   .0       .00        .00      1583.34     1583.34        .00
    2700 SUB                     S SUBCNT'R       600.00   .0       .00        .00       600.00      600.00        .00
                                                --------  ----   -------  -----------  -----------  --------    --------
         TOTAL SUB                              25729.63   5.6  1430.00    1430.00     24389.63    25819.63     -90.00
                                                ==================================================================

                                                --------  ----   -------  -----------  -----------  --------    --------
         REPORT TOTALS                          45972.74  25.0 11513.74   11513.74     34269.12    45782.86     189.88
                                                ==================================================================
```

Figure 13-33b   Cost Control Report After Cost Update—Wood Frame House, page 2

EXERCISES

1. Small Commercial Concrete Block Building—Cost Tracking

Prepare the following updated reports for the small commercial concrete block building located in the Appendix:

1. Cost control report summarized by activity
2. Cost control report detailed by resource

2. Large Commercial Building—Resource Tracking

Prepare the following updated reports for the large commercial building located in the Appendix:

1. Cost control report summarized by activity
2. Cost control report detailed by resource

Appendix

Drawings for Example Problem and Exercises

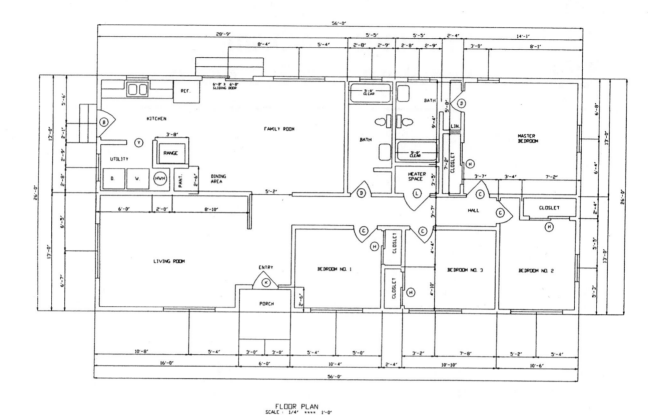

FLOOR PLAN
SCALE : 1/4" ===== 1'-0"

COMPOSITION SHINGLES

VINYL SIDING

ELEVATION
SCALE : 1/4" ===== 1'-0"

Figure A-1 Wood Frame House

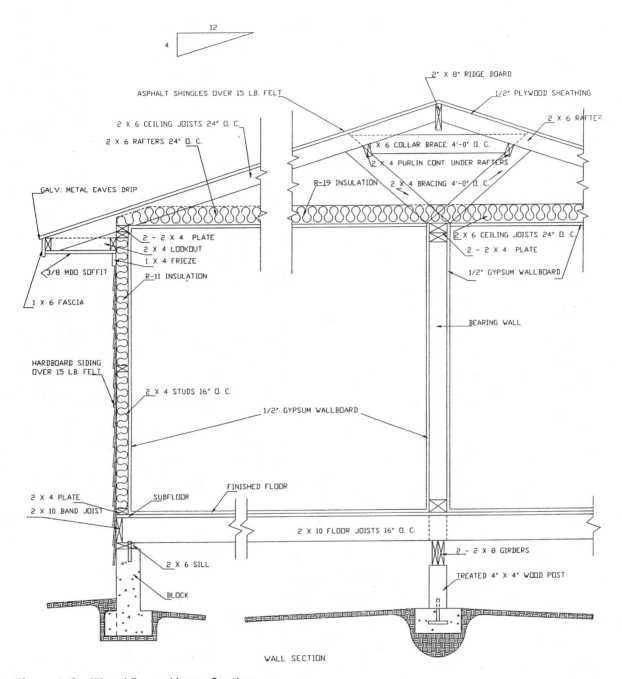

12
4

2' X 8' RIDGE BOARD

ASPHALT SHINGLES OVER 15 LB. FELT

1/2' PLYWOOD SHEATHING

2 X 6 CEILING JOISTS 24' O. C.

2 X 6 RAFTER

2 X 6 RAFTERS 24' O. C.

X 6 COLLAR BRACE 4'-0' O. C.

2 X 4 PURLIN CONT. UNDER RAFTERS

GALV. METAL EAVES DRIP

R-19 INSULATION   2 X 4 BRACING 4'-0' O. C.

2 - 2 X 4 PLATE

2 X 6 CEILING JOISTS 24' O. C.

2 X 4 LOOKOUT

2 - 2 X 4 PLATE

1 X 4 FRIEZE

3/8 MDO SOFFIT

1/2' GYPSUM WALLBOARD

R-11 INSULATION

1 X 6 FASCIA

BEARING WALL

HARDBOARD SIDING
OVER 15 LB. FELT

2 X 4 STUDS 16' O. C.

1/2' GYPSUM WALLBOARD

FINISHED FLOOR

2 X 4 PLATE   SUBFLOOR

2 X 10 BAND JOIST

2 X 10 FLOOR JOISTS 16' O. C.

2 - 2 X 8 GIRDERS

2 X 6 SILL

TREATED 4' X 4' WOOD POST

BLOCK

WALL SECTION

Figure A-2 Wood Frame House Section

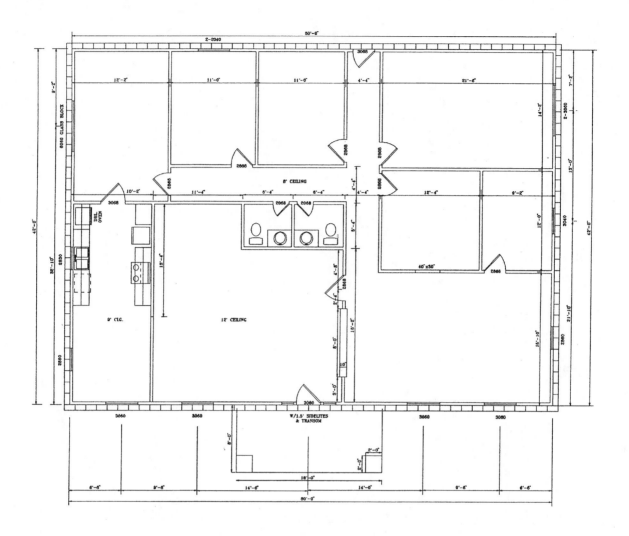

FLOOR PLAN
SCALE : 1/4" ==== 1'-0"

Figure A-3 Small Commercial Concrete Block Building—Floor Plan

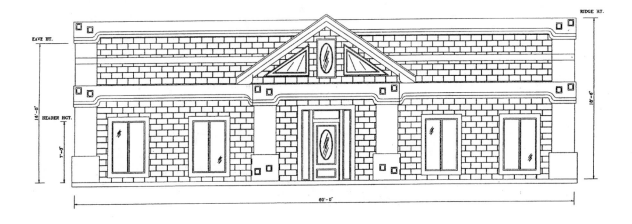

FRONT ELEVATION

SCALE : 1/4" ==== 1'-0"

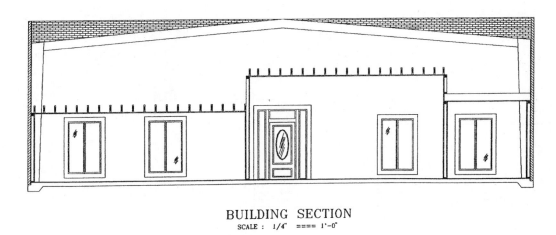

BUILDING SECTION

SCALE : 1/4" ==== 1'-0"

Figure A-4 Small Commercial Concrete Block Building

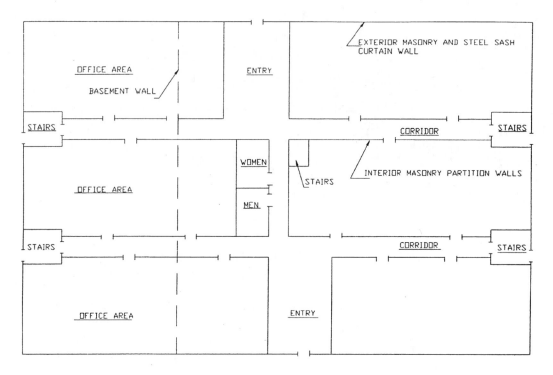

FLOOR PLAN

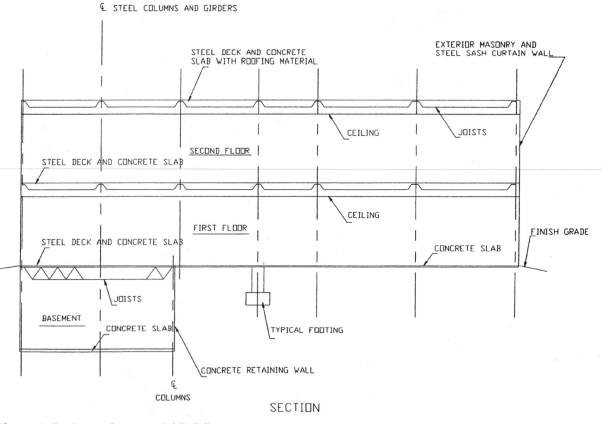

SECTION

Figure A-5 Large Commercial Building

Index